Studies in Systems, Decision and Control

Volume 334

Series Editor

Janusz Kacprzyk, Systems Research Institute, Polish Academy of Sciences, Warsaw, Poland

The series "Studies in Systems, Decision and Control" (SSDC) covers both new developments and advances, as well as the state of the art, in the various areas of broadly perceived systems, decision making and control–quickly, up to date and with a high quality. The intent is to cover the theory, applications, and perspectives on the state of the art and future developments relevant to systems, decision making, control, complex processes and related areas, as embedded in the fields of engineering, computer science, physics, economics, social and life sciences, as well as the paradigms and methodologies behind them. The series contains monographs, textbooks, lecture notes and edited volumes in systems, decision making and control spanning the areas of Cyber-Physical Systems, Autonomous Systems, Sensor Networks, Control Systems, Energy Systems, Automotive Systems, Biological Systems, Vehicular Networking and Connected Vehicles, Aerospace Systems, Automation, Manufacturing, Smart Grids, Nonlinear Systems, Power Systems, Robotics, Social Systems, Economic Systems and other. Of particular value to both the contributors and the readership are the short publication timeframe and the world-wide distribution and exposure which enable both a wide and rapid dissemination of research output.

Indexed by SCOPUS, DBLP, WTI Frankfurt eG, zbMATH, SCImago.

All books published in the series are submitted for consideration in Web of Science.

More information about this series at http://www.springer.com/series/13304

Muhammad Turki Alshurideh ·
Aboul Ella Hassanien · Ra'ed Masa'deh
Editors

The Effect of Coronavirus Disease (COVID-19) on Business Intelligence

 Springer

Editors
Muhammad Turki Alshurideh 🆔
Department of Management
College of Business Administration
University of Sharjah
Sharjah, United Arab Emirates

Department of Marketing
School of Business
The University of Jordan
Amman, Jordan

Ra'ed Masa'deh 🆔
The University of Jordan
Aqaba, Jordan

Aboul Ella Hassanien
Faculty of Computer and Artificial
Intellgence
Cairo University, Cairo, Egypt

ISSN 2198-4182   ISSN 2198-4190 (electronic)
Studies in Systems, Decision and Control
ISBN 978-3-030-67153-2   ISBN 978-3-030-67151-8 (eBook)
https://doi.org/10.1007/978-3-030-67151-8

This Springer imprint is published by the registered company Springer Nature Switzerland AG
The registered company address is: Gewerbestrasse 11, 6330 Cham, Switzerland

Preface

A coronavirus pandemic is one of the most significant events that have recently happened to the whole world. The outbreak started in Wuhan, China, in 2019 and soon spread across the globe, affecting the people's health and paralyzing numerous life spheres. The pandemic has introduced new phenomena, including social distancing, self-isolation, and quarantine, and it is challenging to state whether these preventive measures will eventually disappear in the near future. That is why millions of people from various countries suffer from the pandemic and its consequences. According to the information above, it is not a surprise that COVID-19 has affected every sphere of life, and business is not an exception. It is so because governments took appropriate measures to minimize social contacts to prevent the disease from spreading. This volume discusses through 22 chapters the impact of the COVID-19 on business intelligent systems. The chapters discuss the effect of threat control management strategies on number infected by COVID-19 and how the Business Intelligence in the New Start-up performs during COVID-19. It also discusses e-learning for students and the university branding image that supports community college students in light of the Corona pandemic from offline learning to online learning and understanding the effects of cognitive biases on financial decision-making during the norm the pandemic COVID-19 in some Arab Emirates. This book also discuss the Impact of COVID-19 on Human Resource Practices and the Effect of Service Recovery Justices' Strategies.

This book targets high-quality scientific research papers with applications, including intelligent data analysis, e-payment/m-payment systems and e-learning, business innovation and entrepreneurship and corporate governance and social media and their impact during and after COVID-19.

Finally, editors of this book would like to thank all the authors for their studies and contributions. Editors also would like to encourage the reader to explore and expand the knowledge in order to create their implementations according to their necessities.

Sharjah, United Arab Emirates/Amman, Jordan   Muhammad Turki Alshurideh
Cairo, Egypt   Aboul Ella Hassanien
Aqaba, Jordan   Ra'ed Masa'deh
December 2019

Contents

Intelligent Data Analysis

Role of Intelligent Technology in Crises Management: Systematic Litrature Review the Case of Corona Virus

Maha Alkhaffaf⊙ and **Hesham Almomani**

Abstract This paper's focus is to novel the use of intelligent technology in Crises management, more specifically, Corona Virus Disease (COVID-19) resilience and preparedness in China. This research article aims at identifying the role of digital technology and specifically the role of intelligent technology, big data, artificial intelligence, intelligent applications, and reporting system in crisis management, by analyzing and systematically comparing previous studies to use the above four technological factors in previous case studies and compare them with the case of China when confronting Corona Virus. This research's motivation is the successful crisis management exhibited by China during the Corona Virus pandemic that struck the world in late 2019. Literature about the role of intelligent technology is used to develop a model consisting of factors that directly affect crisis management. The study found a significant role of intelligent technology in decreasing the spread of crises in China's case, This Chinese success in tackling Corona Virus Disease should be studied in terms of intelligent technology that used to limit its spread and other systems or countries.

Keywords Intelligent technology · Crises management · Coronavirus crises

1 Introduction

Technology has entered human life and affected even the way we deal with disasters. In many cases, technology is the easiest part. It has become clear from how much-developed countries were able to use technology to improve every aspect of life [1]. On the other hand, the world is also affected by disasters, whether natural or

M. Alkhaffaf (✉)
MIS Department, World Islamic Sciences University, 11947, Amman 1101, Jordan
e-mail: maha.khafaf@wise.edu.jo

H. Almomani
Industrial Engineering, The Hashemite University, 330127, Zarqa 13133, Jordan
e-mail: heshamalmomani@hu.edu.jo

© The Author(s), under exclusive license to Springer Nature Switzerland AG 2021
M. T. Alshurideh et al. (eds.), *The Effect of Coronavirus Disease (COVID-19) on Business Intelligence*, Studies in Systems, Decision and Control 334,
https://doi.org/10.1007/978-3-030-67151-8_1

3

from human actions. The question that arises is if new technology, including its artificial intelligence and smart applications, is developed to create, quickly, and react to disasters that may hinder societies and countries' ability to adapt and survive amidst their new circumstances. The answer is yes [2]. Many scientific studies contributed to the introduction of intelligent computer science, smart applications, and digital technology in the field of disaster adamant. Hence, many studies demonstrate the role of digital technology in the degree of readiness and preparedness to respond to disasters, such as [3], who realized that digital items, like any other collection of materials, must be protected from the crisis. Further caused by technical or organizational failure. Speaking in the same field, [4] also found that there is a requirement for digitizing resources to ensure real-time responses, appropriate information retrieval, and expediting undertaking and response preparation for disasters. All of these are indications of research indicate the topic and role of digital technology in preparing for disaster and crisis response in all of its forms. What is happening today in China and the world from the spread of the Corona Virus is a biological catastrophe that claimed many people's lives. This is why we have chosen this case in our research paper, highlighting how China has used smart technology, including big data, smart applications, and artificial intelligence, to respond to this incredibly dangerous disease.

The impact of recent technological developments has had a tremendous impact on how we manage disasters. These developments have changed how countries and governments collect information.

About disasters, as well as how they prepare and react to disasters. However, governments and agencies are struggling to coordinate practical disaster and crisis relief programs. Many tools and technological techniques can be used to help in dealing with disasters such as the Corona Virus including, artificial intelligence [AI], machine learning [ML], and natural language processing [NLP], all of them which will be discussed in this research article. The paper will also outline their use in solving and preparing for disasters and crises, with examples de-rived from the new Corona Virus pandemic in China.

2 Research Methodology

This research-based on the idea of the Corona Virus outbreak in China, and for this, the scientific. The method used here came with a foe.us on the academic. Ease. A case study is considered one of the most reliable ways to solve problems that may occur in the future [5]. Our research relied on two sources of information: primary and secondary sources. The primary sources were by relying on access to previous studies in the field of smart technology and its application in disaster and crisis management in the fields of natural disasters and human-made disasters. They provide various references adopted in more than one country and deal with these catastrophes using modern smart technology. After this stage, we compared the previous studies in the field of disaster and crisis management between countries

Fig. 1 The role of digital technology in crisis management

on the one hand and between the intelligent technologies tools used on the other hand. The secondary source of information relied upon in our research includes a case study. Obtain the information required for the state of China in dealing with the Corona Virus; official media sources were referred to, as they continually announced the latest Developments of the crisis there. More specifically, we used official articles published about the use of intelligent technology during the Corona Virus outbreak in China. To ensure accuracy and legitimacy, we used official Chinese reports reviewed by the World Health Organization (WHO).

3   Research Model

Figure 1 shows the research model that explores how intelligent technology factors were received in previous scientific studies. These past uses were then applied to the case of the 2019 Corona Virus pandemic in China to test their success.

4   Literature Review

The Role of Intelligent Technology in Disaster and Crisis Management In the following section, smart technologies and their tools studied under the four elements presented in the study model. Firstly, we cover each element's use and impact during crisis Management, followed by their definitions and uses in past studies [6, 7].

4.1 Artificial Intelligent

It is obvious now that researchers have different understandings of artificial intelligence. For example, Pie Wang defines AI as "building computer systems that are similar to the human mind; they have very different ideas on where this similarity should be". Furthermore, [8] explain that artificial intelligence is an umbrella term for describing a range of technologies and methods, such as natural language processing, data mining, machine learning, and neural networks and algorithm. This study inspired by the work of [9], which explored the benefits of artificial intelligence in crisis response through technologies such as ontology and semantic web, robotics, and multi-agent systems [9]. On the other hand, researchers [7] took a very different approach when studying this topic. They conducted a systematic literature review of the data posted by victim's affected by disasters and crises in the past. They focused on the initial analysis of artificial intelligence applications to process social media big data for efficient disaster and crisis management. The study proved that advances in technology and the analysis of big data of crises de-rived from social media could be quite challenging. This is due to the overwhelming amount of data incurred by society's response posted to social media within minutes of a crisis occurrence [7] Not only are affected individuals using social media platforms, such as Twitter and Facebook, to express themselves but emergency response organizations who want to communicate with the public about the crisis, are also using the same platforms, hence producing An overwhelming amount of data to sort through [10].

Moreover, [11] proposed a hybrid crowdsourcing and real-time ma-chine learning solution to process large volumes of data for disaster and crisis response in a sensitive manner [12]. Conducted a very similar study, which focuses on the use of artificial intelligence during natural disasters, especially in flooding, to improve flood resilience and preparedness. The study delves into the Building Resilience into Risk Management (BRIM) project. The authors focused on the approach to big data through Machine Learning (ML) as an Artificial Intelligence tool. They gathered big data from previous mass floods to produce various ML models that helped create a classification system for floods. Here is the meeting point with the previous study, which talks about the use of big data, its relationship with Artificial Intelligence, and how they can both be used to improve the efficiency of responses to natural disasters. This study reveals the possibility of using our first factor in the study model as a form of smart technology in the face of disaster crisis management. While [12]. studied the use of Artificial Intelligence and its impact (including the learning of machine) on managing natural disasters, our study revolves around a global pandemic known as the Corona Virus or COVID-19. The two circumstances can be compared, but it is decisive and essential to recognize the unpredictability of a virus, spreads at un-precedent speed in a transparent and dangerously lethal manner. This inevitably changes the response to a crisis that cannot be seen or monitored as natural disasters can be [13] used a different approach, which involved focusing only on previous studies that dealt with the role of Artificial Intelligence in crisis and disaster management, in an organized manner. They titled their study "Artificial Intelligence in Disaster Risk Communication: A Systematic Literature Review" [14]. Their research highlighted the various developments that occurred in the field of Business Intelligence in modern-day. They focus on the role of technology learning machines in developing Artificial Intelligence services for the department of crisis and disaster management. After studying the literature written in the field of Artificial Intelligence systematically, they were able to conclude that there are two broad areas that modern researchers focus on: Firstly, the prediction and monitoring of early warning signs, and secondly, information extraction and classification for situational awareness. To conclude, it became apparent through our joint re-search; these two areas of the study prove how AI can be effectively used during a disaster or crisis.

4.2 Big Data

After studying and analyzing many papers in the field of intelligent technology and modern technology, we found, especially in very recent studies, that the subject of big data comes in very often in this field. The proliferation of cameras, websites, and sensors in our world today resulted in useful big data. Cameras, for example, produce an ample amount of big data when recording. So do sensors that track the workday of employees through their entry and exit. References [15, 16] the question arises, where do these institutions store all of the big data that they retrieve daily?

Put, 'big data' is defined as data with a considerable size; it is a collection of data that grows exponentially with time [17]. Its sheer size makes it complicated enough to require non-traditional data management tools to store or process all of the collected [18]. The literature written on this topic reveals that many scholars discuss the role of big data in transforming everyday life, organizations, and mainly the way that the Surrounding environment is dealt with.

"Big Data is the Information asset characterized by such a High Volume, Velocity, and Variety" [17]. The question here is: how does big data relate to the topic of disaster management? During a crisis or disaster, big data is accumulated consisting of personal, medical, patient, survivor, and geographic information. Big data produces valuable information on which to act and provide practical response efforts [19], via crowdsourcing and improved situational awareness [20]. Also studied how big data affected crisis and disaster management with a research paper presented by him and his colleagues at a scientific conference in Bangladesh. This particular study was very similar to the last, as it concluded that big data generated from different sources such as social media, sensors, networks, etc., is catching the attention of crisis managers and researchers.

Most importantly, these studies unveiled the active role of big data collected from varying sources in all phases of disaster management, which will effectively lead to a decrease in Joss of life and property damage during a disaster. In Malaysia, [21], also studied the big data analytics framework for natural disaster management. They explore the use of the big data analytics approach to form strategic decision-making in government. Their research found that big data is one of the tools used to deal with natural disasters because big data is a part of the system. Nevertheless, they explain that big data integrated into other tools that can manage the continuously changing data revolving around weather conditions and water-based natural disasters. What is remarkable in this study is that it provided a framework that contains a group of spirits that help researchers in the field of big data, to classify the stages of dealing with data. This, in turn, may help decision-makers in the government deal with natural disasters. The three stages identified were: data acquisition, data computation, and data interpretation [21]. Reference [22] Also realized that big data provides a new approach to natural disaster management, contributing to the analysis, visualization, and prediction of natural disasters. This study is very similar to our study in that it reviewed previous studies in an organized system but only focused on the role of big data in crisis management.

In comparison, our study focused on a set of smart technologies such as big data and smart applications within disaster and crisis management, as a result of this study and as other Studies indicate Ushahidi, is an example of an open-source crisis-mapping software that creates a database of geo-tagged and time-stamped reports gathered via email, SMS, or tweets. It builds a comprehensive, real-time picture of what is happening on the ground. This platform has become widely spread worldwide, as many countries use it to obtain important information while responding to disasters [11] However, the challenge that organizations and governments face is ensuring that they have a robust digital foundation. This is integral in enabling them to invest in, test, and scale technological solutions before disasters, so they are prepared when they need it the most [17].

4.3   Intelligent Application

After extensive research of previous studies in intelligent applications [23], we did not find studies that specifically address the impact of the intelligent application on crisis management. For example, an article written by Imran, [10], explores artificial intelligence and intelligent technology and their relationship with and uses in disaster management, but neglect to mention intelligent application [24]. This confirms the assumption that the researchers of that paper believed that China was a pioneer in developing intelligent applications to deal with a crisis, like Corona Virus (COVID-19). Moreover, this study recognized that existing studies in intelligent technology did not cover smart applications. Instead, they committed to identifying how to take advantage of big data, artificial intelligence, and other factors in crisis management. They also tended to focus just on natural disasters that took place in various countries around the world, such as Malaysia, Bangladesh, and the United States. In the next part, like the previous parts, we reviewed the procedural definitions mentioned in the scientific articles for each term. First, we present the definition of artificial intelligence, and then we review studies in the same field, along with their impact on disaster and crisis management. Surprisingly, despite all of our research, we did not find a clear definition of intelligent applications.

To get a better understanding of the term, we explored its two components separately, Intelligent and application. The word intelligence is defined by [24] as "the ability to use memory, knowledge, experience, understanding, reasoning, imagination, and judgment to solve problem s and adapt to new situations." Application-defined as "a piece of software designed and intended to fulfill a specific purpose; an application program is part of the main system's package. Web browsers, word processors, games, utilities, and email programs, are all applications" [25] the word application mainly used because each program package has an exact purpose and application for the user [16]. Applications are intended as major goals of the software system since they are made of low and medium-level programs.

Moreover, inherent in the operating system [15]. Through the individual definitions, we conclude that intelligent applications defined as a piece of software designed to perform a specific target in an intelligent way, where it uses knowledge, memory, experience, and reasoning to solve problems. If we apply this definition to developing intelligent applications for crisis management, what are the problems that can be solved through smart applications? We identified them as reporting on disasters, dealing with them in a fast and efficient manner, and contacting the authorities responsible for handling such disasters. When looking at the case of the Corona Virus crisis, what applications have been developed to deal with a crisis of this nature? The following case study talks about China's development of such smart applications that smartly and effectively aid in dealing with an enemy virus that does not discriminate as it destroys human Jives.

5   The Case of Corona Virus in China

5.1   The Use of Artificial Intelligent

The use of Artificial Intelligence in the Corona Virus Crisis made a clear and significant impact on confronting and limiting the disease. The primary way in which this was achievable, was with robots and other technological devices, listed below:

1. Robots used to test more than ten people at once.
2. Robots used to deliver orders to avoid human contact.
3. Robots used to sterilize the air and certain surfaces.
4. Robots used to detect those that were infected/injured and provide audible advice to passers-by.
5. Drones sent to monitor neighborhoods, streets, and citizens.
6. Advanced helmets were used to remotely measure the temperature of passers-by, to identify whether or not they were infected.

These technologies are all examples of Artificial Intelligence tools that used to combat and control the spread of the Corona Virus in China.

5.2   The Use of Big Data

It is well known that the Chinese government uses advanced technology to monitor its citizens, including about 30 million cameras across all of China. The size of the data that produced solely from their cameras provides a perfect example of *big data.* These cameras mentioned above use a modern technology that enables them to recognize feelings, and now identify people carrying the Corona Virus, along with their names, medical identification numbers, and their exact locations. It is evident how prominent of a role digital technology can play when attempting to manage and eradicate disasters and crises. China relies on big data through smart algorithms that combine each citizen's health file, criminal file, and his/her map of travel through public transport. Therefore, this use of smart technology will enable the government to identify all of the people who have had any contact with the disease, and then quarantine them. In return, this will ensure they do not spread the disease to other people, hence, flattening the curve.

5.3   The Use of Intelligent Application

Many companies, such as Alibaba, have developed a QR code linked to government data that identifies a present citizen's health status through the use of colors (red, yellow, and green). This code is required to appear in most public transportation stations and retail stores. Similarly, their government has been able to develop applications that are capable of knowing the whereabouts of people suspected of being infected in the ocean by entering their name and national identification number.

5.4   The Use of Reporting System

The government called on Chinese citizens to report their relatives, neighbors, or friends, if they suspect that they are hiding the disease or if they have some symptoms, by sending digital messages or other methods of reporting. Crises management depends on disaster nature itself and consists of many interrelated dimensions, where all of them centered better management of crises and disaster events and put more emphasis on the expected impacts and effects. O, coronavirus crises management, the internal viewpoint dominated by the crisis manager and leadership, crisis effects derived from Ebola and other epidemic case studies, and On the complexities of risk control and dynamics, complexity and the technology used within the organizations as main factors for successful crises management [26, 27]. Furthermore, the stakeholders' perceptions of organizational reputation, trust, and legitimacy [28]. Succeed in the external perspective too where embrace the stakeholder's interactions with other worlds Countries avoid, stop, solve and overcome the grow out or evolving of a crisis and used as opportunities for integration [28, 29]. The profound implications such as anger of the stakeholders and Decrease or eliminate the number of breaches to curfew. Leaders and health mangers took the actions immediate aftermath of a crisis. Chinese researchers review the core concepts from each perspective and they reduced the likelihood of a crisis, they successfully limit damage to their citizens in a short time, decreased the number of casualties or death per time unit, enable health sector limited capacity for such events and improve organizational preparedness to deal better and absorbing the shock or increased number of patients, not only this but building new field large hospitals with large capacities, some closed after stabilizing the crises and be under control, their organization.

Furthermore, people's ability to learn from the crisis management techniques all over the world the was fast and under control resulted in decreasing the likelihood of crisis reoccurrence, people and government "mindfulness, " cognitive and behavioral process actions please enable them to adapt and accommodate promptly within a reasonable short time via changes in culture and structure to prevent the health and economic systems breakdowns [30, 31]. Chinese organizations believed to work as high-reliability organizations as declared by many researchers [32–34] the availability and use of organizational resources were available too.

The organization may incorporate and use the pre-planned crises strategies quickly enough to meet all expected and predictable dimensions of the crises or disaster emerging events, circumstances, and proceedings, and this will improve the speed and quality of the crises risk response actions and decrease the organizational probability of failures or losses to the minimum. For the financial and economic sector, Fast actions reported from china has a limited effect on financial performance, organization reputation, and improve the continuance and survival rate [34, 35].

6 Conclusion and Future Studies

After studying and analyzing the previous studies mentioned in the field of supporting Intelligent technology for disaster management, Four elements were identified for smart technology that affects disaster management (artificial intelligence, big data, smart applications, and reporting system), In the beginning, a case study of China were analyzed in the face of the Corona Virus crisis, and from these four factors were identified above. Then the systematic approach was used to study and track previous studies in these four areas. The result was the availability of many studies that show the degree of the contribution of artificial intelligence in disaster and crisis management [36, 37]. Furthermore, many other studies emerged from it, too, where big data contributed to crisis and disaster management. However, in intelligent applications, we did not find studies that show a direct impact on crisis management

[38]. Here, China was a pioneer in designing and building intelligent applications to limit the number of people affected by the disease, control their movement, and spread infection. As for the studies mentioned in the impact of the reporting system on crisis and disaster management, they were non-existent, and the reason is recent studies on artificial intelligence and smart technology on the subject of this reporting system, implicitly [39, 40].

For this reason, this study was not able to obtain studies of the role of the reporting system in crisis management in particular. [41, 42] Second: This study proposes future studies on the role of other intelligent information technologies in crisis management: Are there other tools, techniques, and technologies that can be used to track and solve the crisis before and after its occurrence? As for the future studies proposed from this research paper, it was the first; the question that appears in the minds here, What is the time required to develop this type of computer application and whether the IT institutions that have developed these applications had prior knowledge of the possibility of the emergence of the virus before spreading so that it has time to develop these smart systems to track and prevent the spread of the disease.

Accusing China of exporting the Corona crises abroad is not accurate; they were more receptive. less defensive, to criticism, China's international business reputation at stake and has many problems with the USA on many economic issues, China first began fighting COVID-19 at the mainland time ago without serious help from other major countries, according to European Council on Foreign Relations: "This crisis will be a decisive instant of EU-China future relationships - for good or bad remains to be seen." Finally, the pandemic crises should be appropriately addressed based on facts and not on political readings ". Study limitations included, Lack of previous research studies on the Corona Pandemic, the Insufficient sample size, Limited access to data and time constraints, The implication of this research that Heterogeneous impacts unleashed by the outbreak of COVID-19 are hurting, Industrial production and trade, Manufacturing, economies on all levels besides the human life-threatening for all peoples.

References

1. Hanandeh, F., Shannaq, M., Alkhaffaf, M.: Using data mining techniques with open source software to evaluate the various factors affecting academic performance: a case study of students in the faculty of information technology. Int. J. Open Source Softw. Process. 7(2) (2016)
2. Alkhaffaf, M., Makahleh, F.: Investigating the role of planned behavior in knowledge-sharing the mediating role of behavioral intention in Jordanian Commercial Banks. Eur. J. Sci. Res. 148(4), 425–439 (2018)
3. Palme, G.: Emergency and Disaster Preparedness and Response. Can It Play a Role in Disaster Preparedness? OCLC Western, vol. 14, Number 4. Notes from the Field, pp. 20–22
4. Lama, V., Jain, S.: Digitization of disaster management: a multimedia ontological approach. In: Proceedings of Third International Conference on ICTCS 2017. In book: Information and Communication Technology for Competitive Strategies (2019)
5. Sekaran, U.: Research Methods for Business A Skill-Building Approach, 4th edn. Southern Illinois University at Carbondale (2003)

6. Di Martino, G., Iodice, A., Riccio, D., Ruello, G.: A novel approach for disaster monitoring: fractal models and tools. IEEE Trans. Geosci. Remote Sens. **45**, 1559–1570. [CrossRef] 1 (2007)
7. Nunavath, V., Goodwin, M.: The role of artificial intelligence in social media big data analytics for disaster management-initial results of a systematic literature review. In: 2018 5th Conference Proceeding on Information and Communication Technologies for Disaster Management (ICT-DM), Sendai, Japan, 2018, pp. 1–4 (2018)
8. Zawacki-Richter, O., Marín, V.I., Bond, M. et al.: Systematic review of research on artificial intelligence applications in higher education – where are the educators? Int. J. Educ. Technol. High Educ. **16**, 39 (2019). https://doi.org/10.1186/s41239-019-0171-0
9. Khalil, M., Abdel-Aziz, M., Taymour T., Abdel-Badeeh, M.: The role of artificial intelligence technologies in crisis response (2008). Available from: https://www.researchgate.net/pub lication/1735352_The_Role_of_Artificial_Intelligence_Technologies_in_Crisis_Response. Accessed 27 Mar 2020
10. Imran, M., Alam, F., Ofli, F., Aupetit, M.: Enabling Rapid Disaster Response Using Artificial Intelligence and Social Media. Qatar Computing Research Institute (2017)
11. Ofli, F., Meier, P., Imran, M.: Combining Human Computing and Machine Learning to Make Sense of Big (Aerial) Data for Disaster Response projects. AIDR: Artificial Intelligence for Digital Response (2016). https://doi.org/10.1089/big.2014.0064
12. Saravi, S., Kalawsky, R., Joannou, D., Casado, M.R., Fu, G., Meng F.:Use of Artificial Intelligence to Improve Resilience and Preparedness Against Adverse Flood Events. Water **11**, 973. https://doi.org/10.3390/w11050973. 39 (2019)
13. Ogie, R.I., Castilla, J.R., Clarke, R.J.: Artificial intelligence in disaster risk communication: a systematic literature review. In: Conference Proceeding of the International Conference on Information and Communication Technologies for Disaster Management (ICT-DM) (2018)
14. Imran, M., Castillo, C., Lucas J.: AIDR: artificial intelligence for disaster response. In: Conference Proceeding: The 23rd international conference on World Wide Web Seoul, Korea, April 2014 (2014)
15. Arslan, M., Roxin, A., Cruz, C., Ginhac, D.: A review on applications of big data for disaster management. In: Proceedings of the 2017 13th Conference proceeding: on Signal-Image Technology & Internet-Based Systems (SITIS), Jaipur, India, 4–7 Dec 2017, pp. 370–375 (2017) . [CrossRef] 2. S
16. Alkhaffaf, M.: Investigating the mediation role of information technology (IT) components in the relationship between knowledge management (KM) processes and decision-making. Int. J. Bus. Manag. (2018)
17. De Mauro, A., Greco, M., Grimaldi, M.: A formal definition of Big Data based on its essential features. Lib. Rev. **65**(3), 122–135 (2016)
18. Alkhaffaf, M., Almomani, H., Abujamous, F.: Decision support system for disaster management proposed conceptual framework. Int. J. Manag. Appl. Sci. **6**(4) (2020)
19. Akter, S., Wamba, S.F.: Big data and disaster management: a systematic review and agenda for future research. Ann. Oper. Res. **283**, 939–959 (2019). https://doi.org/10.1007/s10479-017-2584-2
20. Rahman, S.H., Di, L., Ul-Zannat, E.: The role of big data in disaster management. In: Proceeding of Conference Proceeding: on Disaster Risk Mitigation, At: Dhaka, Bangladesh (2017)
21. Abdulla, M.F., Ibrahim, M., Zulkifli, H.: Big Data analytics framework. In: Natural Disaster Management in Malaysia, Conference proceeding: 2nd International Conference on Internet of Things, Big Data and Security (2017). https://doi.org/10.5220/0006367204060411
22. Yu, M., Yang, C.: Big Data in natural disaster management: a review. Geosci. Rev. MDPA, **8**, 165 (2018). https://doi.org/10.3390/geosciences8050165
23. Kouns, J., Minoli, D.: Information Technology Risk Management in Enterprise Environments: A Review of Industry Practices and a Practical Guide to Risk Management Teams. Wiley, New York (2011)
24. Legg, Sh. Hutter, M.: A Collection of Definitions of Intelligence. Technical report. ADISA. 7-07. (PDF) A Collection of Definitions of Intelligence (2007). Available from: https://www.res

earchgate.net/publication/1895883_A_Collection_of_Definitions_of_Intelligence. Accessed 26 Mar 2020
25. Trim, Peter R.J.: Disaster management and the role of the intelligence and security services. Disaster Prev. Manag. **12**(1), 6–15 (2003). https://doi.org/10.1108/09653560310463801
26. Gephart, Jr., R.P., Van Maanen, J., Oberlechner, T.: Organizations and risk in late modernity. Organ. Stud. **30**, 141–155 (2009)
27. Starbuck, W.H., Milliken, F.J.: Challenger: fine tuning the odds until something breaks. J. Manage. Stud. **25**, 319–340 (1988)
28. Coombs, W.T., Holladay, S.J.: The negative communication dynamic: exploring the impact of stakeholder effect on behavioral intentions. J. Commun. Manag. **11**, 300–312 (2007)
29. Pfarrer, M.D., Smith, K.G., Bartol, K.M., Khanin, D.M., Zhang, X.: Coming forward: the effects of social and regulatory forces on the voluntary restatement of earnings subsequent to wrongdoing. Organ. Sci. **19**, 386–403 (2008)
30. Weick, K.E., Sutcliffe, K.: Managing the Unexpected: Assuring High Performance in an Age of Complexity. Jossey Bass, San Francisco (2001)
31. Weick, K.E., Sutfcliffe, K.M., Obstfeld, D.: Organizing for high reliability: processes of collective mindfulness. In: Staw, B.M., Cummings, L.L. (eds.) Research in Organizational Behavior, pp. 81–123. JAI Press, Greenwich, CT (1999)
32. Bigley, G.A., Roberts, K.H.: The incident command system: high-reliability organizing for complex and volatile task environments. Acad. Manag. J. **44**, 1281–1299 (2001)
33. Roberts, K.H., Madsen, P., Desai, V.: Organizational sensemaking during crisis. In: Pearson, C.M., Roux C. (eds.) (2007)
34. D'Aveni, R.A., MacMillan, I.C.: Crisis and the content of managerial communications: a study of the focus of attention of top managers in surviving and failing firms. Adm. Sci. Q. **35**, 634–657 (1990)
35. Marcus, A.A., Goodman, R.: Victims and shareholders: the dilemmas of presenting corporate policy during a crisis. Acad. Manag. J. **34**, 281–305 (1991)
36. Coombs, W.T., Holladay, S.J.: Helping crisis managers protect reputational assets. Manag. Commun. Q. **16**, 165–186 (2002)
37. Dufort, Clair, J.A. (eds.) International Handbook of Organizational Crisis Management, pp. 107–122. Thousand Oaks, CA: Sage
38. Muhammad, A., Roxin, M., Cruz, C., Ginhac, D.: A review on applications of Big Data for disaster management. In: The 13th International Conference on Signal Image Technology & Internet Based Systems, Dec 2017, Jaipur, India (2017). ffhal-01678238f
39. Wang, P.: What do you mean by "AI"? Front. Artif. Intell. Appl. **171**(1), 362–373. In: Conference: Proceedings of the 2008 conference on Artificial General Intelligence 2008: Proceedings of the First AGI Conference (2008)
40. Perrow, C.: Normal Accidents: Living with High-Risk Technologies. Princeton, NJ: Princeton University Press (1984). In: Petriglieri, J.L.: Co-creating relationship repair. Admin. Sci. Q., **60**, 518–557 (2015)
41. Weick, K.E.: The collapse of sense making in organizations: the Mann Gulch disaster. Adm. Sci. Q. **38**, 628–652 (1993)
42. Wilson, D.W., Lin, X., Longstreet, P., Sarker, S.: Web 2.0: a definition, literature review, and directions for future research. In: AMCIS 2011 Proceedings—All Submissions. Paper 368 (2011). http://aisel.aisnet.org/amcis2011_submissions/368

Effect of Threat Control Management Strategies on Number Infected by COVID-19

Ahmad Sharieh⑩, **Ruba Abu Khurmah**⑩, **Raja Masadeh**⑩,
Abdullah Alzaqebah⑩, **Nesreen Alsharman**⑩, **and Farah Sharieh**⑩

Abstract As COVID-19 pandemic emerged, quick decisions in response to sudden emergence and rapid spread around the world were required. Strict actions deployed to tackle the COVID-19 pandemic are likely to have prevented millions of additional infections and reduce the number of fatalities during the pandemic. The actions varied from one country or territory to another based on the threat control management (TCM) strategy such as preventing, avoiding, mitigating, and accepting. The TCM involved emergency measures such as isolation, restriction on traveling, closing of nonessential businesses, physical distancing, lockdown and quarantine. Thus, this study introduces a generic dynamic framework that a country can follow to reduce the effects of COVID-19 on the number of infected people and fatalities. Samples of information and data about countries were reviewed, collected, and analyzed. The countries were classified based on the collected data and the curve representing the numbers of infections into countries; with green (flattened the curve and winning), orange (need more actions), and red (failing) colors. The analyses indicate that most of the countries deployed strict actions and applied TCM with

A. Sharieh (✉) · R. A. Khurmah
Computer Science Department, The University of Jordan, Amman 11942, Jordan
e-mail: sharieh@ju.edu.jo

R. A. Khurmah
e-mail: RBA9150539@fgs.ju.edu.jo

R. Masadeh · A. Alzaqebah · N. Alsharman
Computer Science Department, The World Islamic Sciences and Education University, Tabarbour, PO Box 1101, Amman 11947, Jordan
e-mail: raja.masadeh@wise.edu.jo

A. Alzaqebah
e-mail: abdullah.zaqebah@wise.edu.jo

N. Alsharman
e-mail: nesreen.alsharman@wise.edu.jo

F. Sharieh
School of Medicine, The University of Jordan, Amman 11942, Jordan
e-mail: fahmad-jo@hotmail.com

M. T. Alshurideh et al. (eds.), *The Effect of Coronavirus Disease (COVID-19) on Business Intelligence*, Studies in Systems, Decision and Control 334,
https://doi.org/10.1007/978-3-030-67151-8_2

preventing and avoiding strategies were winning countries, apparently, avoiding is the best TCM based on the values of the performance indicators. In conclusion, the generic dynamic framework can be implemented to study the effects of the TCM on the number of infected patients and fatalities caused by the COVID-19. Also, strict actions in response to the pandemic are promising to prevent millions of additional infections and reducing the number of fatalities during the pandemic.

Keywords Avoid · COVID-19 pandemic · Immunity · Mortality · SARS-CoV-2 coronavirus · Strategic plan · Infected patients · Threat control management (TCM) · The world health organization (WHO)

1 Introduction

The Coronavirus (COVID-19) threats all aspects of life over the globe. The problem is to figure what are the actions that lead to prevent the spread of this pandemic and reduce its effects on our life. Thus, every country tries to set strategies to tackle COVID-19 and prepare for this threat before or after it reaches the country. A mission of such a plan is to beat the virus. The main goal is to reduce the effects of the spreading of the virus by reducing the number of infected patients and reducing the mortality rate. The objectives are to have fewer people infected with less complication, to provide treatments to infected people to save their lives and to prevent or avoid the spreading of the virus. The action plan should be dynamic in actions, in policies, and in adapting to better achievement of the aims of the strategic plan.

To help countries in implementing a successful strategy for avoiding the spreading of COVID-19 pandemic, this chapter proposes a generic dynamic framework, based on threat control management, and investigates its effectiveness in tackling the virus. It investigates the effects of set of actions on reducing the number of infected patients, rate of mortality, and number of days to avoid spreading of the COVID-19.

The COVID-19 pandemic enforces hard challenges in our lives worldwide. Plans are required to beat this pandemic and reduce the number of infected patients and mortality rate. The main challenge of COVID-19 pandemic is how the countries can stop the spread of the virus while there are no available treatments or vaccine yet. Most countries have to either try to mitigate the spread of infection with a variety of approaches to infection control or suppress it altogether. Suppression aims to reduce the growth of the epidemic, keep low numbers of cases and sustain this situation for as long as necessary.

Major actions that used for winning or beating COVID-19 pandemic could be summarized as follow: quarantine, wear masks in shared spaces, extensive testing, support medical care, travel restrictions, lockdown, essential services safety, health guidelines, and get everyone on board. There is yet no agreement on an effective and specific treatment.

One of the most challenging things about this virus that symptoms might take in average 5 days to appear; with the possibility of transmission from presymptomatic patients [3, 38]. For management by medications such as Remdesivir, for example, it may reduce the patient's recovery time from 15 to 11 days but did not help reducing the mortality rate [39]. This virus is highly contagious, and its spread varies between people based on adherence to health instructions such as hygiene, social distancing and duration of contact, natural body immunity and ability to resist [40]. A vaccine has not been discovered; and a vaccine usually goes through many steps from discovering until available for manufacturing which takes time. The vaccine should also be given to 70% of the population of earth to develop enough immunity against the virus [41].

In general, the literature indicates that four approaches can be applied to such risk control management of disease threats: accepting, avoiding, mitigating, and transferring. From an operating system point of view in handing deadlock, there are four approaches, also. These are based on avoiding, detection and recovery, ignoring and preventing concepts. These approaches inspire us with an idea to investigate the effects of these approaches on the average of spreading of COVID-19.

The main contributions of this chapter are:

- Propose a framework to manage risk and threats by the COVID-19.
- Investigate the effects of several plan actions and policies deployed by different countries in all continents on the number of infected patients.
- Flag out suitable actions to reduce the effects of the Coronavirus.
- Propose ideas for future studies and predict the spread of COVID-19 virus.

The performance indicators (PIs), which we concentrate on in this study, are the number of infected patients/confirmed cases, weekly average infected patients, the number of days from the first registered infected case until the first peak of the number of infected patients and the rate of mortality.

This chapter is organized into several sections. Section 2 provides a background about COVID-19 and related recent studies. Section 3 presents the proposed framework to tackle COVID-19. Section 4 explains the proposed method. Section 5 presents and discusses the statistical results and their analysis. Finally, Sect. 6 concludes the chapter and suggests possible future directions.

2   Background and Literature Review

This section has the following subsections. Section 2.1 presents a background on COVID-19. Section 2.2 presents sample of the related studies and research conducted on COVID-19.

2.1   A Background on the COVID-19

Coronavirus is a known family of RNA viruses to cause diseases in humans. Some common examples of this family are severe acute respiratory syndrome (SARS) coronavirus, and Middle East respiratory syndrome (MERS) coronavirus [6]. For its proximity to SARS virus genome sequencing and the same receptor, it attacks to get into the body cells. This virus was named as Severe Acute Respiratory Syndrome Coronavirus 2 (SARS-CoV-2) [7]. SARS-CoV-2 was first recognized as a cause of pneumonia (lung inflammation) cases in Wuhan city of China. According to the initial investigations, it is believed that SARS2-Cov-2 is originated in bats then got transmitted, with a yet unknown intermediate carrier, to humans and gained the ability to transmit between people. The first outbreak at china has been tied to a seafood market where people got affected and was later shut down for disinfecting and further investigations [6, 7]. On January 2020, the World Health Organization (WHO) has declared this as a public health emergency, then announced COVID-19 (coronavirus disease 2019) a pandemic on March 2020, after confirmed cases started to rise rapidly in several countries [8, 9].

COVID-19 has a wide spectrum of symptoms, from mild disease to critically ill patients. The main symptoms are fever, dry cough, fatigue, loss of taste and smell, and other symptoms [15]. In severe cases that require hospitalizations, lung involvement is more prevalent [16]. It has been recorded that up to 80% of patients have a mild disease that would only require self-isolation and home care [10]. Asymptomatic infections (patients who get the infection but show no symptoms) so far have been documented but their prevalence and significance in the transmission of the disease are still being studied [14].

As for the diagnosis of COVID-19, patients who are residing or traveled from countries with community transmission within the last 14 days, people in close contact with confirmed or suspected cases, and those who are showing symptoms of the disease could get tested. The diagnosis of COVID-19 is done primarily through reverse transcription-polymerase chain reaction (RT-PCR) to detect the nucleic acid of SARS-CoV-2 [17]. The specimens for the test are collected from upper respiratory tract samples and there have been different types of tests with different degrees of sensitivity. Upon confirmation of the diagnosis, patients who don't show symptoms or have mild symptoms are asked to self-isolate and monitor their symptoms frequently. If they started to show a worsening of symptoms, they are asked to seek medical attention immediately. However, patients that are high risk and show more severe symptoms are usually admitted to hospitals for further evaluation and treatment. The incubation period, the time between a patient gets exposed to the virus to the time showing the first symptom of the disease, is an important factor when studying and tracing any infection. It is thought that SARS-CoV-2 has an incubation period that ranges between 2 days and 14 days with a median of 5 days [22]. It is important to note that there are specific risk factors for patients to develop a severe illness, one of which has been significant is the age of a patient. According to a study done in New York, the majority of patients admitted to hospitals of confirmed COVID-19 were above 44 years old [12]. Moreover, some comorbidities have shown to cause more rapid and serious deterioration of the patient's status. Some of these comorbidities are cardiovascular diseases, diabetes mellitus, hypertension, chronic lung diseases, smoking, obesity, and other chronic diseases that might compromise the immune system of a patient [12]. Mostly, these severe cases were hospitalized and needed at least mechanical ventilation support or treatment with one or more of proposed medications such as Remdesivir, hydroxychloroquine, or other interventions through clinical trials [13, 18, 19]. The death rate has been reported differently in many countries and worse outcomes were related to older patients that received invasive mechanical ventilation management [11, 13].

COVID19 is primarily spread from person to person. A person can become infected by coming into close contact (about 6 feet or two meters) with a person who has COVID19. He/she can become infected from respiratory droplets when an infected person coughs, sneezes, or talks. He/she may also be able to get it by touching a surface or object that has the virus on it, and then by touching his/her mouth, nose, or eyes [20]. Other means of transmission such as airborne (where smaller particles of the disease remain suspended in air) and other body secretions like stool have been studied but showed less significant role in transmission. As a result of these conclusions, healthcare workers who work near COVID19 patients were recommended to have personal protective equipment (surgical mask, face shield, gloves…etc.) on while treating these patients [21]. Upon collecting enough data on the main way of transmission and the incubation period, WHO announced a few steps that can stop the secondary transmission of the virus between people of one community [23]. Those measures include but not limited to the emphasis on the essential role of hand hygiene such as frequent hand washing and the use of hand sanitizers. Also, social distancing included keeping a 6 ft (two meters) distance between people and eliminating all social events with a large number of people. The importance of wearing a face-covering in public where social distancing is difficult to perform also was recommended to stop the spread from symptomatic and presymptomatic patients [25].

Due to the current lack of COVID-19 specific treatment or vaccine, many countries around the world had to also take some measures to stop the widespread of the disease and minimize its effect on the health system with better patients' outcomes and less mortality rates. Those measures included, social distancing, stay at home orders, shut down of airports as well as test and contact tracing [24]. As the pandemic started to reach more countries with larger community spread, further interest surfaced in the transmission means of the virus and what measures could be taken to decrease the impact of the virus on the health systems and avoid higher death rates from the disease.

2.2 Related Studies and Researches on the COVID-19

Many researchers introduced several studies of COVID-19 that taken into consideration only one criterion such as lockdown, social distancing, travel restriction, testing, etc. While this research considered seventeen criteria including travel restrictions, social distancing, restaurants/bars official closure, widespread testing and lockdowns.

The researchers [29] introduced a precise prediction model of the COVID-19. This model is based on Verhulst equation which is capable of fitting data with a high regression coefficient, expect all infections as well as infection peak day. Verhulst equation was employed to fit data of COVID-19 that spread in China, Italy and Spain.

Based on the research [29], the infection peak was on March 23 in Italy and on March 29 in Spain. In addition, the effects of both total and partial lockdowns were studied. However, the infected people and the new infections' rate at the beginning of the lockdowns play significant roles in the spread of infection. More precisely, the model is a significant tool that is used to predict the spread of the disease and gives some important evidences about the major factors which impact the spread of COVID-19 and quantifies the influences of both total and partial lockdowns.

The researchers [30] studied the policy of the optimal lockdown for scheme which controls the number of deaths while reducing the costs' output of the lockdown. The policy relies on the rate of susceptible and infected population. The adversity of the optimal lockdown relies on the gradient of the death rate taken into consideration the infected and, the presence of antibody testing. Moreover, the study analyzed the policy of test-tracing and quarantine (TTQ) and it found that TTQ is complementary to a lockdown. The structures of social connection define the infection spread and, in the absence of vaccines, controlling these structures through large-scale social spacing measures seems to be the most efficient mitigation measure. Thus, the researchers [31] employed age-structured SIR model with matrices of social communication acquired from Bayesian imputation and surveys in order to study COVID-19 disease's progress in India. R0, which is the basic reproductive ratio and its generalization time are calculated based on the structure of social communication, case data and age distribution factors. Then, they investigated the effect of social distancing measures, closed schools, lockdown, work from home, and their efficiency in relation of the duration of implementation. Lockdown for three weeks is insufficient to prohibit resurgence and instead of that, protocols of persistent lockdown with League relaxation are proposed. Predictions are supplied for minimizing in age structured morbidity and deaths as an outcome of these measurements. The research [31] emphasized the significance of structures of social distancing and age in evaluating the country specific influence of mitigatory social distancing.

The widespread testing is a key strategy to stop an outbreak transmission that must be stopped [32, 37]. This leads to additional social cost if the test is not accurate and to isolate additional people who are not infected. However, it will still stop the outbreak. There are ways for testing: self-reporting and diagnosis, contact tracing, lockdown—geographic community identification, massive specific testing, and targeted random sampling. In the beginning of the outbreak, community isolation was an imperative, and was used in China to remarkable success augmenting. In Wuhan, a massive more traditional contact tracing effort (670,000 people) [33] because there was a severe limit to the number of tests that could be done. Subsequently in South Korea, a lockdown was implemented [32] and [33], and there was indication that the outbreak was under control [34].

On March 17, 2020, all the external borders of European Union are closed which is the first time that happened in order to prevent spreading of Coronavirus. For this reason, all governments and authorities around the world implemented huge travel restrictions and control of borders to reduce the outbreak of this global disease. However, the accurate impacts of travel restrictions on the disease outbreak dynamics stay unknown.

The researchers [35] merged a model of global network mobility with a model of local epidemiology in order to simulate and conclude the outbreak dynamics and control of Coronavirus all over Europe. The study connected the mobility model to statistics of passenger air travel and inspect epidemiology mode utilizing the number of infected cases for each country. The suggested simulation [35] displayed that mobility networks of air travel could conclude the emerging global spreading at early phases of the outbreak. Substantially, the proposed model can help political decision making and determining the suitable strategy from total or partial lockdown and current travel restrictions.

3 A Framework for Managing COVID-19

This section presents the proposed framework to tackle COVID-19. Section 3.1 presents a diagram to show the processes of dynamic calibrating of the taken strategies and actions. Section 3.2 explains the adaptation of risk control management on tackling COVID-19 threats, and sect. 3.3 presents some actions to execute the strategic plan to beat the coronavirus by an authority or a country.

3.1 Adapted Framework

Figure 1 shows a diagram of the proposed framework to tackle COVID-19. A country or an authority deploys a strategic plan and actions to stop spreading of the virus and handle its effects. The strategy or actions can be based on operating system techniques in handling deadlocked processes and threats risk control management as shown in Fig. 2. Also, on selected actions such as acting quickly, isolation, travel and transporting restrictions, massive testing, face masks and other protection clothes, social distancing and reopening. Adapting one or more of these strategies or actions will reduce the values of the PIs. A dynamic adaptation of plans and actions should be applied based on the values of the number of infected cases, rate of mortality and time of implementing actions to control the virus.

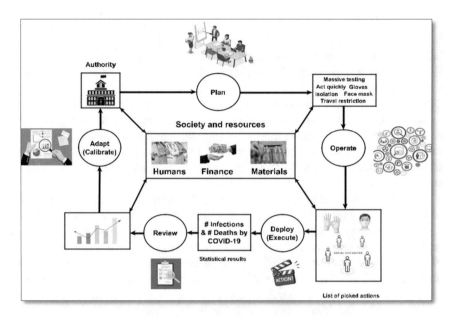

Fig. 1 A dynamic framework followed by a country to beat the COVID-19

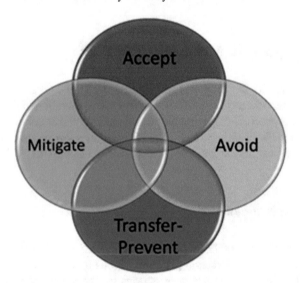

Fig. 2 Threat (risk) control management strategies

In Fig. 1, a country or an authority plans to deploy a strategy and action(s). The authority will activate its operation plan. Then, it will select actions. The actions will be executed based on the target processes and available resources. After that, the authority will review its progress and achievements based on the deployed strategies and actions and the values of the PIs. The authority will calibrate its measures and repeat the processes until it beats COVID-19.

The self-study report that will come from the cycle of this frame will have four main components based on the ADRI approach. In ADRI, the A stands for the approach in policies, procedures, instructions, action strategies, and bylaw stetted by the authority to tackle COVID-19. The D stands for the deployment and execution of the operation plan and action plan set to fight the virus. The R stands for the collecting and analysis of the results of the previous two steps in the number of infected people, fatalities, and how long it took to flatten the curve. Finally, the I stands for the improvement, adaptive, and calibrating of the strategies and actions to reduce the values of the PIs.

An operating system performs one or more of four different methods to handle the deadlock: Ignorance or no setting, Avoidance, Prevention and Detection, and Recovery. For space limitation of the size of this chapter, these are shortly explained in Table 1. Because the spreading of COVID-19 is an unusual situation, predicting the occurrence of deadlock state is preparing the handling strategies and algorithms to cope with what exhausts the system.

3.2 Threat Control Management (TCM)

Four basic strategies are used to control the risk that results from vulnerability [1]. The country management shall decide to accept, mitigate, transfer, exploit or avoid the COVID-19 risks. There are two main factors that help lead this decision by authorities. Those factors are; the likelihood of occurrence and impact. In general, there are three parts of risks: event, consequence, and impact [42]. The event of COVID-19 infects human individuals has devastating consequences and lead to severe impacts. COVID-19 risks can be measured by estimating the likelihood of the event happening and the impact it has on our life. An event of an individual being tested positive or has symptoms of COVID-19 leads to the spreading of COVID-19 among uninfected people, which will increase the number of infected people and the mortality rate. The increase in these numbers has many effects on all aspects of life including effects on services such as delivery, effects on health and safety such as loss of lives, effects on local and international environment and effects on financial aspect such as loss of money and supports for living. Since COVID-19 has high likelihood to occur, high impacts and high risks, it requires very active management to take decisions [4]. It requires active management to deal with the three parts of COVID-19 risks: event, consequence, and impact.

Table 1 Mapping deadlock methods and threat control management strategies

Deadlock handling methods	Threat control management (TCM)	Degree of commitment
Avoidance (keep in safe state) Do not guarantee a resource to any request by a process if it leads to an unsafe state	**Avoid (AV-TCM) (safe-guards)** Deploy actions that lead to avoid the spreading of COVID-19, infection, and death cases	Strict and tough execution for authority actions ($AC1$; $AC2,\ldots ACn$) with high commitment from the community to execute individual actions ($I1, I2, \ldots In$)
Preventing Processes are ordered in a way that the four necessary conditions will not be satisfied simultaneously	**Transfer-Prevent (TR-TCM)** COVID-19 can be transferred through contacts. So, prevent transferring entities that lead to contacts and spread of the virus. The authority applies a resource allocation policy in which the spreading is impossible to arise	Strict execution of the authority actions on specified sample of the society with high commitment from them to execute individual actions
Detection and recovery An operating system should detect if there is a process or more in a deadlock state and then recover the system	**Mitigate (MI-TCM)** Make something less severe or less unpleasant. The authority simply acts to reduce the effect of the virus. The authority manages the risk of COVID-19 spreading through its system internal controls, treat patients and to have less number of death cases	Soft execution for authority actions with intermediate commitment from community to execute individual actions
Ignorance The system will be built as if there will be no deadlock	**Accept (AC-TCM)** This is risk retention, where the authority accepts certain levels of risks and losses that may subsequently arise. It is sometimes referred to as herd or society immunity. The authority, in this case, informs itself and understands the consequences and accepts the risk without control or mitigation	Lack of clarity in the authority actions with low commitment from community to execute individual actions

Figure 2 shows the four threats or risk control management strategies. The way the circles are intersected indicates that an authority can deploy steps in none of these strategies, one strategy, a combination steps of two, three, or all four strategies. To investigate the effects of a followed strategy and actions taken by a country or an authority, we assume that the independent variables, in this study, are (S) set of strategies and (C) set of actions. The dependent variables are assumed to be (E) number of recorded infected cases by COVID-19, (F) rate of fatalities to the number of infected (mortality rate), and (P) the period it takes from the date of the first recorded case until the date it reaches first highest number of recorded infected cases (peak).

Algorithm 1 A dynamic algorithm to beat COVID-19

Input: Strategic plan, Set of actions, and Numberofcases
Output: Curve of COVID-19 number of infected people and death cases, PIs
-Initialize all variables that affect COVID-19 number of cases in a country: Political, geographical, environmental and economical factors; Days
-Set the date of the first announced case in the country. -Set the date of the highest registered number of cases (peak) in the country.
   While Numberofcases > 0 and IPs are not satisfied do
      Prepare a strategic plan (add or delete actions); Pick
      up a list of actions (AC1; AC2,...,ACn);
      Deploy actions;
      Review actions;
        if IP s are notsatisfied then
        Adapt and calibrate actions;
      End if
      Days = Days + 1;
   End while

Avoid threat control management (AV-TCM); which means an authority can apply safeguards (avoidance) that eliminate or reduce the remaining uncontrolled risks for the vulnerability. In avoidance, there are steps taken to prevent the coronavirus from infecting people and try to prevent death cases. In this, the authority shall take an action not to accept the spread of COVID-19. Mitigate threat Control Management (MI-TCM): means make something less severe, less serious, less painful, and/or less unpleasant. For example, getting a lot of sleep, drinking fluids, medications, or the use of respiratory mechanics can mitigate the effects of the virus. In this risk control management, the authority takes actions to manage the risk of avoiding spreading through its system internal controls. Transfer (TR-TCM): COVID-19 can be transferred through contacts. In this, the authority takes steps to shift the loss to another party such as outreaches or someone else, or other areas or outside the countries or transfer people from outside the country and let the virus spread among them to have a society or herd immunity. This cannot be taken for COVID-19 these days. In this case, the authority may replace the transfer with transfer-prevent; where it prevents transferable of the virus from one place to another by using vaccine of uninfected areas or using actions such as isolation, quarantine and lockdown. Accept (AC-TCM): This is risk retention where the authority accepts certain levels of risks and losses that may subsequently arise. It is sometimes referred to as herd or society immunity. The authority, in this case, informs itself and understands the consequences and accepts the risk without control or mitigation. The authority may adopt certain or no practices to minimize the spreading, the number of infected people, rate of mortality, and period to flatten the accumulative number of confirmed cases or death cases.

3.3   Executable Actions

Some actions can be taken to win the battle against COVID-19 [5]. These actions can be classified as actions of authority and actions of individuals. The following list presents some of actions that have been taken by the authorities to cope with COVID-19:

- AC1: Act quickly. Immediate responses and acts by the country speed the process going back to normal.
- AC2: Set facilities to isolate infected patients from their families, classmates, roommates, groups, …etc.
- AC3: Strict Travel Restrictions. Impose travel restriction on areas with infected patients and outward travel to avoid infecting others.
- AC4: Massive amount of testing. This will help in identifying infected individuals separating them from the community.
- AC5: Wearing face masks and gloves is highly effective in preventing infections.
- AC6: Practicing social distancing. Stay away from crowded areas and keep enough distance between nearest neighbors.
- AC7: Keep away those with comorbidities. Do not introduce the elderly, overweight, immunocompromised…etc. to general public.
- AC8: Do not reopen too early to avoid triggering of the risk growth again.
- AC9: Vaccine against the disease.
- AC10: Use respiratory mechanical support.

The following list presents some of actions that have been taken by the individuals to cope with COVID-19:

- I1: Wash hands; and avoid touching eyes, nose, and mouth.
- I2: Wear face masks, gloves, lab coat, glasses; and avoid touching contaminated surfaces.
- I3: Practice social distancing.
- I4: Cover cough and sneezes coming from mouth and nose with tissue.
- I5: Clean and disinfect touched surfaces including tables, door locks, and other sharable things.
- I6: Getting a lot of sleep and drinking fluid can mitigate the effects of the virus.
- I7: Ask for testing and inform if any symptoms of COVID-19 start to appear, such as [5] are fever, dry cough, tiredness, aches and pains, sore throat, diarrhea, conjunctivitis, headache, loss of taste or smell, a skin rash or discoloration of fingers or toes, difficulty breathing or shortness of breath, chest pain or pressure.

Table 1 shows a mapping between strategies and actions. Assume that an authority deploys steps in one or more of the set of strategies S = {Avoid; Transfer prevent; ignore; Detect and recover; Mitigate; Accept} and/or one or more actions in the set of actions C = {AC1: quick response; AC2: face mask; AC3; ACn}. Furthermore, assume that E is the number of infected patients, F is the rate of mortality, and P is the period to reach peak of number of patients. Then, each of E, F, and P can be expressed as a function of S and C. These functions are shown in Eqs. 1–3.

$$E = f(S \, and/or \, C) \tag{1}$$

$$F = h(S \, and/or \, C) \tag{2}$$

$$P = g(S \, and/or \, C) \tag{3}$$

Thus, the proposed framework as sketched in Fig. 1, the processes as expressed in the generic algorithm (Algorithm 1), and the mathematical functions expressed in Eqs. 1–3 are bases of the methodology to investigate the effects of COVID-19.

4 Methodology

This section explains the methodology adapted in this study. Section 4.1 briefly explain the methodology and Sect. 4.2 provide a representation of the number of infected people.

4.1 Proposed Methodology

To investigate the effects of the strategies performed by an authority to tackle COVID-19, the following steps were conducted: collecting information about COVID-19; collecting information about the strategies and actions taken to tackle the virus; proposing an adapted dynamic framework and a generic algorithm to implement the framework. Assume that the sets of strategies and actions as independent variables and the number of infected people, number of fatalities, and period between the date of the first infected case until the date of the first peak in number of infected patients as dependent variables. Eventually, discussing and analyzing the relationships between the dependent and independent variables.

Literature was reviewed to check the strategies and actions were taken for TCM of COVID-19. The strategies include avoid, prevent, mitigate, accept, detect and actions as listed in Sect. 3.3. The collected data was taken from 4. It is published by the European Centre for Disease Prevention and Control (ECDC). The ECDC Epidemic Intelligence team continually collects and refines the data based on the daily reports from international organizations. This data includes the number of deaths and number of infected cases over the world. Because of the dynamic nature of the pandemic and for the accuracy of the recorded data, the dataset is frequently monitored and updated. The data, includes the name of each country or territory, the continents of 210 territories, available population as reported in 2018, date of the first registered case, date of last registered number of cases which is set to May 23rd

2020 in this manuscript, and daily number of cases and fatalities. Section 3 presents the framework, the algorithm, set of strategies and set of actions.

The slopes of the segment lines, represented the weekly average of the number of confirmed cases, were computed. Based on the shape of segment lines and its slopes, the territories were classified into three levels. A territory with increasing daily number of infected people was assumed to have the curve with red color. A territory with increasing number that reached a peak and started to decrease was assumed to have the curve with orange color. A territory with increasing number that reached a peak and started to decrease until it reached 0 or close to 0, was assumed to have the green curve. Samples of different countries were selected to investigate the effects of the strategies on PIs. The samples include; 10 territories of each color group and sample of territories of the highest mortality rates.

4.2   Representation of Number of Infected People

The numbers of confirmed cases can be represented daily, weekly, or per specific number of days. Charts were drawn for each territory to show the number of confirmed cases and number of confirmed deaths. We draw the chart by taking the average cases per week. A daily accumulative of the number of confirmed cases curve can be plotted for each territory. The countries were classified into win (green lines), fail (red lines), and need more actions (orange lines). The countries with the green color were considered successful in tackling COVID-19. The countries with the red color were considered fail to deal with COVID-19. Those countries with the orange color need actions to better deal with COVID-19.

Figures 3, 4, 5, and 6 show the names of the territories, each with its curve of segment lines for the numbers of confirmed cases in different continents. A curve of segment lines shows the average number of cases against time for a specified country or territory. Mathematically, the colors of curves can be produced by finding the slopes of COVID-19 segment lines of the country before the peak point and after the peak point. In the figures, the Y-axis represents the average number of cases per week and the X-axis represents the numbers of weeks. The slope between the peak point $(x_p; y_p)$ and the last point $(x_l; y_l)$ has a significant role in classifying each country. Equation 4 is the equation of Slope_after_peak. If the angle of the Slope_after_peakis ≥ 135, then a country, with taking into account other considerations, might be classified as Win (green-color). But, if the angle is <135 and >90, this implies that a country needs to take more actions to enhance its epidemic situation so it can be classified as Need-actions (orange-color). If the angle is ≤ 90, this implies that number of infected cases is still not decreasing, there is no clear peak point, and the country needs to take effective actions to control its epidemic situation; in this case, it can be classified as Fail (red-color).

$$Slope_{after_{peak}} = SLOPE\left(Y_p : Y_I; X_p : X_I\right) \tag{4}$$

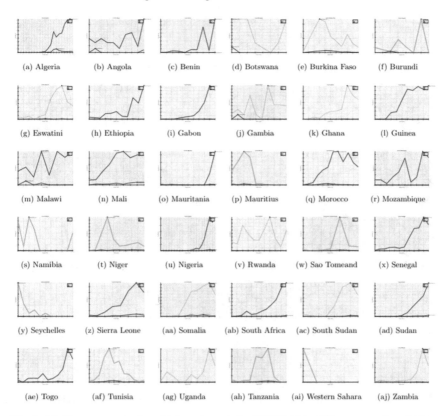

Fig. 3 Weekly average number of infected people by COVID-19 in different countries of Africa continent

4.2.1 Fail-Countries Actions to Beat COVID-19

Table 2 shows sample of territories that had red curves. In Pakistan, for example, the government did not respond early. The image was complicated in the beginning and the capacity of the health sector towards the outbreak was unpredictable. There was an announcement for a soft lockdown. Public relief was focused on the most vulnerable regions of the country [43]. Weakness of the health system, the political situation in Kashmir and the economic instability of the country have all affected COVID-19 outbreak and actions taken towards it [2].

4.2.2 Win-Countries Actions to Beat COVID-19

Table 3 shows list of countries with values of independent variables and dependent variables for sample territories with green curves. For example, on February 27, the Australian government triggered its emergency response to COVID-19, designating

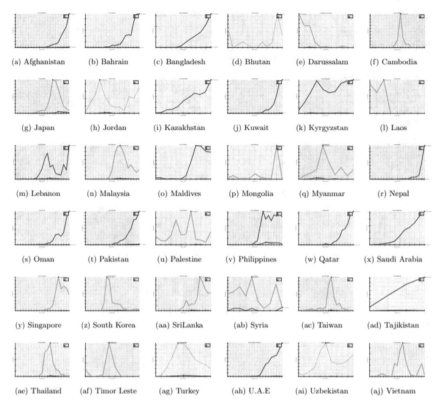

Fig. 4 Weekly average number of infected people by COVID-19 in different countries of Asia continent

it a global pandemic far earlier than the WHO. Their modeling involved making assumptions about how the Coronavirus transmits. They utilized early data from China and other countries which helped to understand how the virus behaves. In addition, the modeling considered the effect of various levels of social distancing and isolating. The initial modeling [27] demonstrated an uncontrolled outbreak scenario with a peak of daily Intensive Care Unit (ICU) beds need to reach 35,000 beds, which exceeds the country's capacity of 7000 ICU beds. By applying quarantine, the need was minimized to 17,000 ICU beds at its peak; which was still above expanded capacity. In case of applying quarantine, isolation and social distancing daily, the need of ICU beds was minimized to below 5000. Thus, the Australian Government directly began to improve possible scenarios by adapting relatively strict measures on February 27 [26]. Firstly, they announced that anyone traveling to Australia must quarantine themselves for 14 days, regardless of history of exposure to the virus. Also, they closed Australia's borders [27]. Moreover, they aimed to reduce the number of people who get infected with coronavirus, manage the request on their health systems and assist to manage their own risk and the risk to the community. After taking these

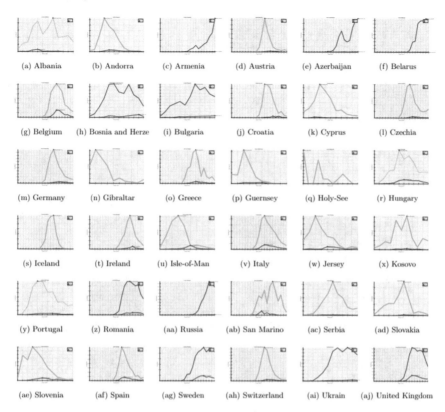

Fig. 5 Weekly average number of infected people by COVID-19 in different countries of Europe continent

measures, by March 30, they recorded 4300 confirmed cases with only 17 deaths. It is worth noticing that all the territories of Oceania had green curves.

4.2.3 Need-Actions Countries Actions to Beat COVID-19

Table 4 shows sample of territories with orange curves. For example, on March 12, the Hungarian government announced a state of emergency and curfew. Moreover, it declared closing of schools as well as other safety recommendations. The international travel was halted on 17 March. However, this decision did not affect private travel within the European Union, goods flow or returning travelers.

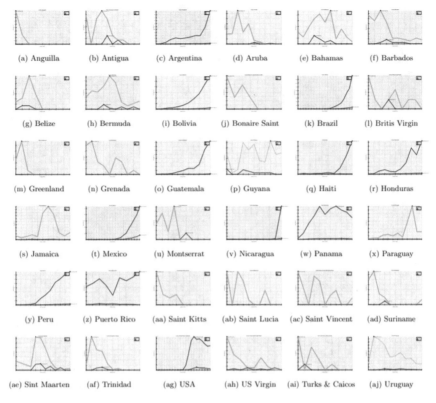

Fig. 6 Weekly average number of infected people by COVID-19 in different countries of America continent

5 Results and Analysis

Tables 2, 3 and 4 show some statistical results for territories with red-color, green-color and orange color, respectively. Each table is divided into two parts. The first part includes the country, population size, actions that executed by the authority of the country, degree of commitment, and the applied TCM strategy. The second part shows the number of patients infected by COVID-19, average number of patients per day, the ratio of the number of patients to the population size, the number of fatalities, the mortality ratio and the period of time to reach a peak during the period from Dec. 31, 2019 until May 23, 2020. Average number of patients per day is calculated by dividing the number of patients into the duration of the epidemic in a country. The duration of the epidemic in a country is the period from the date of the first recorded cases in a country to the date of the last recorded confirmed cases in this study, (23/5/2020). Thus, each country will have different epidemic period. Mortality is calculated by dividing the number of deaths caused by COVID-19 in a country into the number of patients.

Table 2 Sample of countries with red-color and values of independent and dependent variables

Independent variables					Dependent variables					
Country	Pop-size as in 2018	Executed actions	Degree of commitment	TCM	#Patients	Avg. patients/days	#Patients: pop-size	#Fatalities	Mortality	Period to reach peak
Nigeria	195,874,740	AC2, AC3, AC6	Strict	AV-TCM	7261⁻	96.81⁻	0.00004⁻	221⁻	0.030⁻	71⁺
Pakistan	212,215,030	AC2	Soft	MI-TCM	52,437⁺	639.48⁺	0.00025⁺	1101⁻	0.021⁻	85⁺
Russia	144,478,050	AC1, AC2	Strict	AV-TCM	326,448⁺	2888.92⁺	0.00226⁺	3249⁺	0.010⁻	100⁺
Saudi Arabia	33,699,947	AC2, AC3	Strict	AV-TCM	67,719+	857.20+	0.00201⁺	364⁻	0.005⁻	76⁺
Yemen	28,498,687	NA*	Low	AC-TCM	204⁻	8.5⁻	0.00001⁻	33⁻	0.162⁺	21⁻
Ukrania	44,622,516	AC2	Soft	MI-TCM	20,148⁻	283.77⁺	0.00045⁺	588⁻	0.029⁻	13⁻
India	1,352,617,328	AC1–AC10	Strict on some regions	TR-TCM	125,101⁺	1097.38⁺	0.00009⁻	3720⁺	0.030⁻	114⁺
Mexico	126,190,788	NA*	Low	AC-TCM	62,527⁺	976.98⁺	0.00050⁺	6989⁺	0.112⁺	63⁺
Sweden	10,183,175	NA*	Low	AC-TCM	32,809⁺	290.35⁺	0.00322⁺	3925⁺	0.120⁺	85⁺
USA	327,167,434	NA*	Low	AC-TCM	1,601,434⁺	12,914.79⁺	0.00489⁺	96,007⁺	0.060≈	87⁺

NA*: No clear specific action, +: higher than the global number or average, −: less than the global number or average, =: equal to the global number or average, ≈: approximately equal to the global number or average

Table 3 Sample of countries with green-color and values of independent and dependent variables

Independent variables					Dependent variables					
Country	Pop-size as in 2018	Executed actions	Degree of commitment	TCM	#Patients	Avg. patients/days	#Patients: pop-size	#Fatalities	Mortality	Period to reach peak
China	1,392,730,000	AC1, AC2, AC3, AC6	Strict	AV-TCM	84,081+	579.869+	0.000060−	4638+	0.055−	44−
Australia	24,992,369	AC3	Strict	AV-TCM	7095−	48.93−	0.00028+	101−	0.0142−	83+
Taiwan	23,780,452	AC1, AC2, AC3, AC6	Strict	AV-TCM	441−	3.084−	0.0000018−	7−	0.0159−	79+
Vietnam	95,540,395	AC1, AC2, AC3, AC6	Strict	AV-TCM	324−	2.29−	0.000003−	0−	0−	83+
Japan	126,529,100	AC1, AC2, AC3, AC6	Strict	AV-TCM	16536−	114.04−	0.00013−	808−	0.049−	103+
Switzerland	8,516,543	AC1, AC2, AC3, AC7, AC8	Strict	AV-TCM	30,624−	211.2+	0.00359+	1637+	0.053−	88+
Greenland	56,025	AC1, AC2, AC3, AC7	Strict	AV-TCM	11−	0.16923−	0.0001−	0−	0−	8−
Jamaica	2,934,855	AC1, AC2, AC3, AC7	Strict	AV-TCM	544−	7.5−	0.0001−	9−	0.0165−	46−
Iceland	353,574	AC1, AC2, AC3, AC7, AC8	Strict	AV-TCM	1803−	12.43−	0.005+	10−	0.0055−	94+
Tunisia	11,565,204	AC1, AC2, AC3, AC7, AC8	Strict	AV-TCM	1048−	13.78−	0.00009−	47−	0.044−	31−

NA*: No clear specific action, +: higher than the global number or average, −: less than the global number or average, =: equal to the global number or average, ≈: approximately equal to the global number or average

Table 4 Sample of countries with orange-color and values of independent and dependent variables

	Independent variables				Dependent variables					
Country	Pop-size as in 2018	Executed actions	Degree of commitment	TCM	#Patients	Avg. patients/days	#Patients: pop-size	#Fatalities	Mortality	Period to reach peak
Albania	2,866,376	AC1, AC2, AC3, AC5, AC8	Strict	AV-TCM	981^-	12.90^-	0.00034^+	31^-	0.0316^-	48^-
Hungary	9,768,785	AC3, AC5, AC6	Soft	MI-TCM	3713^-	47^-	0.00038^+	470^-	0.127^+	37^-
Jordan	9,956,011	AC1, AC2, AC3, AC4, AC5, AC6	Strict	AV-TCM	700^-	9.72^-	0.00007^-	9^-	0.0128^-	25^-
Latvia	1,926,542	AC1, AC2, AC3, AC6	Soft	MI-TCM	1030^-	13.205^-	0.00053^+	22^-	0.0214^-	29^-
Panama	4,176,873	AC2, AC3, AC4	Strict	AV-TCM	$10{,}267^-$	136.89^-	0.00246^+	295^-	0.029^-	54^-
Portugal	10,281,762	AC1, AC2, AC3, AC6	Strict	AV-TCM	$30{,}200^+$	366.21^+	0.00294^+	1289^-	0.0427^-	40^-
Singapore	5,638,676	AC10	Soft	MI-TCM	$30{,}426^+$	482.95^+	0.00540^+	23^-	0.001^-	89^+
Somalia	15,008,154	AC2, AC3	Soft	MI-TCM	1594^-	23.44^-	0.00011^+	61^-	0.038^-	58^-
Uganda	42,723,139	AC1, AC2, AC3, AC6	Strict	AV-TCM	264^-	4.19^-	0.000001^-	0	0	56^-

(continued)

Table 4 (continued)

Independent variables					Dependent variables					
Country	Pop-size as in 2018	Executed actions	Degree of commitment	TCM	#Patients	Avg. patients/days	#Patients: pop-size	#Fatalities	Mortality	Period to reach peak
Uzbekistan	32,955,400	AC2, AC3, AC6	Strict	AV-TCM	3078⁻	44.609⁻	0.00009⁻	13⁻	0.0042⁻	28⁻

NA*: No clear specific action

+: higher than the global number or average

−: less than the global number or average

=: equal to the global number or average

≈: approximately equal to the global number or average

All these statistics for each country or territory are compared against the global numbers and ratios which are calculated in the same way but with respect to all counties in the world. Therefore, the signs "+", "−", "=" and "≈" are used to indicate the status of the country in comparison with global ratios. They mean respectively above, under, equal or approximately equal to the global number/ratio. Based on the used dataset the following statistics are found; the number of studied countries = 210 territories. The average number of patients in the world = 24,645.124 patients/territory. The global average number of patients during (31/12/2019-23/5/2020) = 35,940.81 patients/day. The global average number of patients in the world/duration of the epidemic = 171.150 patients/day. The ratio between the number of confirmed patients and the world population = 5,175,476/3,323,259,632 = 0.00010. So, there is one infected patient per every 10,000 people. The average deaths in the world = 1609.710 deaths/country. The global mortality rate = 0.065 (deaths/infected patients). The average number of days to reach a peak per country = 58.43 days.

There are 103 territories out of 210 territories estimated with green curve, 26 with orange curve and 81 with red curve during the 144 days' period. Thus, 49% of the territories had strategies that were successful to flatten the curve of their infected people; and almost 39% of the territories could not offer an effective strategy to beat COVID-19 in the given period.

Table 2 shows a sample of territories deployed different TCM with mitigating (MI-TCM), accepting (AC-TCM), avoiding AV-TCM, or transfer TR-TVM. They executed different limited actions including AC2 or not specific clear ones. These strategies, actions and commitments lead, in general, to have higher values of PIs; in average number of patients per territory, in average number of patients per day, in ratio of number of patients to population, and in number of days to reach a peak of the number of infected cases compared to other territories. Most of these territories deployed limited number of actions which may explain why they have red curves. Table 3 shows a sample of territories with green color of their curves. It is clear from the table that all territories, except Australia, deployed avoid TCM with AC1, AC2, AC3, AC4 and AC6, and with a strict level of commitment. The majority of these territories had PIs values less than global average except in the number of days to reach the peak points. Thus, a country will have a high chance to flatten its curves if it at least deploys avoidance of spreading the COVID-19 pandemic strategy in its TCM with actions AC1, AC2, AC3 and AC6; and strict commitment to individuals' actions of their communities. Table 4 shows a sample of countries with orange-color. Some territories deployed avoid TCM and some deployed mitigation TCM. The majority executed AC1, AC2, and AC3 with strict or soft degree of commitment to individuals' actions of their communities. The majority of the dependent variables had values below the global average values except in the ratio between the number of confirmed infected people and the population size. These countries are close to green curve and need more actions to flatten their curves of the number of patients and fatalities.

In computing mortality rates, Europe has 9.59% of the world population, but had 40.3% of the infected cases and 49.76% of total deaths. America has (North, 4.73% and South, 5.53%) 10.36% of the world population but has 30.9% of the infected cases and 40% of total deaths of the world. Both America and Europe have 80.8% of the total infected people and 91% of the total deaths, but with 20.44% of the world population. In the other side, Asia has 9% of infected patients and 9% of the deaths, and 59.54% of the world population. These numbers and percentages indicate that some countries did not deploy sufficient actions and strategies to reduce the number of patients infected by COVID-19. Some of the European countries like Italy did not responded quickly, some applied the MI-TCM as the case of United-Kingdom, and some have weak health facilities. The United States of America, for example, at the beginning did not have a clear policy to tackle the virus. However, China deployed actions such as lockdown and it controlled the threats of COVID-19, even though it has the highest population in the world. Oceania, for example, has 0.55% and almost all its territories have green curves. Australia, for example, responded early before the virus spread with control in the country. These results support our intended goal to show that there are effects of taking strategies such as avoiding and preventing, and it will reduce the number of infected patients by the virus and enable territory to flatten its curve of number of the patients. France, Italy, United Kingdom, Spain and Mexico had high mortality rates than the other countries and higher than the global one. Most of these countries have number of infected people ratio to the size of the population larger than the global one. For example, French government failed for weeks to take decisive action to impose strict (no (AC1 and AC6)) measures or promote large-scale testing (no AC7). Moreover, the average age of an infected person in Germany is 49 years old, compared to about 62.5 years old in France [45]. Italy is one of the worst nations affected by COVID-19 pandemic. Some reasons could be because they have elder people (no AC7), Italians find it difficult to commit to social distancing as it's widely known as a tourists' attraction (no AC6), it is a country with dense population, and attracts businesses from China and other countries (no AC10). Italy and Spain share several characteristics such as fair weather, highly social citizens, and large elderly populations. Unfortunately, overconfidence was the main feature that shared by the governments of both countries in the critical days before drastic effects of COVID-19 happened. Moreover, Spanish and Italian governments underestimated how rapid the virus could transmit and how fast it could pressure their systems of healthcare to the edge of breakdown [36]. As of March 30, France had over 37,568 confirmed cases and 2606 fatalities [28]. Spain became the most virus-infected nation after U.S. and Italy before May 23. It has taken the similar actions as France. Thus, it had over 85,000 confirmed cases and 7300 deaths [27]. Locking down early is an important to stop transmission of Coronavirus and its link with keeping the infection rate number down. The reason for the UK's high death toll was because it didn't lock down quickly enough [44]. In addition, the UK government has been criticized for not allowing large-scale testing early in March [44].

6   Conclusions and Limitations of a Study

There are different actions were taken by the governments and community to tackle the spread of COVID-19. This chapter reviews several important actions and their effects on the number of infected people and number of deaths. We proposed a conceptual model to study the cases as a generic dynamic framework and presented the logical model in an algorithm. The effects of ten authority actions and seven individual actions on the number of infected people, rate of mortality and number of days to reach the maximum number of confirmed infected case are reported in 210 territories. The information is collected from the literature, reports and news about the territories and the statistics of infected people and deaths. The implementation of the proposed algorithm in Spreadsheet and MATLAB helped in drawing the curves and classify the territories. The weekly average number of infected people and deaths were represented as connected continuous linear spline functions (curves). Based on the shape of these functions, 210 territories where classified as those with green, orange and red curves.

In general, it is found that those countries deployed several actions such as quick response, lockdown, quarantine, traveling restriction, perform more testing, wearing face masks, apply social distancing, isolate the elderly, delay reopening, and set sufficient health equipment with more commitment from the communities win the battle. For countries that did not seriously apply some of these measures had high number of infected people and high rates of mortality; and still could not flatten their curves. Thus, following the highlighted strategies and actions in the proposed dynamic framework will have very promising effects in health, social and economic aspects of life globally. This is due to preventing the COVID-19 pandemic spreading, reducing the mortality rate and shortening the period to control the effects of the virus.

There are few limitations of this study. There is lack of reported measures about the virus in all studied countries. There is limited number of pages to report all the finding of this study. The framework needs long period to be verified and validated the proposed framework for more measures.

This work has established bases for many more ideas for future research. The framework can be implemented to be a generic system to deal with information and produce more informative curves and statistical results. There is a chance to study the effects of the actions taken by all territories not samples only. The dependent variables can be interpolated to figure the growth and predict the future behavior of COVID-19.

References

1. Booker, F.: Developing effective risk management strategies to protect your organization. Risk Manag. **5**, 27–32 (2005)
2. Almeida, Olson, Pakistan's Looming Coronavirus Crisis, April, 2020. https://www.usip.org/publications/2020/04/pakistans-looming-coronaviruscrisis

3. "Nigeria: COVID-19 Cases On the Rise", March, 2020. https://www.hrw.org/news/2020/03/25/nigeria-covid-19-cases-rise
4. Gasmi, A., et al.: Individual risk management strategy and potential therapeutic options for the COVID-19 pandemic. Clin. Immunol. 108409 (2020)
5. Jewell: Everything you should know about the 2019 coronavirus and COVID-19. June, 2020. https://www.healthline.com/health/coronavirus-covid-19
6. A Pneumonia Outbreak Associated with a New Coronavirus of Probable Bat Origin. Peng Zhou. Nature. **579**(7798), 270–273 (2020). https://doi.org/10.1038/s41586-020-2012-7. Epub 2020 Feb 3
7. Karim, A., et al.: Knowledge and attitude towards COVID-19 in Bangladesh: population-level estimation and a comparison. Nature **5**(4), 536–544 (2020)
8. Wang, C., et al.: A novel coronavirus outbreak of global health concern. The Lancet **395**(10223), 470–473 (2020)
9. World Health Organization: Coronavirus disease 2019 (COVID-19) Situation Report-51, 11 March 2020. 2020. 51. March 11 2020
10. Wu, Z., McGoogan, J.M.: Characteristics of and important lessons from the coronavirus disease 2019 (COVID-19) outbreak in China: summary of a report of 72,314 cases from the Chinese Center for Disease Control and Prevention. JAMA **323**(13), 1239–1242 (2020)
11. Hopkins: Mortality Analyses Report, last updated, June 12, 2020. https://coronavirus.jhu.edu/data/mortality
12. Petrilli, C.M., et al.: Factors associated with hospital admission and critical illness among 5279 people with coronavirus disease 2019 in New York City: prospective cohort study. BMJ 369 (2020)
13. Richardson, S., et al.: Presenting characteristics. Comorbidities, and Outcomes Among 5700 (2020)
14. Kimball, A., et al.: Asymptomatic and presymptomatic SARS-CoV-2 infections in residents of a long-term care skilled nursing facility—King County, Washington, March 2020. Morb. Mortal. Wky. Rep. **69**(13), 377 (2020)
15. Altwairqi, R.G., Aljuaid, S.M., Alqahtani, A.S.: Effect of tonsillectomy on humeral and cellular immunity: a systematic review of published studies from 2009 to 2019. Eur. Arch. Oto-Rhino-Laryngol., pp. 1–7 (2019)
16. Wang, D., et al.: Clinical characteristics of 138 hospitalized patients with 2019 novel coronavirus–infected pneumonia in Wuhan, China. JAMA 323(11), 1061–1069 (2020)
17. Bhimraj, A., et al.: Infectious diseases Society of America guidelines on the treatment and management of patients with COVID-19. Clin. Infect. Dis. (2020)
18. Beigel, J.H., et al.: Remdesivir for the treatment of Covid-19—preliminary report. N. Engl. J. Med. (2020)
19. Geleris, J., et al.: Observational study of hydroxychloroquine in hospitalized patients with Covid-19. N. Engl. J. Med. (2020)
20. Ong, S.W.X., Tan, Y.K., Chia, P.Y.: Air, surface environmental, and personal protective equipment contamination by severe acute respiratory coronavirus (SARS-CoV-2) from a symptomatic patient. JAMA (2020)
21. World Health Organization: Coronavirus disease (COVID-19) outbreak: rights, roles and responsibilities of health workers, including key considerations for occupational safety and health: interim guidance, 19 March 2020. No. WHO/2019-nCov/HCW_advice/2020.2. World Health Organization, 2020
22. Lauer, S.A., Grantz, K.H.: Qifang Bi, Forrest K Jones, Qulu Zheng, Hannah R Meredith, Andrew S Azman, Nicholas G Reich, and Justin Lessler. The incubation period of coronavirus disease 2019 (covid-19) from publicly reported confirmed cases: estimation and application. Ann. Intern. Med. 3 (2020)
23. Wang, Y., et al.: Reduction of secondary transmission of SARS-CoV-2 in households by face mask use, disinfection and social distancing: a cohort study in Beijing, China. BMJ Glob. Health 5(5) (2020)

24. Lyu, Wei, Wehby, George L.: Comparison of estimated rates of coronavirus disease 2019 (COVID-19) in border counties in Iowa without a stay-at-home order and border counties in Illinois with a stay-at-home order. JAMA Netw. Open **3**(5), e2011102–e2011102 (2020)
25. Adams, J.: Recommendation regarding the use of cloth face coverings, especially in areas of significant community-based transmission. (2020)
26. Kim, H.: These countries are doing the best and worst jobs fighting coronavirus, last updated, 17 March 2020. https://www.usatoday.com/story/news/world/2020/03/17/coronavirus-how-countries-across-globeresponding-covid-19/5065867002/
27. Australia Government: Coronavirus (COVID-19) in Australia–Pandemic Health Intelligence Plan, last updated, June 12, 2020. https://www.health.gov.au/res;ources/publications/corona virus-covid-19-inaustralia-pandemic-health-intelligence-plan
28. Lancet Public Health: The French response to COVID-19: intrinsic difficulties at the interface of science, public health, and policy, last updated, May, 2020. https://www.ncbi.nlm.nih.gov/pmc/articles/PMC7141446/. https://www.euractiv.com/section/coronavirus/short_news/norwayupdate-covid-19/
29. Sanchez-Caballero, S., Selles, M.A., Peydro, M.A., Perez-Bernabeu, E.: An efficient COVID-19 prediction model validated with the cases of China, Italy and Spain: total or partial lockdowns? J. Clin. Med. **9**(5), 1547 (2020)
30. Alvarez, F.E., Argente, D., Lippi, F.: A Simple Planning Problem for Covid-19 Lockdown (No. w26981). National Bureau of Economic Research (2020)
31. Singh, R., Adhikari, R.: Age-structured impact of social distancing on the COVID-19 epidemic in India (2020). arXiv preprint arXiv:2003.12055
32. Shen, C., Bar-Yam, Y.: Massive Testing Can Stop the Coronavirus Outbreak. New England Complex Systems Institute (March 6, 2020)
33. WHO-China: Report of the WHO-China Joint Mission on Coronavirus Disease 2019 (COVID-19), February 2020. https://www.who.int/docs/default-source/coronaviruse/who-china-joint-mission-on-covid-19-final-report.pdf
34. China Goes Door to Door in Wuhan, Seeking Infections, February 19, 2020. https://www.cou rthousenews.com/china-goes-door-to-door-in-wuhanseeking-infections/
35. Linka, K., Peirlinck, M., Sahli Costabal, F., Kuhl, E.: Outbreak dynamics of COVID-19 in Europe and the effect of travel restrictions. Comput. Methods Biomech. Biomed. Eng. 1–8 (2020)
36. COVID-19: What went wrong in Italy and Spain? February 2020. https://www.aa.com.tr/en/europe/covid-19-what-went-wrong-in-italy-and-spain/1797461
37. Guo, Y.-R., et al.: The origin, transmission and clinical therapies on coronavirus disease 2019 (COVID-19) outbreak–an update on the status. Mil. Med. Res. **7**(1), 1–10 (2020)
38. Lauer, S.A., et al.: The incubation period of coronavirus disease 2019 (COVID-19) from publicly reported confirmed cases: estimation and application. Ann. Intern. Med. **172**(9), 577–582 (2020)
39. Cao, B., et al.: A trial of lopinavir–ritonavir in adults hospitalized with severe Covid-19. N. Engl. J. Med. (2020)
40. Howard, J., et al.: Face masks against COVID-19: an evidence review (2020)
41. Yang, Z., et al.: A DNA vaccine induces SARS coronavirus neutralization and protective immunity in mice. Nature **428**(6982), 561–564 (2004)
42. Renn, O.: Three decades of risk research: accomplishments and new challenges. J. Risk Res. **1**(1), 49–71 (1998)
43. Jihan, A.: Mexico's government under pressure over coronavirus response, last updated, 17 Mar 2020. https://www.aljazeera.com/news/2020/05/8000-excess-deaths-mexico-city-corona virus-rages-study-200526011843496.html
44. Euronews: Coronavirus: Why is the UK's COVID-19 death toll higher than other EU coun-tries? Last updated: 06/05/2020. https://www.euronews.com/2020/05/06/coronavirus-why-is-the-uk-s-covid-19-death-toll-higherthan-other-eu-countries
45. Ward: Why France has 4 times as many coronavirus deaths as Germany. Vox, https://www.vox.com/2020/4/17/21223915/coronavirus-germany-france-casesdeath-rate

The Effect of Lockdown on Coronavirus (COVID-19) Outbreak Prevention in Jordan: A Data-Driven Analysis

Hussam N. Fakhouri[ID]**, Faten Hamad, Raja Masadeh, Sandi N. Fakhouri, Amal Akour, and Tamara Al-Daghastani**

Abstract COVID-19 appears in Wuhan in December 2019 and spread all around the world including Jordan. Jordan set successful example in dealing with covid-19 crisis, with its low number of deaths, increased recovery cases and decreased cases of infections in comparison with Middle East and surrounding countries. It succeeded in eliminating the virus outbreak throughout the Kingdom in record time. This paper first describes the stages of Jordan's handling of the Corona pandemic (COVID-19) and how the pandemic controlling has been impacted by governmental and medical efforts, common cooperation, between all organizations in Jordan. Then it studies the effect of lockdown on coronavirus (COVID-19) outbreak prevention in Jordan and provides a data-driven analysis. After that, it proposes a lockdown model to prevent the spread of the virus).

Keywords COVID-19 pandemic · Coronavirus · Treatments · Jordan · Lockdown

H. N. Fakhouri (✉)
Visual Design, The University of Jordan, Amman, Jordan
e-mail: h.fakhouri@ju.edu.jo

F. Hamad
Library and Information, The University of Jordan, Amman, Jordan
e-mail: f.hamad@ju.edu.jo

R. Masadeh
Computer Science Department, The World Islamic Sciences and Education University, Tabarbour, PO Box 1101, Amman 11947, Jordan
e-mail: raja.masadeh@wise.edu.jo

S. N. Fakhouri
Computer Science, The University of Jordan, Amman, Jordan
e-mail: sandi_fakhouri@yahoo.com

A. Akour
Biopharmaceutics and Clinical Pharmacy, The University of Jordan, Amman, Jordan
e-mail: a.akour@ju.edu.jo

T. Al-Daghastani
Al-Salt College for Human Sciences, Al-Balqa Applied University, Al-Salt, Jordan
e-mail: t.aldaghastani@bau.edu.jo

43

1 Introduction

A novel coronavirus (COVID-19), which originated from Wuhan, China, and classified as pandemic by WHO [1]. It has spread to almost all countries worldwide. Up to 3 May 2020, there have been 3,356,205 confirmed cases of COVID-19, including 238,730 deaths, reported to WHO [1].

Most of countries, including Jordan, have taken special procedure to prevent the spread of corona virus COVID-19 pandemic. These procedures varied from one country to another. Jordan procedure included full shutdown of the country when the number of cases was less than 20 [2]. However as soon as the first case was reported in Jordan to be infected with COVID-19 on March 2, 2020. Jordan responded very fast by announcing a full shutdown of the country on March 15 after thirteen days from announcing the first case [1, 2].

Jordan procedures started early by temporarily banned travellers from countries which have COVID-19 outbreak such as China, South Korea, and Iran, starting from February 24. Each person entering Jordanian border crossings and airports was examined, with mandatory chest and throat checks, as well as temperature checks. Moreover, 14-day quarantine for anyone with a positive test results. Later, on March 17th, all incoming and outgoing flights to Jordan have stopped [2, 3].

Before banning the flights, about 5050 people had returned from other countries and all of them had been put in quarantine for 14 days in hotels. This quarantine was very beneficial as some of the later positively confirmed cases in Jordan were from the people who had returned from outside [4]. This step helped to stop the spreading of COVID-19 in Jordan and to ensure their safety and the safety of other citizens.

Jordan procedure to prevent the outbreak of COVID-19 was very serious including issuing the defence orders to apply the full lockdown [3, 4]. To prevent the spread of COVID-19 between Jordan cities as isolation of each city has been applied and city to city transfer banned especially to Irbid city, where some infected people with coronavirus has been reported. Even inside Irbid city, it has been isolated from its related villages.

Furthermore, Jordan has applied building isolation, in case the building reports approved a patient of COVID-19, the authority announces an isolation of the building from other neighbouring buildings and perform random tests to people who live in the same neighbourhood. Moreover, in case more cases are reported as positive, the neighbourhood is isolated.

Because of the full lockdown reduced the possibility of spreading COVID-19 by human-to-human transmission in Jordan, it is very importance to evaluate the isolation and full lockdown effect on the prevention for COVID-19 transmissibility and to forecast the probable size of the pandemic when following similar procedure in the future.

Accordingly, this study estimated the spreading factor (SF) of COVID-19 when full lockdown and isolation are applied and made a prediction of daily new cases for the next thirty days.

Beside the transmission rate, three factors have been studied to see the effect of lockdown on COVID-19, which are number of critical cases, number of infected cases, and number of deaths.

This study outlines the first steps in the treatment of the Corona Virus Disease in Jordan (COVID-19) and how government and medical activities, mutual collaboration between all organizations in Jordan, affected corona virus pandemic controlling. Moreover, it investigates the impact of corona-virus lock-down in Jordan and provides an analysis driven by data. Then a lock-down mechanism is suggested to avoid the replication of the virus.

2   Related Work

Many researchers presented several studies of the effect of lockdown on coronavirus (COVID-19) outbreak prevention.

The number of confirmed infected were 105,792 cases till March 31, 2020 in Italy. However, the number of fatalities was 15,726 which demonstrate the impact of the epidemic on the country. Thus, Italian government implemented a lock-down on March 9. Based on this view, the study [5] evaluated Italy's situation and possible effects in case lockdown continue for two months. Data of COVID-19 infected people were extracted from the website of Italian Health Ministry including cases of registered and recovered from February 15 to March 31. The forecasts were made with 93.75% accuracy for the models of registered case and, 84.4% accuracy for the models of recovered case. While, the results of the study [6] provided a timeline of the efficiency of the implemented closure, which is important for many countries that have followed Italy in applying similar measures.

Indian government implemented full lockdown on the country in order to decrease transmission of COVID-19 in its country in case the absence of effective vaccine or therapy. A novel mathematical model on COVID-19 is introduced [7] which incorporated the impact of lockdown, to research the influence of social distancing measure. The proposed model is validated to the reported cases' data from five various states and overall India. The researchers assessed diverse epidemiologically significant parameters and the number of basic reproduction (R0). By merging the model of mechanistic mathematical with diverse models of statistical forecast, the researchers projected reported cases in six positions for the duration 17 May 2020, till 31 May 2020. A global sensitivity analysis is performed to correlate two epidemiological parameters about the effect of closure as well as on R0. The outcomes indicated that closure will be effective in those positions where there is a higher incidence of symptoms in the population. Moreover, a comprehensive testing of COVID-19 is required on a large scale to decrease community infection.

On March 17, 2020, all the external borders of European Union are closed which is the first time that happened in order to prevent spreading of corona virus. Thus, all authorities around the world announced enormous restrictions of travel and border surveillance in order to decrease the spreading of this universal disease. The study [8] integrated a global model of network mobility with a local model of epidemiology in order imitate and foretell COVID-19 outbreak dynamics and control across Europe. The researchers linked the mobility model to statistics of air travel for passenger and test the model of epidemiology utilizing the number of infected patients for each country. The imitations displayed that air travel networks can foretell the emerging global epidemic pattern in the early stages of an outbreak.

French government implemented full lockdown on March 16, 2020 to stop spreading the COVID-19 disease that increasing in the country. Moreover, it announced the suspension of unnecessary economic and educational activities, keeping retailers of mainly food and healthcare organizations. The researchers [9] improved the SARS-CoV-2 transmission model spatially, age-structured and organized, which able to reproduce the pre-close dynamics of the disease in each of the 13 regions of the country. Based on this model, the study assessed, at the levels of regional and national, the overall number of hospitalizations, admission of ICU, and requirements of hospital beds and, hospital fatalities that may have been prohibited by this huge and unprecedented intervention in the country.

3 Stages of COVID-19 in Jordan

The first case of Corona was recorded in Jordan, which had been arrived on March 2020 by a citizen from Italy by a man who displayed symptoms of the disease 16 days after returning from Italy to Jordan [2], he and his family members were quarantined by the Ministry of Health for laboratory testing.

That is not surprised at the arrival of the virus, as the speed of its spread is very high all-around world countries and the nature of the disease makes controlling it somewhat complicated as it is for other countries.

Elaborate arrangements put in place an automatic step by step to get out of this crisis with minimal damage and losses. Thus, the Kingdom of Jordan resorted to measures that were described as a difficult beginning, after which it became a role model for the most capable and strong countries in the world with certificates that are repeated daily through the international and Arab media, and through meetings and statements from Arab and international leaders and decision makers.

Talking about this success, inevitably, did not come from a vacuum, or by self-testimony, but by international and Arab media, discussing this aspect with details which confirm that Jordan has become a role model, and a positive example that has not been defeated by the conditions imposed on it, once again asserting that it is a country that has long endured harsh conditions, and under his wise leadership, he was able to overcome them. Not only were these media articles telling about Jordan and its successes, but huge interactions via social media were followed by admiration and appreciation for this unique case of distinction.

Jordan is now producing medical equipment and will be able within 10 days to produce one million muzzles daily, and it currently produces a quarter of a million facial masks daily, as well as producing two thousand every day, in addition to increasing the numbers of epidemiological investigation teams [2].

Moreover, the authorities imposed a comprehensive curfew for 48 h ending on Sunday morning every weekend beginning on Friday, while epidemiological survey teams conducted hundreds of random checks in various cities [2].

The great efforts that the country is working on to face coronavirus, the government is working with exceptional and distinct efforts to stop spreading this virus. Furthermore, people of society and its youth are working to support these efforts by adhering to all government decisions that aim to preserve the lives of people of this country by adhering to and following health instructions and guidelines from its sources official.

Jordan today records a global position in managing this crisis despite the limited resources of Jordan, but directives had a rapid impact in issuing a defense order and a set of decisions, such as closing the borders by land, air and sea.

In addition to providing hotels and care for all those coming from abroad for the purposes of quarantine in order to protect them and protect society, although the cost is high, the state has proven that citizen health is the most important in everything.

Jordanian citizen proved his great role through his awareness, commitment, and the size of the social solidarity that considered as a gift to help the neighbor, the chaste families, and the abundance of guidance on the communication platforms for fear for each other as one body and spirit of unity, what distinguishes this stage is that there is a real achievement for the meaning of the state of institutions and the size of the bid The state provides specialized institutions, and the proof was a successful and integrated management of this crisis through the National Center for Security and Crisis Management and its cadres with diverse and important experiences that support government decisions and direct management of the crisis in its various aspects.

Nowadays, Jordan is producing medical equipment and will be able within 10 days to produce one million face masks daily, and it currently produces a quarter of a million muzzles daily, as well as producing two thousand every day, in addition to increasing the numbers of epidemiological investigation teams.

3.1 Activating Defense Law

The government has taken many preventive measures, including issuing defense orders, and the defense order included a ban on the movement of people and their roaming in all regions of the Kingdom, and the reports issued pursuant to the penalties for violators with a fine of not less than 100 Jordanian dinars (JOD) and not more than 500 JOD if the violation was for the first time imprisonment for a period not more than a year or a fine of not less than 100 JOD and not more than 500 JOD, or both penalties in the event of repetition [3]. What is permitted is a fine of not less than 1000 JOD, and the closure of its place for a period of 14 days for each of the opening of his shop from unauthorized, imprisonment for a period not exceeding 3 months or a fine of 3000 JOD, or both [3].

In accordance with Defense Order No. 4 [3], a national fund was established, to support efforts to combat the Corona epidemic, and in view of the exceptional circumstances the kingdom is going through to confront the Corona virus epidemic. Furthermore, to face its economic and social effects on the Kingdom, and to support the efforts made by the government to confront those conditions.

The validity of all periods and dates stipulated in the applicable legislation shall be suspended, whether they are periods of limitation, lapse or lack of hearing a case or periods for taking any action of litigation in all types of courts in the Kingdom, public prosecution departments, arbitration bodies, implementation departments, wages authority and any council of conciliation, mediation and disciplinary councils and others who exercise competences similar to those of these councils, even if these periods are periods that the suspension does not apply to.

All periods of time and deadlines necessary to take any action with any ministry, government department, official public institution or public institution shall be ceased to apply in accordance with any legislation in force including the social security corporation, the lands and survey department, the companies control department, the customs department, the greater Amman municipality and the municipalities, including the prescribed periods for non-segregating agencies.

The validity of the deadlines for submitting public and private sales tax returns that must be filed during the official holiday scheduled during the period of work in the Defense Law No. (13) of 1992 is suspended [3].

The suspension does not include the deadlines and deadlines set to fulfill the financial obligations owed to state departments, public and official institutions and public institutions.

3.2 Partial and Total Curfews

The partial curfew was imposed from six o'clock in the evening to ten in the morning from the day following the days from Sunday to Thursday [2, 3]. The main goal of the curfew is to protect the lives and health of Jordanians, to make every possible effort to take the necessary measures to reduce the negative economic effects on operators, private sector companies and workers in it, and for the purposes of enabling the economy to recover after the end of the current crisis, while studying the possibility of a gradual opening and operation of economic sectors in accordance with safety and public health regulations and national priorities.

3.3 Minimum Work Organization and Remote Work Activation

All workers in private sector institutions and establishments or any other entity subject to the Labor Law are entitled to their usual wages for the period from March 18, 2020 to March 31, 2020 provided that none of the workers in the sectors excluded from the Cabinet's decision to disrupt the additional fee for Their work during that period, unless they are assigned to overtime in accordance with the provisions of Article (59) of Labor Law No. (8) for the year 1996. Specify the sectors, institutions or establishments excluded from the private sector or from any other entity subject to the Labor Law from the decision to suspend and that will be authorized to work after obtaining approval from the Minister of Industry and Trade and the Minister of Labor and Health and the competent minister collectively [2–4].

Considering the prevailing exceptional circumstances, a defense order No. (7) for the year 2020 was issued, with the aim of preventing the spread of the Corona epidemic [2–4]. More precisely, to ensure the continuation of the educational process, and enabling it to rely on non-traditional methods of education, through modern electronic means, and as of the date of the cabinet's decision, the prime minister informed that the ministries, official departments, public institutions and bodies have suspended their work, including educational institutions, and for the purpose of directing them to work, and to return to study there as usual.

According to the decisions issued by the official authorities, defense order decided the following [3, 10, 11]; education institutions that work in the kingdom in accordance with the provisions of the education act, unconventional methods and methods of education, and the various forms of evaluation of student achievement that are carried out by electronic means, or distance education as accepted methods that are approved in all governmental and private educational institutions within the Kingdom only.

Initially, Information Systems play a crucial role for organizational competitive advantage [12, 13]. Non-traditional education, or distance education, is a regular and accepted actual study of all the goals stipulated in the legislation related to educational institutions in accordance with the provisions of the education law, including the period prescribed for the current academic year 2019/2020 [10, 11]. Thus, necessary to implement the provisions of this defense order on private government educational institutions.

For the purposes of adopting non-traditional means in the form of studying and conducting evaluation, they apply to institutes, vocational training institutions and centers that are not subject to the education act. With regard to higher education institutions [11], such as universities, community colleges, and intermediate colleges that operate inside the Kingdom only in accordance with the provisions of the higher education law: unconventional methods and methods of education that are carried out by electronic means or distance education are adopted, and are considered acceptable for all purposes stipulated in the legislation related to higher education institutions including the prescribed period for the academic year 2019/2020.

The higher education council is empowered to take the decision that deems convenient regarding the extension of the duration of both second and summer semesters of the academic year 2019/2020 over the specific period for each of them in higher education institutions [11].

Electronic classes are used as an alternative to the weekly lecture or seminar for the purposes of evaluating credit hours in all subjects during the remainder of the academic year 2019/2020.

The provisions stipulated in the instructions applied in institutions of higher education related to attendance, including absence and deprivation, or those related to student warning and dismissal due to a low cumulative average in the second academic semester of the academic year 2019/2020, during the period from the date of the cabinet's decision and the prime minister's notification to disable the ministries, official departments and public institutions and bodies work for the purpose of directing them to work and returning to study there as usual [11].

Regarding to dean's councils in all universities in Jordan to modify the mechanism of calculating grades and their proportions for the second semester of the academic year 2019/2020 and submitting them to the council of higher education for approval according to the following: the student can adopt his mark for any subject or adopt successful or failure principle [11].

For the purposes of implementing the provisions of this defense order and until the return to study as usual, the student's interest shall prevail and the texts of the instructions in force in the institutions of higher general education shall be interpreted in his favor.

4   Stages of Jordan's Handling of the Corona Crisis (COVID-19)

Jordan passed three stages to handle the COVID-19 crisis. First stage was the immediate response phase. In this phase, all institutions, the crisis management center and the government are responded to this crisis and the directives from the first day. Thus, the issue of the health of the Jordanian citizen was not delayed, the immediate response phase was started, and the transition took place after weeks to the stage of adjustment and interdependence until the beginning of May which represent the second stage. Third stage was the recovery stage. These stages reviewed because they presented significant lessons and current and upcoming procedures as shown in Fig. 1.

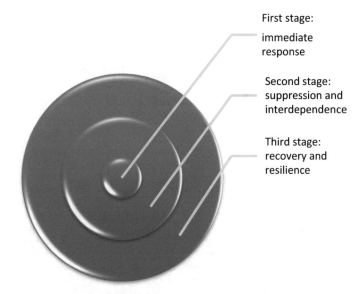

First stage: immediate response

Second stage: suppression and interdependence

Third stage: recovery and resilience

Fig. 1 Stages of Jordan's handling of the COVID-19 crisis

5 Immediate Response

The stage of the immediate response started with three axes, as Jordan started early with the complete banning and enforcement of defense orders and providing the basics for the citizens in the first days. Moreover, mission was undertaken jointly by all concerned authorities to take the necessary action, and then the mechanisms of random examination, epidemiological investigation and partial closure were developed, where quick decisions were made in these mechanisms. The development of mechanisms for quarantine and treatment, as needed a rapid response to all these requirements, which had a major impact in limiting the spread of this epidemic. Since the beginning of the spread of the COVID-19, the three hypothetical scenarios for the spread of this epidemic and the rate of infection by the Jordanian government have been elaborated and clarified and that determines the situation based on the measures taken by governments to confront this epidemic. However, Fig. 2 shows the steps taken in the immediate response stage.

The three scenarios for the spread of the epidemic and the prevalence rate were laid out according to the graph that shows the ability of the health system to absorb this epidemic and the number of expected cases according to the number of days. Figure 3a illustrates the first scenario which displayed more than 600 casualties arrive daily and will lead to a critical impairment of the absorptive capacity of Jordan's health institutions.

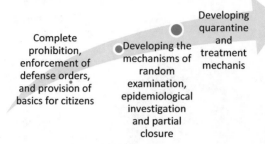

Fig. 2 The steps taken in the immediate response stage

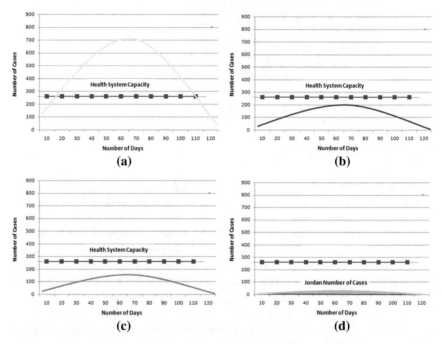

Fig. 3 Scenarios for the spread of the epidemic and the prevalence rate in world countries and Jordan

The second scenario that was studied, was with a different curve less than the first in the top of 200 infected per day as shown in Fig. 3b, which is better. However, what has been sought is a third path that is being worked out in conjunction with the awareness of the Jordanian citizen and the ability in the health sector and the various state agencies, which is the third track that arrives in the severity of the limit is 150 cases per day as illustrated in Fig. 3c, leaving an important margin in the health sector to accommodate other conditions.

Table 1 Number of deaths and number of COVID-19 cases for the 3 paths and Jordan situation

Cases	Number of deaths	Number of covid19 cases
1st path 3-a	2000	26,780
2nd path 3-b	698	9561
3rd path 3-c	310	4246
Jordan situation to 2nd of May	465	9

As for the current situation of preparing Cases in Jordan, according to the best graph of all the scenarios studied, which reflects where Jordan reached compared to these three tracks located in different countries of the world, which represents a greater achievement for Jordan and Jordanians, which was achieved as a result of the measures taken as shown in Fig. 3d, what is done and what is accomplished in hospitals and in all facilities of public life, this achievement is recorded for Jordan under his wise leadership and is registered for all Jordanian citizens who adhered to the instructions issued to manage the crisis as a delegate and the foundations of prevention and social divergence, this level of awareness is what has achieved this path, which is practically curve, is a model for other countries.

The number of possible cases and deaths for each of the previous paths if 1st path 3-a is taken, the number of cases would have been 26,780 and the number of deaths is 2000. However, for the second path 3-b, the number of cases would have been 9561 and the number of deaths is 698, while the 3rd path 3-c which was expected that the ability of the medical sector to be 4246 to prepare cases of COVID-19 and 465 to prepare deaths, but the numbers that arrived in Jordan were less than the 3rd path 3-c as the number of infections until 5th of May is 456 and the number of deaths is 9 deaths as shown in Table 1.

5.1 Adaptation and Solidarity Stage

This is the stage of the immediate response in Jordan, and in the second half of April until the beginning of May as shown in Table 2, it entered the phase of adjustment and interdependence, adaptation because this epidemic that the world is poisoned by the new natural situation requires adaptation to these new conditions on our world with appropriate procedures for solidarity as well. Several packages of measures taken by Jordan, including learning and working remotely, came wonderfully, feedback and all notes were received for development in the field of distance education, and also work remotely through more workers were able to conduct their tasks remotely and from their homes without the need to be in the work and in occupations that do not require the presence of cadres in the field and at work sites for the medical field example, the gradual conquest of the sectors geographically within the provinces and subject to public health and safety with the possibility of re-closure and the ban

Table 2 The Jordanian model for dealing with the epidemic and daily COVID-19 number of cases

Date	Number of cases	Growth rate (%)	Date	Number of cases	Growth rate (%)
2/3/2020	1	0.2	8/4/2020	5	1.0
15/3/2021	12	25.8	9/4/2020	14	3.0
16/3/2021	16	34.4	10/4/2020	0	0
17/3/2021	11	23.6	11/4/2020	9	1.9
18/3/2021	16	34.4	12/4/2020	8	1.7
19/3/2021	13	27.9	13/4/2020	2	0.4
20/3/2022	15	32.2	14/4/2020	6	1.2
21/3/2022	15	32.2	15/4/2021	4	0.8
22/3/2020	13	35.6	16/4/2021	1	0.2
23/3/2020	15	32.2	17/4/2021	5	1.0
24/3/2020	26	5.5	18/4/2021	6	1.2
25/3/2020	19	40.8	19/4/2021	4	0.8
26/3/2020	40	86.0	20/4/2022	8	1.7
27/3/2020	23	04.9	21/4/2022	3	0.6
28/3/2021	11	23.6	22/4/2020	7	15.0
29/3/2022	13	02.7	23/4/2020	2	0.4
30/3/2020	9	01.9	24/4/2020	4	0.8
31/3/2020	6	1.2	25/4/2020	3	0.6
1/4/2020	4	0.8	26/4/2020	3	0.6
2/4/2020	21	4.5	27/4/2020	2	0.4
3/4/2020	11	2.3	28/4/2021	0	0
4/4/2020	13	27.9	29/4/2022	2	0.4
5/4/2020	22	47.3	30/4/2020	2	0.4
6/4/2020	4	0.8	1/5/2020	6	1.2

in the event of renewed spread of the virus and this process is ongoing mitigation as shown in Figs. 4 and 5.

This first package of adaptation and the second package are characterized by financial and economic measures because it is a big economic impact. The first line of defense is to postpone the payments of fees, tax, customs, social security, water and electricity because current expenditures on companies and the private sector in general should have been reduced so that he can overcome this crisis. There is awareness that there are other measures, so the central bank has instituted a wide range of measures that increase facilities from banks to companies and small institutions, and towards reducing interest on them and rescheduling payments. These are the adaptation measures that we have started and acclimatization with the new situation.

Solidarity is taking measures including expanding the coverage of the beneficiaries of the National Aid Fund to include daily workers at a rate of 155,000 new families entered and the goal is to enter 200,000 families, expanding the coverage

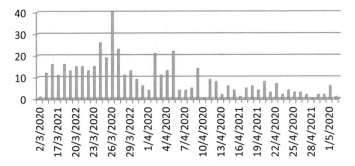

Fig. 4 The Jordanian model for dealing with the epidemic

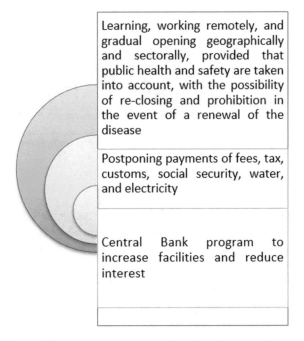

Learning, working remotely, and gradual opening geographically and sectorally, provided that public health and safety are taken into account, with the possibility of re-closing and prohibition in the event of a renewal of the disease

Postponing payments of fees, tax, customs, social security, water, and electricity

Central Bank program to increase facilities and reduce interest

Fig. 5 Adaptation procedures

of beneficiaries of social security to include the new unemployed at a rate of 11,000 new facilities, and by the efforts of everyone and the private sector, a nationwide fund has been created. The balance reached 81 million dinars in donations from Jordanian citizens. Donations were received in millions of some individuals and institutions in solidarity with the community in this crisis and the application of the highest foundations of governance by the government in managing this fund as illustrated in Fig. 6.

Fig. 6 Solidarity procedures

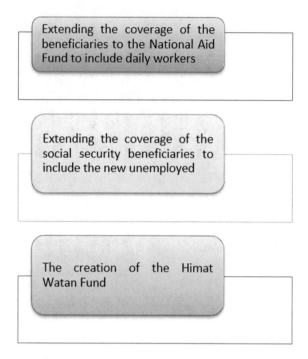

5.2 Recovery and Resilience

This is the stage of adjustment and solidarity that Jordan has started, and Jordan will continue with, and the other stage that Jordan has also begun to cooperate with the private sector is recovery and resilience. The defense line for institutions, establishments and companies that will not be able to recover has started. The measures taken to recover started by creating an investment fund that contributes with promising companies capable of Follow-up if an investment is pumped into its capital and this matter is studied in partnership between the central bank, banks and the private sector in general and with the contribution of the Jordanian government, in addition to specific measures, each sector is affected differently, so the sectors (For tourism, the largest sector affected, land and air transport, industries other than food and drug industries, trade, construction and other services) all require specific measures that need to be developed to the ground as displayed in Fig. 7.

As a continuation of the prevention measures, the process of transition to a new reality and the process of electronic transformation and delivery services, the expansion of programs Employment for young people in order to avoid high unemployment significantly.

Furthermore, prevention requires a national plan for strategic storage by building a national database for strategic stocks and increasing self-reliance in energy production and food production and the development of the drug sector and tools and medical devices. The productive sector, especially the industrial sector related to equipment,

Fig. 7 Recover procedure

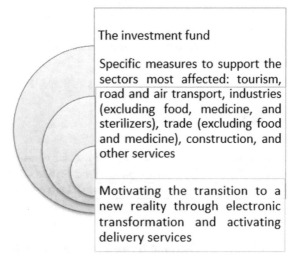

The investment fund

Specific measures to support the sectors most affected: tourism, road and air transport, industries (excluding food, medicine, and sterilizers), trade (excluding food and medicine), construction, and other services

Motivating the transition to a new reality through electronic transformation and activating delivery services

Fig. 8 Strength procedures

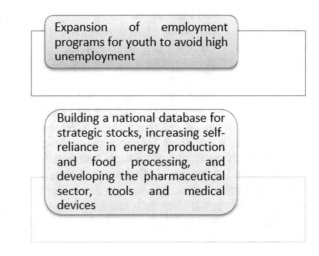

Expansion of employment programs for youth to avoid high unemployment

Building a national database for strategic stocks, increasing self-reliance in energy production and food processing, and developing the pharmaceutical sector, tools and medical devices

medicine and medical equipment, has proven its efficiency not only at the level of Jordan but also at the world level as shown in Fig. 8.

6 Methods

A confirmed case of COVID-19 infection was defined as a case with a positive result using PCR test. Suspected case was defined as a case with symptoms of COVID-19 infection but not confirmed by PCR test.

Serial interval was defined as the duration between symptom onset of the primary case and symptom onset of the secondary in a transmission chain [14]. R0 was defined as the expected number of secondary cases that one primary case will generate in a susceptible population [15].

7 Data Source

All the data were captured from the official website (Ministry of Health, Jordan, 2020 and WHO) [1, 2] that reported the situation of COVID-19 infection in Jordan. The data for the development of the model were updated to May 3, 2020.

8 Model Development and Statistical Analysis

This study developed a model for dealing with pandemic situation according to the successful procedure followed in Jordan country. The first step starts by isolating the country from the surrounding infected countries that the pandemic by banning the traveling. After that we examined the situation inside the country if there is reported cases the country will go into lockdown and the people recently returned from other countries will be in quarantine for determined days according to the pandemic. After that we followed a procedure in isolation according to the state then the city then the street and finally the building that has infected person and the people in contact with him. Until reaching a point to control all the cases and then start removing the isolation and the quarantine from street then the city then the state.

This model has proven its efficiency in Jordan country since its one of the best country that dealt with coronavirus pandemic and the number of reported cases is very low as the number of deaths is low too for months from the start of the pandemic until writing this research is July-2020. The model is shown in Fig. 9 for dealing with pandemic situations.

9 Conclusion

Jordan has succeeded until the beginning of May 2020 in the repercussions of the COVID-19 virus through collective action to prevent, limit, and treat this disease. The state of the law that is implemented on everyone according to the defense law that was issued to redress the crisis Corona, the country of production that unleashes the energies of all, and the state of interdependence that protects everyone, so this is Jordan, the model for all countries in the world in crisis management. The Kingdom of Jordan has proven to be one of the best medical system and crisis management in

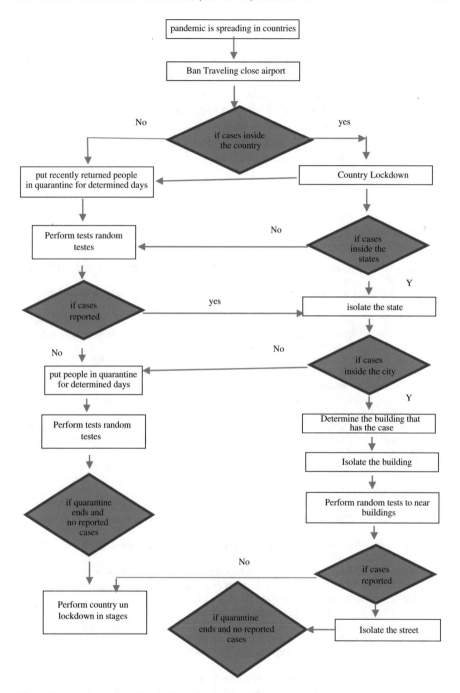

Fig. 9 Proposed model for dealing with pandemic situations

the world, and the countries of the world must go to Jordan after finishing the virus to study its crisis management mechanism.

Jordan found itself facing intertwined conditions and lifestyles in its entirety that it had not previously experienced, as was the case with the rest of the world as a result of the spread of the new Corona virus, in which many of the details that seemed in their beginning dangerous as they were emerging. Soon this small country, with limited capabilities and great actions, directed directly from the crisis management center, was able to draw up a plan to confront this virus, and indeed this global enemy, took a proactive dimension to many of the stages of the spread of this epidemic and strive to reduce it.

References

1. World Health Organization, Jordan: Last updated, June 29, 2020 http://www.emro.who.int/countries/jor/index.html
2. Ministry of Health: The Hashemite Kingdom of Jordan, last updated, July 16, 2020. https://corona.moh.gov.jo/en
3. KPMG: Jordan: government and institution measures in response to COVID-19, last updated, May 14, 2020, https://home.kpmg/xx/en/home/insights/2020/04/jordan-government-and-institution-measures-in-response-to-covid.html
4. Jordan: Free Speech Threats Under Covid-19 Response, last updated, May 5, 2020, https://www.hrw.org/news/2020/05/05/jordan-free-speech-threats-under-covid-19-response
5. Chintalapudi, N., Battineni, G., Amenta, F.: COVID-19 disease outbreak forecasting of registered and recovered cases after sixty day lockdown in Italy: a data driven model approach. J. Microbiol., Immunol. Infect. (2020)
6. Guzzetta, G., Riccardo, F., Marziano, V., Poletti, P., Trentini, F., Bella, A., … Boros, S.: The impact of a nation-wide lockdown on COVID-19 transmissibility in Italy (2020). arXiv preprint arXiv:2004.12338
7. Sardar, T., Nadim, S.S., Rana, S., Chattopadhyay, J.: Assessment of lockdown effect in some states and overall India: a predictive mathematical study on COVID-19 outbreak. Chaos, Solitons & Fractals 110078 (2020)
8. Linka, K., Peirlinck, M., Sahli Costabal, F., Kuhl, E.: Outbreak dynamics of COVID-19 in Europe and the effect of travel restrictions. Comput. Methods Biomech. Biomed. Eng. 1–8 (2020)
9. Roux, J., Massonnaud, C., Crépey, P.: COVID-19: One-month impact of the French lockdown on the epidemic burden. medRxiv (2020)
10. The Ministry of Education of Jordan: Last updated, July 17, 2020. http://www.moe.gov.jo/en
11. The Jordanian Ministry of Higher Education & Scientific Research: Last updated, July 17, 2020. http://www.mohe.gov.jo/en/Pages/default.aspx
12. Altamony, H., Alshurideh, M., Obeidat, B.: Information systems for competitive advantage: implementation of an organisational strategic management process. In: Innovation and Sustainable Competitive Advantage: From Regional Development to World Economies—Proceedings of the 18th International Business Information Management Association Conference, 1, pp. 583–592 (2012)
13. Shannak, R., Obeidat, B., Almajali, D.: Information technology investments: a literature review. In: The 14th IBIMA Conference on Global Business Transformation through Innovation and Knowledge Management: An Academic Perspective, Istanbul-Turkey, 23–24 June, pp. 1356–1368 (2010)

14. Zhang, S., Diao, M., Yu, W., Pei, L., Lin, Z., Chen, D.: Estimation of the reproductive number of novel coronavirus (COVID-19) and the probable outbreak size on the Diamond Princess cruise ship: a data-driven analysis. Int. J. Infect. Dis. **93**, 201–204 (2020)
15. Li, Q., Guan, X., Wu, P., Wang, X., Zhou, L., Tong, Y., … Xing, X.: Early transmission dynamics in Wuhan, China, of novel coronavirus–infected pneumonia. N. Engl. J. Med. (2020)

How the Business Intelligence in the New Startup Performance in UAE During COVID-19: The Mediating Role of Innovativeness

M. T. Nuseir⊙, A. Aljumah⊙, and M. T. Alshurideh⊙

Abstract The current study aims at empirically investigating the impact of the business intelligence on the new start-up performance of UAE during Covid-19. The study also examines the mediating role of innovativeness in the relationship between business intelligence and new start-up performance in UAE. The sample size included distributing the questionnaires to 250 respondents to get the required information for further analyses. 210 questionnaires out of 250 were received, so the response rate of the study was 84%. The data analysis involved the path modeling technique because of the explorative nature of the study. The results indicated that all the paths are significant at a p-value of less than 0.05. The findings of the study will be helpful for policymakers and researchers in formulating the policy concerning the business intelligence, innovation, and start-up performance in UAE.

Keywords Business intelligence · Innovation · Start-up performance in UAE

M. T. Nuseir (✉)
Department of Business Administration, College of Business, al Ain University, Abu Dhabi Campus, P.O. Box 112612, Abu Dhabi, United Arab Emirates
e-mail: mohammed.nuseir@aau.ac.ae

A. Aljumah
Management Department, Emirates College of Technology, Abu Dhabi, P.O. Box 41009, Abu Dhabi, United Arab Emirates
e-mail: aljumah37@gmail.com

M. T. Alshurideh
Department of Marketing, School of Business, The University of Jordan, Amman, Jordan
e-mail: m.alshurideh@ju.edu.jo; malshurideh@sharjah.ac.ae

Department of Management, College of Business Administration, University of Sharjah, Sharjah, United Arab Emirates

1 Background

No term has been widely accepted by scholars to refer to the external and internal intelligence, which are deemed as essential in business decision-making. The Business Intelligence (BI) is considered an umbrella term that covers processes and technologies required for handling information that may enhance the business decision-making [1]. Therefore, business intelligence is a product as well as a process, which comprises those methods that are used by organizations for intelligence and to generate useful information that may help firms thrive and survive in the marketplace. As a product, it serves as information that enables firms to anticipate the behavior of firms' suppliers, customers, competitors, services, products, business environment. and technologies with greater certainty.

The Business intelligence (BI) has gained considerable attention due to its increased accessibility to the information through electronic means. In this process, information is acquired, processed, and presented in a useful form and also forms the basis for organizations' intelligence practices. In addition, the information usage is required to be improved, particularly in the context of the great social and political change taking place around the globe, the rapid technological change, and the increase in the cut-throat competition [2]. However, the increased uncertainty among firms has given rise to the information processing activities [3, 4], which may otherwise jeopardize the firms' survival in the market. Generally, the startups strive to attain a certain position in the market, which is also essential for them to grow and survive. It is noteworthy that small firms are not exactly a smaller version of large firms since both have different structures, management practices, ran an source availability, the competition level and the environmental response in the market they compete [5]. Thus under a highly dynamic, volatile, and competitive business environment, the information must be gathered by firms for improved decision-making. Although it may seem challenging, particularly for the startup firms who are struggling to survive in the market, this information gathering process facilitates the firms' managers to effectively align with their environment and improve the performance of their organization [6, 7]. In addition, firms can use business intelligence (BI) for information gathering and simultaneously contribute to expanding the available pool of knowledge for the managers. Thus, BI can be deemed as important assets for the organizations [8].

In the current study, BI is approached for its characteristics, whereby it is viewed as a construct which covers several dimensions, such as perceived usefulness, an inter-industry analysis, an intra-industry analysis, and BI formality. On the one hand, the perceived usefulness and BI formality are concerned with the internal structure and information usage, whereas the other two aspects are the external aspects. It provides a basic understanding of how intelligence efforts facilitate the decision-making process. Innovativeness is another key concept which is related to the organizational success. This study also studied knowledge gathering through the network learning. Therefore, in order to stimulate the intangible resource acquisition, it is important to interact with actors who possess non-technical or technical knowledge.

Finally, the estimation of the existing relationship between the performance and these constructs is carried out, since all these processes provide assistance to the managers in effectively aligning with the environment and improve the performance [6, 7]. However, limited studies have reported findings related to the business intelligence in small firms. Therefore, this study aims at extending the Business Intelligence (BI) Theory by analyzing different BI aspects in small organizations. Unlike the traditional technological perspective, a managerial perspective considers business intelligence as an important approach [8].

2   COVID 19 and Businesses in the UAE

Globally, all the businesses have experienced the effects of coronavirus (COVID-19). Therefore, business leaders are trying to safely navigate the several interrelated issues, ranging from customers and employees' safety, and reorienting operations to bracing liquidity and cash and maneuvering support programs by the government.

Similar to other governments, The UAE government has also experienced the severity of the upcoming situation and proactively responded by becoming technologically advanced, highly strategic and organized towards slowly and gradually performing activities, as well as following the WHO's advice and observing the developments at the global level. It is well-known that coronavirus has affected the entire world at the end of 2019 and the beginning of 2020. Some countries adopted a denial approach, whereas some were struggling to understand the situation, and others were confused about how they should react. Prior to the announcement of WHO about coronavirus disease as a global pandemic, the UAE government proactively adopted necessary and important steps to deal with risks arising from the COVID-19. Therefore, individuals and businesses in the UAE were assured and felt safe because of the proactive steps taken by their government. Besides, UAE was the first to prepone their Easter/Spring Holidays that were scheduled at the end of March, and announced closing schools and recommended isolation from March 8, 2020, as a necessary action to avoid the spread of coronavirus (COVID-19).

Initially, the office hours and mall timings were also reduced. Subsequently, event organizers were requested to discontinue planning for any forthcoming events and leave them on hold. Besides, the educational and training institutions were ordered to remain and the order was extended to include the means of transport like [9].

After a week or two, the UAE government issued more strict social distancing regulations for all UAE residents, and a lockdown was imposed during the evening hours to perform disinfecting campaigns, followed by a strict lockdown for 24 h till 18th of April. The police introduced a permit system for the residents to purchase the essential home items, such as medicines and food, as well as those who were fighting at the front line, including healthcare professionals, food delivery services, rubbish collection, etc. However, in spite of the permission, people were supposed to wear gloves and masks when leaving their houses and to ensure the compliance of these regulations, the police established numerous checkpoints.

In addition, the government has been issuing regular updates both in English and Arabic through social media platforms. For those who were not abiding by the government's regulations and violated these rules were charged with hefty fines and faced by the legal consequences, such as expelling from the country.

Thus due to strict communications and measurement, no panic purchasing was witnessed in UAE like other countries. In other words, all the residents followed the instructions and guidance by the government in a calm and civilized manner. Some of the essential items, including face masks, were sold out by the mid of the March, which was somehow expected in that situation. However, the UAE region did not face any food shortage or lack of basic amenities. The government repeatedly assured to the UAE residents that pharmacies and supermarkets would be well-stocked, and they should stay home and be safe and should not be concerned about necessities. In a few weeks, everyone shifted from office-based work to working from home station and started to conduct training, meetings, virtual coffees, online services, as well as conferences.

The majority of firms granted paying leaves to their employees; some negotiated with their employees to accept 50% of their salary. At the same time, some chose to send employees on unpaid leave. Unfortunately, some employees were terminated due to COVID-19's impact on the financial condition of some businesses.

In addition, the UAE government has issued a stimulus package for the businesses to generously support them by reducing the cost of starting a business, i.e., 25–98%. This package was announced to facilitate businesses in production, investment, export, import, innovation, and trade, as these will surely go a long way to mitigate the effects of COVI-19 [7]. Moreover, a consolidated platform was also provided by the government, especially for the UAE-based residents who have residence visas but lost their jobs due to COVID-19. Therefore, they are provided with the opportunity to apply through Virtual Labor Marker Portal and companies are directed to publish available vacancies and they were bound to hire the UAE residence visa holders.

In the UAE, most schools are run by private parties who were directly paid by the parents themselves. However, the global pandemic has sabotaged the ability of many parents to pay for their children's school fees. In this regard, several educational groups and school owners realized this problem. They responded by waving off 20–30% of their final term school fee, while others decided to give relief to only those parents who were being affected directly by the pandemic. The utility companies in UAE also cut down water and electricity prices to relieve the impact of COVID-19 on business owners and home.

Businesses and residents in the UAE have been effectively working to fight this global pandemic. In this domain, several businesses in UAE have volunteered to support the economy, such as producing essential equipment, like face masks, financial funding, workforce support to participate in disinfection campaigns and programs, providing free fuel for the equipment that is used in combating the virus, as well as for vehicles, donation of medical equipment, involving hand sanitizers for the police and healthcare staff, and free meals for the volunteers and workers who have been working across the country as part of the national disinfection program.

Further, all private and public hospitals and hotels were used as quarantine centers and as well as for the treatment of COVID-19 patients.

3 Literature Review

3.1 Business Intelligence (BI)

Interestingly, despite the long history of this concept, a significant increase in business intelligence studies has been witnessed in recent years [3, 10]. Business intelligence incorporates different processes and elements from other fields, such as government administration, military and intelligence-driven cultures to some extent [3]. In the context of military, intelligence refers to information gathering about the battlefield environment and the enemy which they would face. In the past, governing bodies have used the military intelligence to achieve their social, political, and economic objectives. In addition, integrating different concepts in the military is not new; rather, business intelligence is one of these concepts used by the military. Generally, business intelligence is generally viewed as an umbrella term that encompasses various processes, technologies and activities to collect, analyze, store, and spread information to achieve improved business decision-making [1]. Several studies have attempted to define and explain this complicated and broad concept, but they could not reach a consensus of proposing a single accepted definition, thus leading to confusing [4]. However, all these studies have integrated the data and information analysis to transform it into useful and condensed managerial knowledge. This concept has been widely studied in the field of management under different contexts [10]. The term BI was also used by some authors as an environmental scanning process, which emphasizes mainly on how the managers scan the organizational environment; At the same time, other scholars described this term as a competitive analysis or intelligence, which focuses on the behavior, weaknesses, and strengths of their competitors [3, 11] and some others refer to it as technological intelligence driven by the technological dynamics [10, 12]. Besides, the same concept was also used with other labels, such as product intelligence, environmental intelligence, customer and marketing intelligence [12]. The business intelligence practice enables to convert the acquired data into useful knowledge, thus leading to effective decision-making to support strategic, operational and tactical decision making at all levels of an organization and ultimately improve the business performance [11, 12]. It is noteworthy that BI not only affects the process of decision-making but also has its influence on the organizational actors that are involved in the practices, i.e., how knowledge is created, shared and how it is perceived. As a consequence, it can be referred to as a cultural dimension [1, 13].

Described BI, and it is a key to perceived benefits as the improvement in business processes, better decisions, and supporting the strategic business objectives.

3.2 Innovativeness

Innovativeness shows the tendency of a firm to support and engage in experimentation, creative processes and new ideas, which result in the creation of new products, technological processes and services [14]. In the entrepreneurship context, innovativeness is the 'newness' level that firms implement in the market. However, after successfully establishing, the start-ups compete with their market competitors by providing their customers with new benefits or by bringing significant improvements in the existing benefits [15]. Based on reviewing the literature review, it is indicated that an innovative performance is a key driver to stimulate other organizational performance aspects and also encourage implementing the learning dynamics in the organization [16]. In this regard, two perspectives of the firm's innovativeness exist, whereby according to the first perspective, innovativeness can be used as a behavioral construct, which explains the firm's rate of adopting innovative ideas and products.

On the contrary, the second perspective denotes the willingness of a firm to change [17]. The innovative capacity of a firm can efficiently utilize the available resources and bring improvements in its potential value and efficiency. In addition, it also helps in developing new intangible assets within the companies. In their study, [18] identified the innovation ability as a determining factor to succeed and survive in the market.

However, more innovativeness may stimulate the value creation and enable firms to develop new abilities and capabilities to be able to address the customer's needs and achieve a better performance and profitability under rapidly changing, competitive, and increasingly complex environment [17, 19]. Numerous prior studies have also reported the innovation capability as a key firm performance determinant [3, 17]. Firms with the innovative capability of creating new technologies and products are likely to achieve a better and improved economic performance [20].

3.3 Startup Performance

Performance refers to the capability of achieving the firm's objectives. Various perspectives of organizational performance exist, such as periods (short versus long term), criteria (employees versus shareholder, profit versus market share), etc. [21]. The previous studies have proposed three different approaches to measure the performance in an organization, namely, (1) the financial performance, (2) the operational and the financial performance dimensions, and (3) the organizational effectiveness. The first approach is the financial performance, which is an outcome-based performance indicator. However, it is a limited concept to assess the business's performance. The second approach covers the operational and financial performance dimensions by integrating non-financial measures, such as introducing new products, the market share, the internal process outcomes, and marketing effectiveness. All these operational factors contribute to the firm's financial performance, whereas the third and the broadest approach to conceptualizing performance is the organizational effectiveness. The commonly used measures to determine the overall effectiveness or organizational effectiveness include the perceived overall performance, the achievement of goals, reputation, and the firm's survival.

4 Hypotheses Development

The business's success level can be understood by analyzing the firm's performance. Therefore, each organization aims at analyzing its performance arising from its internal business processes [5]. A few prior studies [1, 22] have successfully established a linkage between the business performance and the business intelligence (BI result or knowledge). However, there is still limited research concerning the performance and business intelligence relationship in new ventures. Therefore, this interrelationship is suggested to be important due to the actions and the consequent changes that are made by firms to act in accordance with the environmental changes and new opportunities [20]. In an attempt to achieve the superior performance, firms face several challenges. They are required to continually renew their capabilities and resources in response to environmental changes and also manage an existing set of competences in terms of recent success [23]. Another challenge mentioned in the entrepreneurship research is to develop an understanding of the startup performance. Based on this discussion, the following hypothesis is hypothesized.

H1 Business intelligence has a significant impact on the startup performance.

Innovativeness refers to what extent a firm embraces and engages in experimentation, creativity, and new ideas resulting in the creation of new processes, products, and services, and is considered a cultural aspect of a firm, which helps it succeed and survive under an uncertain business environment [14, 17]. In their study, [3] view innovation as an activity that can be controlled by the management, whereas [23] view it as a process of engaging in creative processes and experimentation, which may lead to technological processes, new products, and services. Firms' actions can also be driven and contingent on external factors, such as the government's legislation, customer's demand, and competitors' actions [3]. Thus [17] mentioned that the relationship between the organizational learning and innovation is of significant importance, as well as the existing prior knowledge helps develop an understanding of the market conditions, the creation of new products and ideas, and new technology [19]. They further argued that a higher innovation capacity improves the capability of a firm to quickly and effectively respond against the environmental challenges. In another study [24] assert that businesses that possess the ability to exploit and transform knowledge explain their innovation level, such as the ability of businesses to introduce new products and problem-solving methods to respond to the market demand in a minimum possible time. The literature explains the nature of the relationship between the innovativeness and BI. Hence, business intelligence concerns the use of information for better decision-making is likely to affect a firm's innovative actions. In addition, a better access to information does not necessarily result in increased efficiency and business performance. Rather, how this information is used by the organizations is of prime importance [4]. Thus, it can be hypothesized that the information acquisition and using that information in a better way that can positively affect the innovativeness. Based on this discussion, the following hypothesis is hypothesized.

H2   A positive relationship exists among innovativeness in startups and characteristics of business intelligence (BI).

Therefore, innovativeness is assumed to facilitate in pursuing new opportunities by providing a range of new services or products in the market. The successful achievement of these activities will likely enhance and affect the firm's performance [3]. In addition, firms with a greater innovative capability would be capable to effectively respond and satisfy the customer's needs and develop those desirable capabilities which enable them gain a superior profitability and a better performance. During the past few years, the effects of the innovation aspects on the performance have been gaining more attention among scholars [24]. Therefore, the relationship between the two can be anticipated.

H3   A positive relationship exists between startup performance and innovativeness.
H4   Innovativeness mediates the relationship between business intelligence and start-up performance.

5   Method and Measurements

In this study, based on a survey, we adopted the cross-sectional data and developed a questionnaire to gather the required information in order to meet the objective of the present study [25]. By using a seven-point Likert scale, the questionnaire items were observed. The sample size of the present study was 250 distributed to the targeted respondents to get the required information for further analyses. Only 210 questionnaires out of 250 were received, so the response rate of this study was 84%.

For data analyses, the path modeling technique was required because of the explorative nature of the study. According to the available literature, if the research objective is the extension of an existing theory or validation for the prediction of existing relations among different variables, formally the PLS path modeling is the ideal technique.

Many steps were involved in the data analysis process after the collection of the required data through the survey using the Statistical Package for Social Sciences (SPSS) for screening and encoding the data. With the help of this process, it was ensured that for the PLS analysis, the gathered data is appropriate. In the next step, we have estimated the convergent and discriminant validities, the individual item consistencies and the internal consistency reliabilities in the measurement model (MM). Based on following the suggestions of [26], we used the smart PLS, whereas in the third step for the evaluation of the structural model (SM), we performed a method of bootstrapping for 315 cases and 5000 resamples [27]. Thus, we specifically determined the significance of path coefficients, level of R-square, effect size and predictive relevance for the assessment of SM.

5.1 Measures and Variables

To establish the content validity, the present study extracted some measures from prior studies that were related to the same topic. For instance, the business intelligence measures were derived from [7] study which includes 16 measuring items and these characteristics were grouped into four main dimensions, namely (I) formality, (ii) perceived usefulness, (iii) intra-industry comprehensiveness, and (iv) inter-industry analysis. The perceived rating for each item was obtained from the respondents based on the BI practices and startup experience. A Seven-point Likert scale was adopted in this study, ranging from 1 to 7, where 1 denotes strongly disagree and 7 denotes strongly agree. The Cronbach alpha coefficient was used to assess the internal consistency of the scale, whereby 0.88 value was obtained for the coefficient.

A 10-item scale was used to measure innovativeness based on [28]. The rationale for choosing this scale is that it considers all innovative activities rather than only the inputs that are involved in innovative processes. The Cronbach alpha value obtained for this scale was 0.83.

Besides the financial indicators, other indicators were also integrated into this study because some authors reported only limited applicability in the context to startup. In view of [29], if we closely observe the developing phase of these firms, it can be noticed that sustained improvements in the competitive performance of these firms cannot be reflected merely by financial figures, and cannot easily obtain and interpret subject to new ventures. In addition, it is a multidimensional scale that explains a firm's achievement in comparison to its competitors. This comparison approach has been used in other studies [20]. In another question, firms were asked to rate their performance in relation to the above mentioned six items on a scale from 1 to 7, where 1 denotes 'much worse than competitors' and 7 denotes 'much better than competitors.' The Cronbach alpha value obtained for this scale was 0.76. The normality test was conducted for the variables. Based on reviewing the literature, the reference values for skewness (SK) and kurtosis (Ku) must be satisfied to achieve normality. In this regard, $|sk| < 7$ and $|ku| < 2$ were used as the required conditions for normality, whereby all the variables followed these conditions. The variance inflation factor (VIF) test was also performed, and all values were found to be less than 5 with tolerance values above 0.2, which are in line with the required condition. Thus it can be concluded that no collinearity problem was found in this study.

6 Results

The analysis starts in PLS-SEM by the estimation of the outer or the measurement model (MM). In this process, we assessed the reliabilities and validities, which include convergent and discriminant validity following the estimation of the inner model to analyze the path relationships [30–32]. For the assessment of the model's validity, the recent arguments regarding the PLS path modeling inappropriateness were also keeping in view, so we followed the suggestions of [27] and implemented a two-step process for reporting the outcomes of PLS-SEM.

The different criteria were included in the assessment of MM which is important to observe the outer model, in which the content validity, the convergent and discriminant validity, and the internal consistency reliability (ICR) were included though it is recommended to divide these indicators into two or more subgroups to catch the possible effect [33] (Fig. 1; Table 1).

In their study, [34] suggested that for the measurement of the individual item reliability, we must explore the outer loadings for each indicator. In contrast, according to [35] the degree at which items of specific scale measures, the same concept is known as ICR. In this regard, Cronbach alpha coefficient and composite reliability (CR) are commonly used estimators for the assessment of ICR [31, 32]. The estimates obtained from the CR coefficient are less biased compared to the estimates of Cronbach alpha. However, the Cronbach alpha does not reflect the individual item contribution and assumes all items influence equally towards their respective construct [36]. The other reason for not selecting the Cronbach alpha is that it does not estimate the scale reliability efficiently.

It is also recommended that the coefficient of CR considers the various item loadings [32]. Therefore, the reliability coefficient can be easily selected. Different ranges of ICR value if it is greater or equal to 0.70 and is satisfactory and can be acceptable. In contrast, there is no reliability if the CR value is less than 0.60 (Table 2).

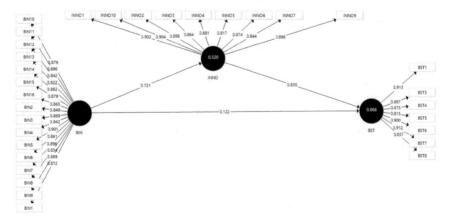

Fig. 1 Measurement model

Table 1 Outer loadings

	BIN	BST	INNO
BIN10	**0.879**	0.635	0.598
BIN11	**0.890**	0.662	0.642
BIN12	**0.842**	0.627	0.638
BIN13	**0.822**	0.630	0.585
BIN14	**0.862**	0.656	0.668
BIN15	**0.879**	0.683	0.686
BIN16	**0.865**	0.629	0.599
BIN2	**0.849**	0.592	0.632
BIN1	**0.872**	0.608	0.608
BIN3	**0.869**	0.594	0.595
BIN4	**0.842**	0.593	0.611
BIN5	**0.901**	0.688	0.679
BIN6	**0.861**	0.604	0.609
BIN7	**0.896**	0.635	0.632
BIN8	**0.834**	0.563	0.543
BIN9	**0.889**	0.650	0.641
BST1	0.694	**0.913**	0.884
BST3	0.631	**0.897**	0.804
BST4	0.639	**0.875**	0.774
BST5	0.573	**0.815**	0.754
BST6	0.636	**0.900**	0.850
BST7	0.668	**0.912**	0.822
BST8	0.623	**0.837**	0.804
INNO1	0.630	0.818	**0.902**
INNO10	0.674	0.844	**0.904**
INNO2	0.636	0.801	**0.898**
INNO3	0.622	0.818	**0.864**
INNO4	0.648	0.796	**0.881**
INNO5	0.579	0.751	**0.817**
INNO6	0.613	0.794	**0.874**
INNO7	0.580	0.792	**0.844**
INNO9	0.693	0.885	**0.896**

Table 2 Reliability

	Cronbach's Alpha	rho_A	Composite Reliability	Average Variance Extracted (AVE)
BIN	0.978	0.978	0.980	0.750
BST	0.951	0.952	0.960	0.773
INNO	0.962	0.963	0.967	0.768

The convergent validity or CV is the extent to which the latent constructs are represented by their specific items and correlate with the other items of a similar construct.

According to [37] view, we can measure the CV via the average variance extracted (AVE). In this way, for all latent contracts of a model, we can compute the AVE. For the enough level of CV for all constructs, the value of AVE should be greater than 0.50, as recommended by [32], While assessing the outer model, the most important criteria are DV. The discriminant validity is the level at which the model particular construct is different from all other latent constructs [34, 35, 37]. Thus for the determination of DV, we compared the correlations among the latent constructs and the square roots of AVE by following the recommendations of [37] (Table 3).

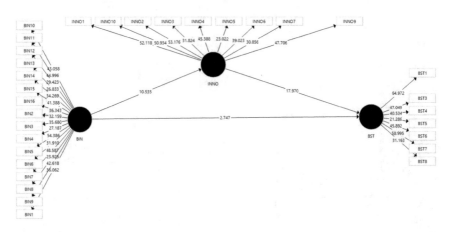

Fig. 2 Structural model

Table 3 Validity

	BIN	BST	INNO
BIN	0.866		
BST	0.727	0.879	
INNO	0.721	0.727	0.876

The results of the structural model are shown in Fig. 2 and Tables 4 and 5. The results indicate that all the paths, namely **BIN → BST, BIN → INNO, and INNO → BST** are significant at p-value less than 0.05.

The results of the mediation analysis are shown in Table 5 which shows that the mediation path, namely **BIN → INNO → BST** is significant at p-value less than 0.05.

For the assessment of SM, the most important measure is the coefficient of determination, which is also known as the value of R-square. In an endogenous variable, the proportional variance is explained by R-square because of the exogenous variables of the model [35] (Fig. 3).

On the endogenous latent constructs, the effects of latent variables were measured by the effect size via R-square. According to the study of [37], if the values of size

Table 4 Direct relationships

	(O)	(M)	(STDEV)	(|O/STDEV|)	P values
BIN -> BST	0.727	0.727	0.064	11.289	0.000
BIN -> INNO	0.721	0.722	0.068	10.535	0.000
INNO -> BST	0.839	0.832	0.047	17.970	0.000

Table 5 Mediation

	(O)	(M)	(STDEV)	(|O/STDEV|)	P values
BIN → INNO → BST	0.605	0.599	0.045	13.519	**0.000**

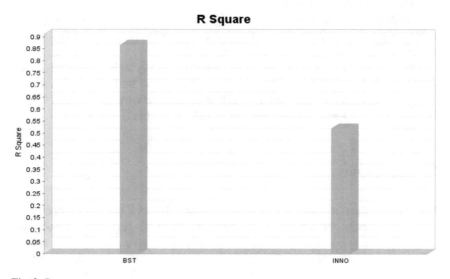

Fig. 3 R-square

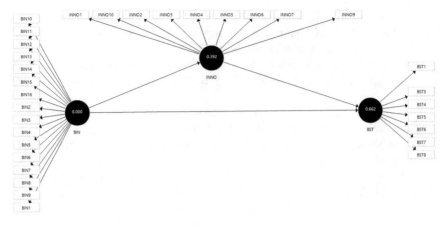

Fig. 4 Q-square

Table 6 Q-square

	SSO	SSE	Q² (=1-SSE/SSO)
BIN	3472.000	3472.000	
BST	1519.000	513.961	0.662
INNO	1953.000	1187.991	0.392

are 0.020, 0.150 and 0.350, it explains the little, moderate and significant effect of the independent variables of SM.

In addition, the quantity predictive weight of the manifest variable is known as indicators predictive relevance in a model. Consequently, if the value of Q-square is greater, the model will be more significant (Fig. 4).

The Q^2 values can be compared to make the mandatory variations in the model. If the value of $Q^2 > 0$ that is a threshold level indicates that the model has predictive relevance and recreates well through a blindfolding procedure, the effectiveness of the predictive relevance can be measured by calculating the Q^2 (Table 6).

7 Conclusions

The results of the study are in line with the Resource-Based View (RBV). According to the RBV Theory, organizations must utilize their human, organizational, and physical assets (intangible & tangible) to develop and sustain a competitive advantage. This theory postulates that firms that possess rare and valuable resources are capable of developing a competitive advantage [20], particularly if their resources cannot be easily substituted or imitated. Correspondingly, in terms of knowledge-based view

(KBV), knowledge is viewed as a company's most valuable resource [38]. The RBV theory provides a theoretical foundation for the Knowledge-based view (KBV) by asserting that a firm's primary factor of production is its knowledge, which helps firms in gaining a competitive advantage. This study was based on the sample size of 250 respondents, whereby only 210 questionnaires out of 250 were received, so the response rate of this study was 84%. For the data analysis, the path modeling technique was required because of the explorative nature of the study. The results indicate that all the paths are significant at a p-value of less than 0.05.

In this study, BI is approached for its characteristics; it is viewed as a construct which covers several dimensions, such as perceived usefulness, an inter-industry analysis, an intra-industry analysis, and BI formality, where the perceived usefulness and BI formality are concerned with the internal structure and information usage, whereas the other two aspects are the external aspects [39–44]. It provides a basic understanding of how intelligence efforts facilitate the decision-making process. Thus, innovativeness is another key concept, which is related to the organizational success. In addition, this study studied knowledge gathering through the network learning. Therefore, in order to stimulate the intangible resource acquisition, it is important to interact with actors who possess non-technical or technical knowledge. Finally, the estimation of the existing relationship between the performance and these constructs is carried out, since all these processes provide assistance to the managers in effectively aligning with the environment and improve the performance [6, 7, 45]. Since limited studies are reported concerning the business intelligence in small firms, this study aims at extending the Business Intelligence (BI) Theory by analyzing different BI aspects in small organizations. Unlike the traditional technological perspective, a managerial perspective considers business intelligence as an important approach [46–48]. More specifically, the cancellation of conferences, events, training, closed air space, and restricted trading naturally caused a substantial impact on the businesses, particularly the small and medium enterprises (SMEs). The Ministry of Economy in the UAE stated that 98% of the total companies in UAE belongs to the SME sector and accounts for 52% of non-oil GDP.

References

1. Wanda, P., Stian, S.: The secret of my success: an exploratory study of business intelligence management in the Norwegian industry. Procedia Comput. Sci. **64**, 240–247 (2015)
2. Geels, F.W.: Low-carbon transition via system reconfiguration? A socio-technical whole system analysis of passenger mobility in Great Britain (1990–2016). Energy Res. Soc. Sci. **46**, 86–102 (2018)
3. Caseiro, N., Coelho, A.: Business intelligence and competitiveness: the mediating role of entrepreneurial orientation. Compet. Rev.: An Int. Bus. J. (2018)
4. Ariyarathna, K., Peter, S.: Business analytics maturity models: a systematic review of literature. Focus 3(10), 4 (2019)
5. Luppi, E., Bolzani, D., Terzieva, L.: Assessment of transversal competences in entrepreneurial education: a literature review and a pilot study. Form@ re-Open Journal per la formazione in rete **19**(2), 251–268 (2019)

6. Knight, G., Czinkota, M., Khan, Z.: International marketing and the migrant-owned enterprise: research propositions: an abstract. Paper presented at the academy of marketing science annual conference (2018)
7. Mishiyi, H., Cumberlege, R., Buys, F.: Challenges facing newly established quantity-surveying firms to secure tenders in South Africa. Acta Structilia **26**(1), 42–60 (2019)
8. Martinez, M.G., Zouaghi, F., Marco, T.G.: What drives business failure? Exploring the role of internal and external knowledge capabilities during the global financial crisis. J. Bus. Res. **98**, 441–449 (2019)
9. Koch, J.: Aiming for Sustainability in Educational Travel-An Empirical Analysis (2020)
10. Santra, K.: Using Structural Equation Modeling To Assess The Budget Hotel Performance In Bali Indonesia. Department Of Management. Bali State Polytechnic, 8, 933–951 (2017)
11. Gudfinnsson, K., Strand, M., Berndtsson, M.: Analyzing business intelligence maturity. J. Decis. Syst. **24**(1), 37–54 (2015)
12. Wyskwarski, M.: Business intelligence-publication analysis using the R language. Zeszyty Naukowe. Organizacja i Zarządzanie/Politechnika Śląska(137 Modernity of Industry and Sciences), pp. 263–274 (2019)
13. Shollo, A., Galliers, R.D.: Towards an understanding of the role of business intelligence systems in organisational knowing. Inf. Syst. J. **26**(4), 339–367 (2016)
14. Steel, P., Saunders, C., Dewald, J.R.: Synergistic impacts of entrepreneurial and learning orientations on performance: a meta-analysis. Paper presented at the academy of management proceedings (2019)
15. Dias, C.S., Ferreira, J.J.: What we (do not) know about research in the strategic management of technological innovation? Innovation **21**(3), 398–420 (2019)
16. Guttentag, D.: Regulating innovation in the collaborative economy: an examination of Airbnb's early legal issues. In: Collaborative Economy and Tourism, pp. 97–128. Springer (2017)
17. Jun, J., Lee, T., & Park, C.: The mediating role of innovativeness and the moderating effects of strategic choice on SME performance. J. Small Bus. Manag. 1–21 (2020)
18. Akgün, A.E., Keskin, H., Kırçovalı, S.Y.: Organizational wisdom practices and firm product innovation. RMS **13**(1), 57–91 (2019)
19. Mennens, K., Van, A., Odekerken, G.: Exploring antecedents of service innovation performance in manufacturing SMEs. Int. Small Bus. J. **36**(5), 500–520 (2018)
20. Hernández, R., Kellermanns, F.W., Fernández, M.C.: A note on the relationships between learning, market, and entrepreneurial orientations in family and nonfamily firms. J. Fam. Bus. Strategy **9**(3), 192–204 (2018)
21. Gerschewski, S., Xiao, S.S.: Beyond financial indicators: an assessment of the measurement of performance for international new ventures. Int. Bus. Rev. **24**(4), 615–629 (2015)
22. Coetzee, B., Mearns, M.: Establishing competitive intelligence process elements in sport performance analysis and coaching: a comparative systematic literature review. Int. J. Inf. Manage. **52**, 102071 (2020)
23. Caseiro, N., Coelho, A.: The influence of business intelligence capacity, network learning and innovativeness on startups performance. J. Innov. Knowl. **4**(3), 139–145 (2019)
24. Liu, W., Atuahene, K.: Enhancing product innovation performance in a dysfunctional competitive environment: The roles of competitive strategies and market-based assets. Ind. Mark. Manage. **73**, 7–20 (2018)
25. Zeffane, R., Melhem, S.B., Baguant, P.: The impact of job satisfaction, trust, gender and supervisor support on perceived organisational performance: an exploratory study in the UAE service sector. Int. J. Bus. Excell. **14**(3), 339–359 (2018)
26. Henseler, J., Ringle, C.M., Sarstedt, M.: A new criterion for assessing discriminant validity in variance-based structural equation modeling. J. Acad. Mark. Sci. **43**(1), 115–135 (2015)
27. Henseler, J., Hubona, G., Ray, P.A.: Using PLS path modeling in new technology research: updated guidelines. Ind. Manage. Data Syst. (2016)
28. Boso, N., Adeleye, I., Donbesuur, F.: Do entrepreneurs always benefit from business failure experience? J. Bus. Res. **98**, 370–379 (2019)

29. Schüler, J.: What do we really know about entrepreneurial orientation? Using quantitative empirical reviews to advance its concept and nomological network (2020)
30. Hafeez, M.H., Basheer, M.F., Rafique, M., Siddiqui, S.H.: Exploring the links between TQM practices, business innovativeness and firm performance: an emerging market perspective. Pak. J. Soc. Sci. (PJSS), **38**(2) (2018)
31. Hair, Hult, G.T.M., Ringle, C.: A Primer on Partial Least Squares Structural Equation Modeling (PLS-SEM). Sage Publications (2016)
32. Hair, Matthews, L.M., Matthews, R.L., Sarstedt, M.: PLS-SEM or CB-SEM: updated guidelines on which method to use. Int. J. Multivar. Data Anal. **1**(2), 107–123 (2017)
33. Hameed, W.U., Basheer, M.F., Iqbal, J., Anwar, A., Ahmad, H.K.: Determinants of Firm's open innovation performance and the role of R & D department: an empirical evidence from Malaysian SME's. J. Glob. Entrep. Res. **8**(1), 29 (2018)
34. Ong, M.H.A., Puteh, F.: Quantitative data analysis: choosing between SPSS, PLS, and AMOS in social science research. Int. Interdiscip. J. Sci. Res. **3**(1), 14–25 (2017)
35. Akter, S., Fosso Wamba, S., Dewan, S.: Why PLS-SEM is suitable for complex modeling? An empirical illustration in big data analytics quality. Prod. Plan. Control **28**(11–12), 1011–1021 (2017)
36. Basheer, M., Siam, M., Awn, A., Hassan, S.: Exploring the role of TQM and supply chain practices for firm supply performance in the presence of information technology capabilities and supply chain technology adoption: a case of textile firms in Pakistan. Uncertain Supply Chain Manage. **7**(2), 275–288 (2019)
37. Naala, M., Nordin, N., Omar, W.: Innovation capability and firm performance relationship: a study of pls-structural equation modeling (Pls-Sem). Int. J. Organ. Bus. Excell. **2**(1), 39–50 (2017)
38. Cerchione, R., Esposito, E.: Using knowledge management systems: a taxonomy of SME strategies. Int. J. Inf. Manage. **37**(1), 1551–1562 (2017)
39. Yousuf, H., Zainal, A.Y., Alshurideh, M., Salloum, S.A.: Artificial intelligence models in power system analysis. In: Artificial Intelligence for Sustainable Development: Theory, Practice and Future Applications, pp. 231–242. Springer, Cham.
40. AlShamsi, M., Salloum, S.A., Alshurideh, M., Abdallah, S. Artificial intelligence and blockchain for transparency in governance. In: Artificial Intelligence for Sustainable Development: Theory, Practice and Future Applications, pp. 219–230. Springer, Cham
41. Alhashmi, S.F., Alshurideh, M., Al Kurdi, B., Salloum, S.A.: A systematic review of the factors affecting the artificial intelligence implementation in the health care sector. In: Joint European-US Workshop on Applications of Invariance in Computer Vision, pp. 37–49. Springer, Cham (2020)
42. Salloum, S.A., Alshurideh, M., Elnagar, A., Shaalan, K.: Mining in educational data: review and future directions. In: Joint European-US Workshop on Applications of Invariance in Computer Vision, pp. 92–102. Springer, Cham (2020)
43. Salloum, S.A., Alshurideh, M., Elnagar, A., Shaalan, K.: Machine learning and deep learning techniques for cybersecurity: a review. In: Joint European-US Workshop on Applications of Invariance in Computer Vision, pp. 50–57. Springer, Cham (2020)
44. AlMehrzi, A., et al.: Investigation of the key internal factors influencing knowledge management, employment, and organisational performance: a qualitative study of the UAE hospitality sector. Int. J. Innov., Creativity Change **14**(1), 1369–1394 (2020)
45. Alshurideh, M., Kurdi, B.A., Shaltoni, A.M., Ghuff, S.S.: Determinants of pro-environmental behaviour in the context of emerging economies. Int. J. Sustain. Soc. **11**(4), 257–277 (2019)
46. Ghannajeh, A.M., AlShurideh, M., Zu'bi, M.F., Abuhamad, A., Rumman, G.A., Suifan, T., Akhorshaideh, A.H.O.: A qualitative analysis of product innovation in Jordan's pharmaceutical sector. Eur. Sci. J. **11**(4), 474–503 (2015)
47. Mehmood, T., Alzoubi, H.M., Alshurideh, M., Al-Gasaymeh, A., Ahmed, G.: Schumpeterian entrepreneurship theory: evolution and relevance. Acad. Entrep. J. **25**(4), 1–10 (2019)
48. Al-Jarrah, I., Al-Zu'bi, M.F., Jaara, O., Alshurideh, M.: Evaluating the impact of financial development on economic growth in Jordan. Int. Res. J. Financ. Econ. **94**, 123–139 (2012)

COVID-19 Outbreak in Jordan: A 150 Days of Successful Response and Re-Open Strategy

Motasem N. Saidan, Radwan A. Al-Weshah, Ra'ed Masa'deh⬡, Hakam Saidan, and Khalid A. Kheirallah

Abstract This study aims to provide analyses and assessment of Jordan's responses and actions in the first 150 days since March 2, 2020. The present study focuses on the emerging cases COVID-19 in Jordan, investigates the socio-economic impacts of COVID-19, and assesses the concise responses and initiatives to lessen such impacts. Analytical method was adopted in the present study using quantitative and qualitative data captured from official sources and news reports. The findings show that Jordan was so far capable to flatten the curve, and effectively controlled the outbreaks within four to six weeks under strict isolation and curfew measures. Jordan was ranked 132nd in the world based on 1136 positive cases. The sources of external transmissions were 10.2% by trucks drivers crossing the borders, and 18.4% by returning expatriates. Several features enabled Jordan to flatten the curve such as closed contacts tracing and testing capabilities, integration their COVID-19 responses and national initiatives with digital technology and Apps utilization, and providing of transparent and up-to-date information and updates through several media about this unprecedented situation. Jordan is attempting to lessen the socio-economic impacts of prolonged

M. N. Saidan (✉)
Chemical Engineering Department, School of Engineering, The University of Jordan, Amman 11942, Jordan
e-mail: m.saidan@gmail.com; m.saidan@ju.edu.jo

R. A. Al-Weshah
Civil Engineering Department, School of Engineering, The University of Jordan, Amman, Jordan
e-mail: weshah11@yahoo.com

R. Masa'deh
School of Business, The University of Jordan, Amman, Jordan
e-mail: r.masadeh@ju.edu.jo

H. Saidan
Jordan Food and Drug Administration, Amman 11181, Jordan
e-mail: hakam.rph@gmail.com

K. A. Kheirallah
Department of Public Health, Medical School, Jordan University of Science and Technology, Ar-Ramtha 21110, Jordan
e-mail: kkheiral@gmail.com

© The Author(s), under exclusive license to Springer Nature Switzerland AG 2021   81
M. T. Alshurideh et al. (eds.), *The Effect of Coronavirus Disease (COVID-19) on Business Intelligence*, Studies in Systems, Decision and Control 334,
https://doi.org/10.1007/978-3-030-67151-8_5

closure of economic sectors on vulnerable population by establishing a fund that tackles the most influenced entities that are informal labor and SMEs. Jordan was exemplified as a successful preparedness and proactive intervention case to combat the unprecedented infectious COVID-19 outbreaks. After 30 days since the lockdown was relaxed, less than 10 cases per day were recorded. Accordingly, Jordan pandemic status is in a moderate risk stage based on the issued re-open strategy and adopted risk assessment matrix.

Keywords Coronavirus · Lockdown · Pandemic · Socio-economic · Digital technology · Jordan

1 Introduction

COVID-19 outbreak was initiated in December 2019, with pneumonia cases of unknown cause, as reported by the Chinese government to the WHO, and then identified to be caused by a novel coronavirus (nCoV) and confirmed by WHO on 12th January 2020 [1, 2]. After the first positive COVID-19 case was declared in Jordan in early of March 2020, Jordan realized that stringent control measures should be implemented as early as possible to reduce the number of daily cases and slow down reaching the peak. This also because Jordan was ranked 128th country out of 178 as it had 1.4 hospital beds per 1000 inhabitants [3]. Jordan has therefore decided to adopt a strategy that flattens the curve in combating the spread of COVID-19. So far, Jordan has controlled several COVID-19 outbreak events, and has been able to stand fast in battling the COVID-19 to date and succeeded in flattening and damping the curve of nCoV spread. This paper presents an analysis and reporting of the rapid actions and responses to COVID-19 emergencies taken by Jordanian government during the first 150 days. IT highlights also the socio-economic impacts of lockdown measures, in addition to the re-open strategy adopted by the government.

2 Methods

2.1 Study Area

Jordan population is 10,203,134 capita in 2020, with 114.93 capita/km^2, while 96.25% of the Jordanians ages are less than 65 years according to [4]. Jordan is home to around 745,192 Refugees while 83.5% of the refugees are living in urban areas and camps [5]. The geographical location of Jordan in the Middle East poised Jordan to impose stringent measures to combat the spread of COVID-19 entirely, with the geometric increase in the daily COVID-19 cases and escalating numbers of confirmed cases in all of Jordan's neighboring countries [6–11].

2.2   Data Collection

The prevalence data of COVID-19 in Jordan was taken from the Ministry of Health in Jordan website that reports the latest information of CVOID-19 in Jordan, and MS Excel was used to build a time-series statistics of the COVID-19 data in Jordan between 2/03/2020–2/07/2020. Moreover, other quantitative and qualitative for data were captured from news and public health reports, official website (WHO), and governmental reports in Jordan.

3   Results

3.1   COVID-19 Outbreak: Statistical Analysis

Jordan has declared its first case on 2nd of March 2020. This was an imported case from Italy. Despite of the fact that no other positive cases were reported until 14th of March 2020, however, on 10th of March, Jordanian government suspended academic institutions (nurseries, kindergartens, schools, colleges, universities, etc.), banned mass gatherings (conferences, workshops, training events, etc.) [12]. Restricting mass gathering is crucial to prevent the spread of COVID-19, since it is transmitted through direct contact [13].

The government also took emergency measures to prevent imported COVID-19 cases from entering Jordan by issuing a decision of closing the airports and borders mainly for travelers and arrivals coming from the surrounding countries (i.e. Lebanon and Syria) and other countries (i.e. France, Germany, and Spain, etc.) that were overwhelmingly hit by COVID-19 [6]. Furthermore, Jordan reduced airline service to Egypt by half, and allowed the overland commercial traffic with Iraq and Syria [12].

Two major COVID-19 outbreaks occurred in Jordan: the first one was caused by a wedding ceremony on 13th of March in northern Jordan, and spread by the bride's father (58-year-old man), who arrived from Spain couple of days prior to the wedding event. The ceremony was attended by 450 persons. On 13th April 2020, 85 persons tested positive (with 76 persons attended the wedding event). Furthermore, 47.4% of those infected were asymptomatic carriers [14]. This large outbreak was controlled and the spread was dampened within four weeks under strict isolation and curfew measures in Irbid City. It is noteworthy, that no new COVID-19 cases connected with wedding events were reported in Jordan to date. [15] showed that the reproduction number (R0) of this wedding party is 5 using mathematical model.

The second major outbreak was on 9th of May till 8th of June. 2020, and caused in Mafraq Governorate by a truck driver, who tested negative on border when he returned to Jordan, but he did not follow self-quarantine signed pledge. Few days later, he started showing symptoms and tested positive. However, he held an Iftar meal event in Ramdan month, invited tens of family members, and accordingly infected

Table 1 Classifications of new cases in Jordan during the first 150 days since early of March 2020 [18]

	Classification	%
A	Imported, i.e., returning expatriates	18.40
B	Imported, i.e., trucks drivers	10.20
C	Part of a known cluster, or contact with a known case	71.22
D	Source unknown	0.18

35 persons. He also moved through several large cities in Jordan and was in contact with more than 140 persons. Ultimately, the contacts tracing activities identified 140 infected persons. This outbreak was a challenging spread cluster due to its spanning into four large cities (the highest populated cities in Jordan), and formed infection of third generation as declared by the Minister of Health [16].

At the time of writing this paper, Jordan was ranked 132nd in the world based on the number of positive cases (1136 positive cases). The active cases are 103 (excluding those more than 100 non-Jordanian trucks drivers who tested positive at the borders and decided to get treated at their origin countries), while the recovered cases are 889, and 10 death cases, as reported by the Ministry of Health (MoH). It is noteworthy to mention that no death case was reported from 2nd of May to 2nd of July, 2020 in Jordan. The death case number 10 was declared on 3rd of July 2020 [17]. The highest number of daily new cases to date (in the first 150 days) were 40 and 38 cases that were recorded on 26th March, and 13th June, 2020, respectively.

Moreover, Jordan adopted and modified the WHO's classification of new cases during the first 150 days since early of March 2020 [18]. Accordingly, among the 1136 reported cases, 10.2% cases were for trucks drivers crossing the borders (Jordanians and non-Jordanians), 18.4% for travelers and arrivals coming back to Jordan, and were quarantined in designated hotels and caravans-complex (see Table 1).

It is noteworthy to mention that the positive-tested trucks drivers forms 2–4% of the daily tested drivers mainly for those who are mainly coming from Saudi Arabia.

Figure 1 shows the total infected, active, recovered, and death cases in Jordan, from March 3 to July 2, 2020, which involved the largest two aforementioned clusters for the COVID-19 outbreak in Jordan, while the third peak cluster started from June 10 to July 2, 2020 (shown in Fig. 1 in red line for the active cases) is a cumulatively added positive-tested arrivals from the airport and borders. Accordingly, the government has declared stopping all air flights for Jordanian Arrivals until 10th of July, 2020 due to the drastically emerging of arrivals' positive-tested cases lately.

3.2 Lockdown Measures and Curfew

As a response to the first large COVID-19 outbreak, the government officially declared a state of emergency, and enacted the National Defense Law, from 17 March

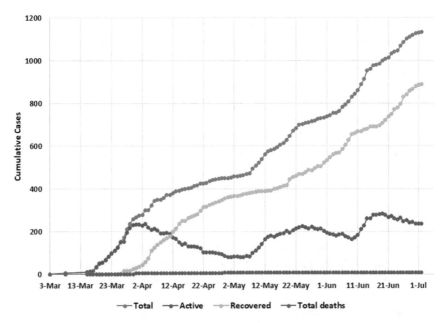

Fig. 1 Total cases, active and recovered cases, and deaths reported daily from March 3 to July 2, 2020 [17]

till April 14. Accordingly, a nationwide curfew was imposed from 6 pm until 10 am the next morning (i.e. sirens were echoed daily at 6 pm to announce the starting of curfew around Jordanian and their neighborhoods), including all public and private sectors, and banning travel between provinces, with exceptions for those with valid permits. In addition, religious activities in mosque and churches are subject to the ban, as well as, banning of gatherings of more than 10 people.

Jordan learnt the national curfew step from China, which had proved its efficiency in curbing the virus spread and containing the pandemic, through isolating the infected group of individuals and practicing social distancing [19]. Therefore, Jordan issued Defense Order No. 3, which controls the movement of people, and thereof a fine between 100 Jordanian Dinars ($141) and 500 Dinars ($705) for the first instance of curfew violation, while seized vehicles are impounded for 30 days in addition to a fine [20]. Reference [20] also announced that 6814 vehicles and 10,874 people violated the 30-day curfew.

As it became clear that the lockdown act caused adverse impacts on the vulnerable population in Jordan, the government eased the closure and reopened the economy on April 30, and accordingly, most sectors were allowed to gradually resume work on 3 May, except academic institution, recreational centers, mass gathering and events, and religious activities in mosques and churches. However, after 6 p.m. curfews and Friday curfews remain in place. Businesses companies are allowed to work at full production capacity with only 75% of its employees. Moreover, only delivery services is allowed to operate.

On June 6, the government has decided to shorten the curfew from midnight until 06:00. Additionally, sports events, worship places, restaurants, nurseries, cafes, and hotels are allowed to reopen with ensuring strict health and preventative measure. Moreover, all limitations of transportations and inter-provincial travel are removed. Because of the increasing CVOID-19 cases among the expatriates returning to Jordan, the government announced the suspension of international commercial flights and the closure of the country's borders from June 20 until at least July 14. This is to ease the burden on the quarantine facilities, reorganize the medical procedures, and keep the COVID-19 cases under control.

3.3 Application of Technology

On global level, COVID-19 is the only pandemic that has utilized the applications of digital technology in pandemic management and response (i.e. surveillance and tracking, contacts tracing, isolation and quarantine, and other health care activities) [21].

Countries such as Taiwan [22], South Korea [23], China [24], Singapore [25], Germany [26], Australia [27], have integrated their COVID-19 responses and national initiatives with digital technology which have in return enabled them to flatten their COVID-19 incidence curves and maintained low mortality rates [28].

In Jordan, four online Apps were deployed by the MoH in order to curb the spread of COVID-19. The three apps can be download from the official MoH's coronavirus website [29]:

(1) CRADAR App used during the corona pandemic crisis to help citizens identify places where mass gatherings or violations of public safety standards or any other practices that increase the likelihood of transmission of the coronavirus.
(2) AMAN App which is simply an exposure detection application to identify and track the close contacts of any positive-tested person.
(3) Bader App is launched to monitor and track the commitment of arrivals returning to Jordan to self-quarantine for 14 days which is now harmonized with an electronic wristband to regulate the self-quarantine at homes for expatriates who accomplished the actual quarantine.
(4) "Sehtak" (Your Health) App is lately launched by the MoH, which provides authentic information and reliable data related to COVID-19 pandemic status in Jordan.

4 Discussions

4.1 Readiness and Institutional Capacity

Ministry of Health was proactive when the COVID-19 outbreak was officially declared by the WHO, and took several quick steps to combat the new epidemic in collaboration with the healthcare system stakeholders in Jordan (i.e. public and royal medical services, private healthcare sector, and the academic medical institutions). One of the first steps taken by the MoH is to enforce health screening at all entry points using thermal scanners to prevent cross-border COVID-19 disease infection amongst expatriates returning from abroad.

The government has formed a National Epidemiological Committee, as a steering committee without authority, and established and mandated a Crisis Management Cell (CMC) in the National Centre for Security and Crisis Management. Moreover, the MoH practiced exclusive role as the national spokesperson of the daily registered COVID-19 cases and other heath indicators related to the pandemic.

CMC announced strict measures when the first COVID-19 outbreak began in the north and resulted in a steady increase in infections spread. These measures imposed mandatory two-week isolation on all arrivals at Amman airport on 17 March. It is noteworthy that Jordan was the first country in the world to isolate arrivals and travelers in hotels in luxury Dead Sea resorts for two weeks at the expense of the government. Subsequently, all academic activities were physically suspended, public gatherings and events were banned, and all borders, ports, and airports were closed. However, the freight and transport flights were left in operation.

Since the outset of the outbreak, Jordan realized that it cannot afford having herd immunity with limited healthcare system in terms of sixteen thousands hospital beds and 1400 ventilators (in addition to another 200 ventilators purchased lately), with occupancy rate might exceed 90% in the public hospitals. Consequently, Jordan has been so far determined, without any compromise, towards adopting the flattening and dampening strategy of the COVID-19 curve, and imposed purposeful measures to achieve that in reality.

The key step taken by the CMC to overcome the spread of COVID-19 was to increase the number of quarantine and isolations rooms by contracting hotels in Amman, Aqaba, and the Dead Sea, with the objective of symptoms monitoring and early detection of COVID-19 cases. For quarantining the expatriates, the hotels were able to host more than twelve locals returning from abroad each 14 days. Other small hotels, nearby the selected hospitals for CVODI-19 treatment in Amman, were contracted to quarantine and isolate mild-cases COVID-19 when needed as an option.

4.2   Contact Tracing, Testing, and Cases Isolation

The MoH's epidemiological taskforce has shown outstanding capabilities in infection cases identification and close-contacts tracing so far. This played a vital role in controlling the spread of the virus through the largest two outbreaks in Jordan. [30] reported that in case of highly effective contact tracing and case isolation, a new COVID-19 outbreak control takes three months to be controlled. However, as shown in Fig. 1, the two largest outbreak in Jordan were controlled in a 45 days for the first outbreak (15 March to 1 May 2020), and 30 days for the second outbreak (7 May to 7 June 2020). In the first and second outbreaks, 85 and 140 positive-tested persons, respectively, were traced, identified, tested, and isolated.

At present 72 epidemiological teams are working in all Jordanian cities, and collecting between six to eight thousands random samples on daily basis. Jordan's testing capacity has been increasing from 30,000 to 420,893 (currently) and would be further enhanced to 900,000 tests. In hot spot area with emerging COVID-19 outbreak, the epidemiological teams are working in support with the police to identify and locate probable exposed and infected contacts, collecting swaps and conducting lab-testing, and eventually imposing a 14-day quarantine in the hospitals.

Currently, only eight private testing labs are authorized to carry out Covid-19 testing. During the curfew and lockdown, sample collection for COVID-19 testing were only done through home visits, while, walk-in testing for COVID-19 is currently permitted. In the public healthcare sector, seven hospitals are authorized to conduct COVID-19 testing in different governorates.

4.3   Economic Impact of COVID-19 on Vulnerable Individuals

Although Jordan has succeeded in flattening the COVID-19 spread curve in the first 150 days since early of March, 2020 to date, however, Jordan has suffered enormous socio-economic losses due to its limited capacity to absorb intermittent national lockdown and prolonged closure of economic sectors. Most of low-and-middle-income countries, which adopted curfew, lockdown strategy, and stay-at-home directive to curb COVID-19 spread, have encountered companies bankruptcy, increasing unemployment, and adverse socio-economic impacts on vulnerable populations livelihood [31].

At the national level, the Society of Banks in Jordan reported that the curfew and closure are expected to have a negative impact on Jordan, so government revenues are expected to fall by 12.5–20%. Jordan's GDP will fall by 3.7%, while inflation will be 0.2%. Moreover, Jordanian exports will fall by 6.6–15.25%, while the imports drop down from 17 to 23.1%, as a result of the imposed curfew and business sector lockdown [32].

Authors in [33] highlighted the impacts of curfew in Jordan, which is mainly on the informal labor and small and medium-sized enterprise (SME). The informal labor represents 52% of the labor in Jordan, most of whom lack social protection, and are at any time at risk of losing the ability to work because of the lack of stability in the working environment, while SMEs forms 95% of the private businesses sector and contributes around 40% of the GDP in Jordan.

The UNDP in Jordan explored the impact of COVID-19 and lockdown measures on households and businesses in Jordan by conducting an online survey of 12,084 respondents. The results revealed that 58.6% of respondents, who were employed before the crisis, have lost their entire income. Moreover, 72.5% of respondents indicated having difficulties covering basic needs (food, medical treatment, rent, etc.) [34].

Hence, and a response to the drastic socio-economic impacts of COVID-19 on the vulnerable population, Jordan issued Defense Order No. 4, which establishes a fund to support the national effort to combat the coronavirus epidemic and to counter its effects. The fund is called *Himmat Watan* (A Nation's Effort). It is allocated for the deposit of cash donations of more than 100,000 Jordan Dinars (JD) ($140,845). The official website for the fund is http://himmatwatan.jo. Until May 30, Himmat Watan Fund has received JD93.5 million since launched on March 31 [35].

4.4 Working Together to Reopen Strategy

The strategy of "working together to reopen" was launched early of June, 2020, and serves as a systematic plan that help keeping business sectors and activities running, while avoiding the emerging of COVID-19 outbreak. The strategy adopts a data-driven approach based on risk assessment matrix considering the public health risk and the economic benefit of reopening of each of the closed economic sectors. As shown in Fig. 2, five risk magnitudes, stages, and classifications are considered in order to evolve from critical risk into low risk. Each stage may take one to two weeks depending on the daily reported COVID-19 cases through 3, 7, and 14 consecutive days for critical; high, medium, and moderate risks; and low risks, respectively. For each risk magnitude, a list of allowed sectors to re-open is stated with exceptions, adjustable levers, and measures.

Currently, Jordan pandemic status is in a moderate risk stage, since the number of positive-tested COVID-19 cases (with internal source of infection) for 14 consecutive days is less than 10 cases per day. Moreover, the ratio of detected cases to the number of random samples is less than 0.5%. Figure 3 shows the allowed sectors to re-open, exceptions, adjustable levers, and measures for current moderate risk stage that Jordan is experiencing since the early of June, 2020.

Critical Risk
- Full lockdown: Doubling the number of cases for 3 consecutive days starting more than local 20 cases (20, 40, 80) 3 foci points of unknown origin of infection OR more than 140 cases in one day OR>3% of test results are positive

High Risk
- Response response: More than 20 local cases per day for 7 consecutive days (MA) OR 2%-3% of test results are positive.

Medium Risk
- Protection: Number of local cases between 10-20 for more than 7 consecutive days (MA) OR 1%-2% of tests results are positive.

Moderate Risk
- Full operations: Number of local cases between 10-20 for more than 7 consecutive days (MA) OR 1%-2% of tests results are positive.
- Full operations: Number of local cases are less than 10 for 7 consecutive days (MA) OR 0.5%-1% of test results are positive.

Low Risk
- Sustainable recovery: No new local cases for 14 consecutive days.

Fig. 2 COVID-19 risk management matrix: Health risk triggers [18]

Allowed Activities	Exceptions	Adjustable Levers	Measures
• Administrative and support service activities • Restaurants and cafes with restrictions • Accommodation and hospitality (hotels, B&B) • Courts • Real estate sector • Touristic sites for domestic tourism • Mining and quarrying • ICT sector (tech companies) • Media sector (including paper-based media) • Publishing houses • Sport clubs and events without spectators, Spas (with restrictions) • Mosques and churches (with restrictions) • Nurseries (with restrictions)	• Schools and kindergartens • Wedding halls and funeral homes • Universities, colleges and institutes • Youth activities • Cinemas • Facilities for organizing parties, festivals, conferences, exhibitions, cultural events and festivals • Oriental Baths • Training and cultural centers • Public parks, game cities and entertainment places • Allowed: regular medical check-ups; allow visit to prisons and care homes • Restricted: seniors > 70 years; individuals with chronic diseases; permit holders from infection hotspot areas, wedding and funeral events	• Curfew: no full-day per week; (6 am-12 am for citizens mobility) (6am-11pm for business) • Travel restrictions: accoring to MoH, MoT and epidemiological committee and crisis cell discretion • Mobility within the regions and between the governorates is permitted, provided that any governorate in which the epidemic appears is closed. • Size of public gatherings: up to 20 individuals • Transportation: all, capacity for public transport: 50%	• Aggressive PCR testing: 70% of testing capacity; increased contact tracing efforts to 100% of the epidemiological investigation team's capacity • Extensive public awareness media campaign; constant updates provided to the public • Allow internal flights • All governmental entities will announce/ activate their own emergency response plans in every stage.

Fig. 3 Moderate risk's allowed sectors to re-open, exceptions, adjustable levers, and measures

5 Conclusions

In the present study, Jordan was exemplified as a successful preparedness and proactive intervention case to combat the unprecedented infectious COVID-19 outbreaks. Jordan implemented effective measures and has been able to stand fast in battling the COVID-19 to date and succeeded in flattening and damping the curve of nCoV spread. While, other countries took three months to control COVID-19 outbreak, Jordan was capable to control the largest two outbreaks in 30–45 days for each. This due to the effective closed contacts tracing and testing capabilities, integration their COVID-19 responses and national initiatives with digital technology and Apps utilization, and providing of transparent and up-to-date information and updates through several media about this unprecedented situation.

Jordan has suffered enormous socio-economic impacts due to its limited capacity to absorb intermittent national lockdown and prolonged closure of economic sectors. Informal labor and SMEs were the most influenced entities. Jordan has issued and adopted a re-open strategy based on risk assessment matrix with several levels of risks stages and magnitude, considering the public health risk and the economic benefit of reopening of each of the closed economic sectors. Accordingly, Jordan pandemic status is in a moderate risk stage.

References

1. Secon, H., Woodward, A., Mosher, D.: Business Insider US. A Comprehensive Timeline of the New Coronavirus Pandemic, from China's first COVID-19 Case to the Present. Business Insider Malaysia (2020). Retrieved from https://www.businessinsider.my/coronavirus-pandemic-tim eline-history-major-events-2020-3?r=US&IR=T. Accessed 25 June 2020
2. World Health Organization: WHO timeline—COVID-19 (2020a). Retrieved from https://www. who.int/news-room/detail/08-04-2020-who-timeline—covid-19. Accessed 19 May 2020
3. Mundi Index: Hospital bed density-country comparison Index Mundi 2019. (2019). https:// www.indexmundi.com/g/r.aspx?c=ha&v=2227. Accessed 23 June 2020
4. Moody's Analytics: Jordan-Economic Indicators (2020). https://www.economy.com/jordan/ indicators. Accessed 21 June 2020
5. UNHCR: UNHCR Jordan Factsheet, November 2019 (2019). https://reliefweb.int/sites/relief web.int/files/resources/72611.pdf. Accessed 17 June 2020
6. Abbara, A., Rayes, D., Fahham, O., Alhiraki, O.A., Khalil, M., Alomar, A., Tarakji, A.: Coronavirus 2019 and health systems affected by protracted conflict: the case of Syria. Int. J. Infect. Dis. **96**, 192–195 (2020)
7. Alahdal, H., Basingab, F., Alotaibi, R.: An analytical study on the awareness, attitude and practice during the COVID-19 pandemic in Riyadh, Saudi Arabia. J. Infect. Public Health (2020). https://doi.org/10.1016/j.jiph.2020.06.015
8. Al-Dmour, H., Masa'deh, R., Salman, A., Abuhashesh, M., Al-Dmour, R.: Influence of social media platforms on public health protection against the COVID-19 pandemic via the mediating effects of public health awareness and behavioral changes: Integrated model. J. Med. Intern. Res. **22**(8) (2020). https://doi.org/10.2196/19996
9. Aziz, P.Y., Hadi, J.M., Sha, A.M., Aziz, S.B., Rahman, H.S., Ahmed, H.A., Abdulla, M.A., Amine, A.S.M.: The strategy for controlling COVID-19 in Kurdistan regional government

(KRG)/Iraq: identification, epidemiology, transmission, treatment, and recovery. Int. J. Surg. Open (2020). https://doi.org/10.1016/j.ijso.2020.06.006

10. Mikhael, E.M., Al-Jumaili, A.A.: Can developing countries face novel coronavirus outbreak alone? The Iraqi situation. Public Health Pract. **1**, 100004 (2020)

11. Mondoweiss: COVID-19 cases double in West Bank's second wave surge (2020). https://mondoweiss.net/2020/06/covid-19-cases-double-in-west-banks-second-wave-surge/. Accessed 1 July 2020

12. The Jordan Times: Schools suspended, borders closed, gatherings banned as gov't responds to continued coronavirus spread (2020a). https://www.jordantimes.com/news/local/schools-suspended-borders-closed-gatherings-banned-govt-responds-continued-coronavirus. Accessed 3 April 2020

13. Chakraborty, I., Maity, P.: COVID-19 outbreak: migration, effects on society, global environment and prevention. Sci. Total Environ. **728**, 138882 (2020)

14. Yusef, D., Hayajneh, W., Awad, S., Momany, S., Khassawneh, B., Samrah, S., et al.: Large outbreak of coronavirus disease among wedding attendees, Jordan. Emerg. Infect. Dis. 2020 Sep [date cited] (2020). https://doi.org/10.3201/eid2609.201469

15. Saidan, M.N., Shbool, M.A., Arabeyyat, O.S., Al-Shihabi, S.T., Al Abdallat, Y., Barghash, M.A., Saidan, H.: Estimation of the probable outbreak size of novel coronavirus (COVID-19) in social gathering events and industrial activities. Int. J. Infect. Dis. (2020). https://doi.org/10.1016/j.ijid.2020.06.105

16. The Jordan Times: 23 new virus cases spanning 4 governorates reported in Jordan on Wednesday (2020b). http://www.jordantimes.com/news/local/23-new-virus-cases-spanning-4-governorates-reported-jordan-Wednesday. Accessed 26 June 2020

17. MoH: Coronavirus cases chart (2020a). https://corona.moh.gov.jo/en

18. World Health Organization. WHO: Operational considerations for case management of COVID-19 in health facility and community. Interim guidance. Geneva (2020b). https://apps.who.int/iris/bitstream/handle/10665/331492/WHO-2019-nCoV-HCF_operations-2020.1-eng.pdf. Accessed 17 May 2020

19. Bernama: Health DG: movement control can boost ministry's fight against Covid-19. The Star (2020). https://www.thestar.com.my/news/nation/2020/03/17/health-dg-movement-control-can-boost-ministry039s-fight-against-covid-19. Accessed 29 May 2020

20. The Jordan Times: Kingdom witnessed 18,000 curfew violations in 30 days—PSD data (2020c). http://jordantimes.com/news/local/kingdom-witnessed-18000-curfew-violations-30-days—psd-data?fbclid=IwAR3uPWK3zMC6XSwvgevxnl5_DlOukdgFoIjhEel6qqxtx7 64jDj-m5UBLQE. Accessed 26 May 2020

21. Ferretti, L., Wymant, C., Kendall, M., et al.: Quantifying SARS-CoV-2 transmission suggests epidemic control with digital contact tracing. Science **368**, eeba6936 (2020)

22. Wang, J., Ng, C.Y., Brook, R.H.: Response to COVID-19 in Taiwan: Big data analytics, new technology, and proactive testing. JAMA (2020). https://doi.org/10.1001/jama.2020.3151

23. Business Insider: South Korea launched wristbands for those breaking quarantine because people were leaving their phones at home to trick government tracking apps (2020). https://www.businessinsider.com/south-korea-wristbands-coronavirus-catchpeople-dodging-tracking-app-2020-4. Accessed 26 May 2020

24. Liu, J.: Deployment of IT in China's fight against the Covid-19 pandemic (2020). https://www.itnonline.com/article/deployment-health-it-china%E2%80%99s-fight-againstcovid-19-pandemic. Accessed 13 June 2020

25. News National: Singapore's coronavirus temperature screening and tracking are leading the way (2020). https://thenewdaily.com.au/news/national/2020/03/19/singapore-coronavirus-temperature-scans/. Accessed 17 May 2020

26. Thomas Reuters: Germany launches new smartwatch application to monitor coronavirus spread (2020). https://www.reuters.com/article/ushealth-coronavirus-germany-tech/germany-launches-smartwatchapp-to-monitor-coronavirus-spread-idUSKBN21P1SS. accessed 26 May 2020

27. The Wall Street Journal: Australia's coronavirus evacuation plan: a tiny island 1000 miles away. (2020). https://www.wsj.com/articles/australias-coronavirus-evacuation-plan-a-tiny-isl and-1-000-milesaway-11580295354. Accessed April 18, 2020

28. Whitelaw, S., Mamas, M.A., Topol, E., Van Spall, H.G.C.: Applications of digital technology in COVID-19 pandemic planning and response. Lancet Dig. Health (2020). https://doi.org/10. 1016/S2589-7500(20)30142-4

29. MoH: Working together to reopen (2020b). https://corona.moh.gov.jo/en/Together-to-Reopen

30. Hellewell, J., Abbott, S., Gimma, A., Bosse, N.I., Jarvis, C.I., Russell, T.W., Munday, J.D., Kucharski, A.J., Edmunds, W.J.: Feasibility of controlling COVID-19 outbreaks by isolation of cases and contacts. Lancet Glob. Health **8**, 488–496 (2020)

31. Petersen, E., Wasserman, S., Lee, S., Ugo, U., Holmesg, A.H., Al-Abri, S., McLellan, S., Blumberg, L., Tambyah, P.: COVID-19–We urgently need to start developing an exit strategy. Int. J. Infect. Dis. **96**, 233–239 (2020)

32. Ammon News: In figures: impacts of Coronavirus on Jordanian Economy. (2020). https://www. ammonnews.net/article/531263. Accessed 27 June 2020

33. Al-Ajlouni, L.: Could COVID-19 Push Jordan to the Edge? Middle East Institute (2020). https:// www.mei.edu/publications/could-covid-19-push-jordan-edge. Accessed 4 June 2020

34. UNDP. (2020). COVID-19 Impact on Households in Jordan: A rapid assessment. May 2020. Amman, Jordan

35. The Jordan Times: Himmat Watan Fund has received JD93.5 m since launch. (2020d). http:// jordantimes.com/news/local/'himmat-watan-fund-has-received-jd935m-launch. Accessed 29 June 2020

Factors Impacts Organization Digital Transformation and Organization Decision Making During Covid19 Pandemic

A. Ahmad ⓘ, M. T. Alshurideh ⓘ, B. H. Al Kurdi ⓘ, and S. A. Salloum ⓘ

Abstract The Digital transformation concept defined as how firms and organizations utilize the power and possibility of technology like cloud computing, artificial intelligence, and machine learning IoT technologies to enhance the business operation and customers' experience or create new business lines and methods. In this study will qualitative research will conduct focusing on the impact of internal and external factors of operational decision making approaches under the digital transformation umbrella during the COVID-19 pandemic. The main results indicated that senior management support and technology infrastructure readiness has a positive impact factor in digital transformation and organization performance as a result, where data breaching threats had a negative impact on organization adoption of digital transformation.

Keywords Digital transformation · Operation decision making · IoT · AI · COVID-19 pandemic

A. Ahmad · M. T. Alshurideh (✉)
Department of Management, College of Business Administration, University of Sharjah, Sharjah, United Arab Emirates
e-mail: malshurideh@sharjah.ac.ae; m.alshurideh@ju.edu.jo

A. Ahmad
e-mail: U19106242@sharjah.ac.ae

M. T. Alshurideh
Department of Marketing, School of Business, The University of Jordan, Amman, Jordan

B. H. Al Kurdi
Department of Management, Faculty of Economics and Administrative Sciences, The Hashemite University, Zarqa, Jordan
e-mail: barween@hu.edu.jo

S. A. Salloum
University of Sharjah, Research Institute of Sciences, University of Sharjah & Engineering, Sharjah, United Arab Emirates
e-mail: ssalloum@sharjah.ac.ae

© The Author(s), under exclusive license to Springer Nature Switzerland AG 2021
M. T. Alshurideh et al. (eds.), *The Effect of Coronavirus Disease (COVID-19) on Business Intelligence*, Studies in Systems, Decision and Control 334, https://doi.org/10.1007/978-3-030-67151-8_6

1 Introduction

The COVID-19 Pandemic will have long lasting impacts and implications for the business in general, where the this pandemic shows clearly the importance of digital transformation, big data, artificial intelligence, and data analytics are probably the most cited math topics in the news around the world, also this outbreak shows the impact on all business to speed up the digital transformation (DI) process and be more innovative to mitigate the crisis impact. Commercial and technical reports usually indicate a high level of success in adopting these methods. In contrast, in the current war the world faces against Covid-19, it is not uncommon to see many references on these technologies. Where the fourth industry revelation almost started and change the old understanding of the business operation and decision-making approaches now, we deal with a very dynamic market, and consumers' behaviors and accelerated curve of technology evolve. The need to understand the implication of the new era becomes a must of organization in order to sustain and growth and survival. Digital transformation (DI) and data-driven organization have become the strategic goal of many organizations and senior management agenda [1, 2], but the amount of literature is little in regards to how organization decision making ability impact. The Definition of Digital transformation as [1] is "used new digital technologies (social media, mobile, analytics or embedded devices) to enable major business improvement such as enhancing users' experience, streaming operation, or creating new business models.

1.1 Research Questions

As we demonstrate earlier about the fourth industrial revolution, dramatic change in the competition disruptive, easy and widespread of technologies, and the changes in customers' behavior change, those factors make the predictions of competitive advantage for organizations more complex and complicated [3]. The research path we will try to demonstrate the time factor role that digital transformation in terms of utilizing the AI-based decision support system, maintaining the competitive advantage of firms on a dynamic market. His study will develop different approaches analyzing quantitative data collected to measure the effects of digital transformation in specific (AIDSS) on organizational survival, growth, and performance at highly dynamics organizations such as technology and the manufacturing industry.

The research question of this study is:

- What the effects of internal and external factors on digital transformation in specific utilizing AI-based decisions support systems on organizational performance under the impact of COVID-19 Pandemic?

1.2 Research Objectives

In the beginning, we need to understand the current situation and impact of COVID-19 pandemic on business, we can catogrise this impact on businesses in three catogeries, first some businesses quickly innovated the overall business models to reduce the impact and keep business running. Second businesses has been highly effected by the COVID-19 due to unpreperness with digital transformation adoption in the business operations. Finally the business has experience major growth and benefits due to redeness and preperness of digital transformation. To elaborate more and get better understanding we need to understand artificial intelligence (AI) before we argue about constrain and issues involved in adopting (AI) and big data in the decision support system (DSS). The literature on (AI) took two directions; the first approach is the trails of simulate human intelligence second approaches of research is artificial life where the focus on how the process of a living system in artificial environments, moreover, both fields have deep implications on leadership decisions making ability cause of AI-based decisions system utilizes both researches approaches in addition of combining communication, computing, and Decisions supporting systems to emulate human intelligence in solving complex unstructured problems. [4] leading to full-fledged automated decision support systems.

1.3 Literature Review

For organizations, the condition requires to stay competitive; it's a must of mangers on operations to have a well understanding of the competition and competitors [5]. In order of organizations made the correct strategy in terms of operations and reflecting it across overall organization activity from design, plan [6–14] to enhance the organization position and competitive edge among competitors by improving the operation process and be more Flexible and fast adopting market changes [15–22]. We need to understand the concept of decision is that there is an 'individual decision-maker facing a choice involving uncertainty about outcomes' [23]. Putting the individual decision-maker at the center of decision-making is the most intuitive approach to studying algorithmic decision-making in organizations (e.g. [24]), where the individual is the recipient of automated decisions or recommendations. For instance, operative decision-makers who provide IT enabled services [25–28]. Study on the implementation of design and evaluations of the Performance Management System, with the context of SME's use. PMS providers targeting SMEs use the empirical natures of PMS to provide proof that cover enablement of their software offering. Also, be considered as a pathway to evaluate and improve existing PMS. PMS doesn't relay on benchmarks to support decisions making along with PMS use. Study shows the interest of business owners and makers of proposed PMS features that were unavailable in existing PMS [29]. Design a new performance modeling system based on data envelopment analysis (DEA) and Artificial Neural Network (ANN) to demonstrate overall business efficiency performance predictions and to be tested underperformance modeling framework empirically. This study focus in advancing research in performance benchmark and modeling by proposing new model can help managers take decisions and solve theory practice issues [30–33], Channing management approaches and overall culture of the organization it's a must since its play major role of successful implementation on any innovative system or concept [34–36].

Where the age group of is a key in this study, significant difference in precipitation found answering the same question about industry 4 concept [37]. Other study shows the deep relationship between human and artificial intelligence decision making algorithms between de-attaching in context of rational distance and attaching to decisions making caused off accidental and infrastructural proximity, imposed engagement, and affective adhesion. Study demonstrates the result of unbalanced relationships as delayed decision and solutions, and manipulation in data, also pointing out the important role of media studies in explaining behavior created during digital transformation in the organization [38], moreover, organization behavior in terms of data drove and business digitalization era where data have driven it has to be an organizational culture. However, the multidisciplinary nature of big data analytics and data science (BDADS) seems to collide with the domain- ant "functional silo design" that characterizes business schools. The scope and breadth of the radical digitally enabled change; they necessitate a global questioning about the nature and structure of business education [39]. Understanding incumbent firms in traditional industries build dynamics capabilities for digital transformation and conceptualized buildings dynamics capabilities as a process of building dynamics capabilities for ongoing strategy renewal will help build to understand the importance of the time factor roles in digital transformation in organization maintenance of their competitive edge [40].

2   Theoretical Background

Explain the disruptive impact of digital transformation on the organization's business model and innovation. Determined the characteristics of digital innovations will use a dynamic capabilities framework to help us understand the nature of strategic change in an organization [41–44]. This framework gives us the ability to describe organizations' capacity. Dynamic capabilities are innovation-based and distinguishable from a firm's operational capabilities, where its conventional help organization in a sense they help a firm in the present by maintaining the status quo, but this leaves the organization vulnerable to environmental change [36, 45–47]. Reference [48] explains that although ordinary capabilities enable the or to perform operational tasks, conventional capabilities in functions such as accounting, human resources management, and sales are now easily replicable because they can be outsourced The literature clarifies the relationship between dynamic capabilities, strategy, and business models argument that a business model is "a reflection of the firm's realized strategy," [49] business models (present or short-term perspective) to face either upcoming or existing contingencies. In other words, dynamic capabilities represent the intermediary between strategy and business models, ensuring the strategic renewal of organizations [50]. Reference [51] emphasizes that firms require a system of dynamic capabilities to orchestrate resources and evolve the business model. Balanced redundancy, requisite variety, and cognitive discretion were the dynamic capabilities that supported the evolution of a news organization's digital business model. For organizations, [52] argues, "In many cases, corporate strategy dictates business model design. At times, however, the arrival of new general-purpose technology (e.g., the Internet) opens opportunities for radically new business models to which corporate strategy must then respond." Building sensing, seizing, and transforming capabilities thus allows a firm to craft future strategy that designs, creates, and refines a defensible business model, guides organizational transformation, and provides a durable source for obtaining a competitive advantage [52].

3 Research Model

Building on organization dynamic capabilities theory [40]. We manage to create a research model highlight the impact and correlation among organizations building dynamic capabilities and business performance, taking into consideration internal and external factors impact, as Fig. 1.

4 Research Propositions

Converting digital transformation insights into decisive action. Utilizing the data outcome of the digital transformation, it's a key now of many organizations where the adoption of data-driven decision start to be one of the key competitive advantages of the firms in the new dynamic nurture of the market.

P1 Testing the impact of the external and internal factors affecting digital transformation where it impacts organization decision ability and expected performance.

5 Methodology

In our study, we followed a qualitative research methodology where we conduct a focus group of operational managers within an organization adopting digital transformation and utilizing data for decision making; the outcome is transcribed, coded, and finally analyzed following [53–56]. By starting to categorize the code, then creating themes defined by the created code finally, gathering all match theme and code in

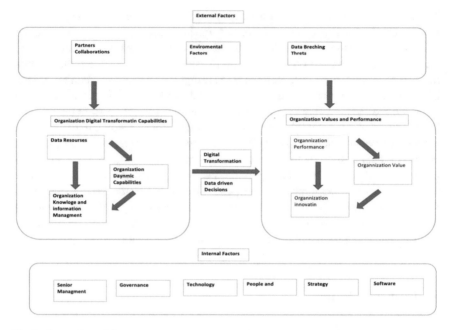

Fig. 1 Research model

order to define main factors affecting the data-driven decision making and the overall impact of digital transformation in organization performance. Table 1 demonstrates the codes of our study factors and focus group.

This search aimed to explores Managers understanding of decision based on data generated of overall digital transformation within their organization. Through a semi-structured interview where the interview question mailed to the participant. As inclusion criteria were a minimum 5 years' experience in operation it's a must. Table 2 demonstrate the Demographics of the participants. The focus group held in virtually in March 5, 2020 due to COVId-19 restriction, all notes are documented and transcribed, coded and themed for the analysis. The six steps used in analysing the collected data followed method commonly used by researchers [47, 56–58].

Table 1 Code of effecting factors

Code	Study items
MS	Senior management support
TI	Technology infrastructure
DT	Data breaching threats

Table 2 Participant demographics

Gender	Male	20
	Female	2
Total participant	Met inclusion criteria	18
Language	English	14
	Arabic	4
Occupations	Operation manager	6
	IT manger	6
	Data security manager	6

Table 3 Frequency counts of managers participate in the factor behavior + or − impact

	Factors					
	MS+	MS−	TI+	TI−	DT+	DT−
Positive behavior	90	0	80	0	20	0
Negative behavior	0	0	0	10	0	70
Total	90	0	80	10	20	70

6 Analysis

Table 3 demonstrate the study factors along with their frequency also the table show the counts of positive and negative responses of the focus group participants.

7 Discussion

The outcomes of the analysis and of Table 3 stated that all two tested internal factors and one external factor (MS, TI and DT) positively impact organization performance in the context of data-driven decision making during COVID-19 Pandmic where total positive was 160, while the negative impact was 110 as shown in Fig. 2.

Based on the theming of focus group discussion about factors that are key to the digital transformation transition, first management support is the main factor where almost all participants agreed about the positive impact of it on organization performance in the context of data-driven decision making specially during the COVID-19 crisis on business during the restrictions and long periods of lockdowns. The second factor is technology infrastructure where the reediness of the organization to adopt digital transformation, and here we talk about the infrastructure where this been a key factors of business operations during the pandemic and allowing the work to be performed remotely the opinion of participants was the technology infrastructures it's a key enabler of transitions and other they argue human skills and acceptance to deal and utilize the digital transformation and business continuity as one main outcome of

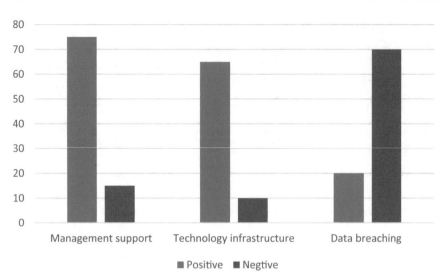

Fig. 2 Impact of the factor (positive and negative incidents)

utilizations. Finally, we analyze the main external factor connected with the digital transformation of the risk of data breaching impact where almost all participants agreed about the negative factors of it, giving many examples about data breaching such as "WannaCry" breaching attack and the loses of organization happened.

8 Managerial Implication

The results and findings of this study have many managerial implications. Where the study give an guidelines of the managers seeking to enhance their business operation an performance utilizing the power of the digital transformation tools this guide focus on factors impact this process of transformations. Study shows the importance of investing on technology infrastructures towards adopting new business operations and decision making tools its play significant roles in business growth and competitive advantage [59–63].

9 Limitation of the Study

One of the major limitation in this study is the size of the sample not large enough due to the COVID-19 pandamic restrictions, more interviews needed to have wider precpective of the impact of these factors on the digital transformation and the business overall performance.

10 Conclusion

Our research studies the impact of internal and external factors affecting the organization's digital transformation process and tries to qualitatively measure the impact of these factors on the overall performance of the organization's ability towards data have driven decision making during COVID-19 pandemic. Three propositions regarding senior management support, technology infrastructure, and data breaching threats factors were tested using focus group discussion as tools of qualitative research. The main results showed that senior management support and technology infrastructure readiness has a positive impact factor in digital transformation and organization performance as a result, where data breaching threats had a negative impact on organization adoption of digital transformation.

References

1. Fitzgerald, M., Kruschwitz, N., Bonnet, D., Welch, M.: Embracing digital technology: a new strategic imperative. MIT Sloan Manag. Rev. **55**(2), 1 (2014)
2. Singh, A., Hess, T.: How chief digital officers promote the digital transformation of their companies. MIS Q. Exec. **16**(1) (2017)
3. McGrath, R.G.: The End of Competitive Advantage: How to Keep Your Strategy Moving as Fast as Your Business. Harvard Business Review Press (2013)
4. Desanctis, G., Gallupe, R.B.: A foundation for the study of group decision support systems. Manage. Sci. **33**(5), 589–609 (1987)
5. Hodgkinson, G.P., Johnson, G.: Exploring the mental models of competitive strategists: the case for a processual approach. J. Manag. Stud. **31**(4), 525–552 (1994)
6. Aboelmaged, M.G.: Harvesting organizational knowledge and innovation practices. Bus. Process Manag. J. (2012)
7. Barnes, D.: Research methods for the empirical investigation of the process of formation of operations strategy. Int. J. Oper. Prod. Manag. (2001)
8. Boer, H., Drejer, A., Minarro-Viseras, E., Baines, T., Sweeney, M.: Key success factors when implementing strategic manufacturing initiatives. Int. J. Oper. Prod. Manag. (2005)
9. Yousuf, H., Zainal, A.Y., Alshurideh, M., Salloum, S.A.: Artificial intelligence models in power system analysis. In: Artificial Intelligence for Sustainable Development: Theory, Practice and Future Applications, pp. 231–242. Springer
10. AlShamsi, M., Salloum, S.A., Alshurideh, M., Abdallah, S.: Artificial intelligence and blockchain for transparency in governance. In: Artificial Intelligence for Sustainable Development: Theory, Practice and Future Applications, pp. 219–230. Springer
11. Hayajneh, N., Suifan, T., Obeidat, B., Abuhashesh, M., Alshurideh, M., Masa'deh, R.: The relationship between organizational changes and job satisfaction through the mediating role of job stress in the Jordanian telecommunication sector. Manage. Sci. Lett. **11**(1), 315–326 (2021)
12. Al Kurdi, B., Alshurideh, M., Salloum, S.: Investigating a theoretical framework for E-learning technology acceptance. Int. J. Electr. Comput. Eng. **10**(6), 6484–6496 (2020)
13. Alhashmi, S.F.S., Alshurideh, M., Al Kurdi, B., Salloum, S.A.: A Systematic Review of the Factors Affecting the Artificial Intelligence Implementation in the Health Care Sector, vol. 1153 AISC (2020)
14. Salloum, S.A., Alshurideh, M., Elnagar, A., Shaalan, K.: Mining in Educational Data: Review and Future Directions, vol. 1153 AISC (2020)

15. Alegre-Vidal, J., Lapiedra-Alcamı, R., Chiva-Gómez, R.: Linking operations strategy and product innovation: an empirical study of Spanish ceramic tile producers. Res. Policy **33**(5), 829–839 (2004)

16. Salloum, S.A., Alshurideh, M., Elnagar, A., Shaalan, K.: Machine Learning and Deep Learning Techniques for Cybersecurity: A Review, vol. 1153 AISC (2020)

17. Fawzi Assad N., Alshurideh, M.: Investment in context of financial reporting quality: a systematic review. WAFFEN-UND Kostumkd. J. **11**(3), 255–286 (2020)

18. Abu Zayyad, H.M., Obeidat, Z.M., Alshurideh, M.T., Abuhashesh, M., Maqableh, M., Masa'deh, R.: Corporate social responsibility and patronage intentions: the mediating effect of brand credibility. J. Mark. Commun. (2020)

19. AlMehrzi, A., Alshurideh, M., Al Kurdi, B.: Investigation of the key internal factors influencing knowledge management, employment, and organisational performance: a qualitative study of the UAE Hospitality sector. Int. J. Innov. Creat. Chang. **14**(1), 1369–1394 (2020)

20. Al-Gasaymeh, A., Almahadin, A., Alshurideh, M., Al-Zoubid, N., Alzoubi, H.: The role of economic freedom in economic growth: evidence from the MENA Region. Int. J. Innov. Creat. Chang. **13**(10), 759–774 (2020)

21. Al Kurdi, B., Alshurideh, M., Al afaishata, T.: Employee retention and organizational performance: evidence from banking industry. Manage. Sci. Lett. **10**(16), 3981–3990 (2020)

22. Kurdi, B., Alshurideh, M., Alnaser, A.: The impact of employee satisfaction on customer satisfaction: theoretical and empirical underpinning. Manage. Sci. Lett. **10**(15), 3561–3570 (2020)

23. Peterson, M.: An Introduction to Decision Theory. Cambridge University Press (2017)

24. Davenport, T.H., Dyché, J.: Big data in big companies. Int. Inst. Anal. **3** (2013)

25. Chae, B.K., Yang, C., Olson, D., Sheu, C.: The impact of advanced analytics and data accuracy on operational performance: a contingent resource based theory (RBT) perspective. Decis. Support Syst. **59**, 119–126 (2014)

26. Alshurideh, M., Gasaymeh, A., Ahmed, G., Alzoubi, H., Kurd, B.: Loyalty program effectiveness: theoretical reviews and practical proofs. Uncertain Supply Chain Manage. **8**(3), 599–612 (2020)

27. Alzoubi, H., Alshurideh, M., Kurdi, B.A., Inairat, M.: Do perceived service value, quality, price fairness and service recovery shape customer satisfaction and delight? A practical study in the service telecommunication context. Uncertain Supply Chain Manage. **8**(3), 579–588 (2020)

28. Aburayya, A., Alshurideh, M., Albqaeen, A., Alawadhi, D., Ayadeh, I.: An investigation of factors affecting patients waiting time in primary health care centers: an assessment study in Dubai. Manag. Sci. Lett. **10**(6), 1265–1276 (2020)

29. Bennett, L.M., Gadlin, H., Marchand, C.: Collaboration Team Science: Field Guide. US Department of Health & Human Services, National Institutes of Health … (2018)

30. Kwon, H.-B., Lee, J., Roh, J.J.: Best performance modeling using complementary DEA-ANN approach. Benchmarking An Int. J. (2016)

31. Alshurideh, M., Salloum, S.A., Al Kurdi, B., Monem, A.A., Shaalan, K.: Understanding the quality determinants that influence the intention to use the mobile learning platforms: a practical study. Int. J. Interact. Mob. Technol. **13**(11), 157–183 (2019)

32. Alshurideh, Do electronic loyalty programs still drive customer choice and repeat purchase behaviour? Int. J. Electron. Cust. Relatsh. Manag. **12**(1), 40–57 (2019)

33. AlShurideh, M., Alsharari, N.M., Al Kurdi, B.: Supply Chain Integration and Customer Relationship Management in the Airline Logistics, vol. 9, no. 02, pp. 392–414 (2019)

34. Mehmood, T., Alzoubi, H., Alshurideh, M., Ahmed, G., Al-Gasaymeh, A.: Schumpeterian entrepreneurship theory: evolution and relevance. Acad. Entrep. J. **25**(4), 1–10 (2019)

35. Alshurideh, M., Al Kurdi, B., Salloum, S.A.: Examining the Main Mobile Learning System Drivers' Effects: A Mix Empirical Examination of Both the Expectation-Confirmation Model (ECM) and the Technology Acceptance Model (TAM), vol. 1058 (2020)

36. Alshurideh, M., Al Kurdi, B., Shaltoni, A.M., Ghuff, S.S.: Determinants of pro-environmental behaviour in the context of emerging economies. Int. J. Sustain. Soc. **11**(4), 257–277 (2019)

37. Mohelska, H., Sokolova, M.: Management approaches for industry 4.0–the organizational culture perspective. Technol. Econ. Dev. Econ. **24**(6), 2225-2240 (2018)
38. Bader, V., Buhr, D.: DIE AMBIVALENZ DES NEUEN
39. Carillo, K.D.A., Galy, N., Guthrie, C., Vanhems, A.: How to turn managers into data-driven decision makers. Bus. Process Manag. J. (2019)
40. Warner, K.S.R., Wäger, M.: Building dynamic capabilities for digital transformation: an ongoing process of strategic renewal. Long Range Plan. **52**(3), 326–349 (2019)
41. Teece, D., Peteraf, M., Leih, S.: Dynamic capabilities and organizational agility: risk, uncertainty, and strategy in the innovation economy. Calif. Manage. Rev. **58**(4), 13–35 (2016)
42. Schilke, O., Hu, S., Helfat, C.E.: Quo vadis, dynamic capabilities? A content-analytic review of the current state of knowledge and recommendations for future research. Acad. Manag. Ann. **12**(1), 390–439 (2018)
43. ELSamen, A., Alshurideh, M.: The impact of internal marketing on internal service quality: a case study in a Jordanian pharmaceutical company. Int. J. Bus. Manage. **7**(19), 84–95 (2012)
44. Alkalha, Z., Al-Zu'bi, Z., Al-Dmour, H., Alshurideh, M., Masa'deh, R.: Investigating the effects of human resource policies on organizational performance: an empirical study on commercial banks operating in Jordan. Eur. J. Econ. Financ. Adm. Sci. **51**(1), 44–64 (2012)
45. Alshraideh, A., Al-Lozi, M., Alshurideh, M.: The impact of training strategy on organizational loyalty via the mediating variables of organizational satisfaction and organizational performance: an empirical study on Jordanian Agricultural Credit Corporation Staff. J. Soc. Sci. **6**, 383–394 (2017)
46. Alshurideh, M., Al Kurdi, B., Vij, A., Obiedat, Z., Naser, A.: Marketing ethics and relationship marketing—an empirical study that measure the effect of ethics practices application on maintaining relationships with customers. Int. Bus. Res. **9**(9), 78–90, 2016
47. Ghannajeh, A., et al.: A qualitative analysis of product innovation in Jordan's pharmaceutical sector. Eur. Sci. J. **11**(4), 474–503 (2015)
48. Teece, D.J.: A dynamic capabilities-based entrepreneurial theory of the multinational enterprise. J. Int. Bus. Stud. **45**(1), 8–37 (2014)
49. DaSilva, C.M., Trkman, P.: Business model: what it is and what it is not. Long Range Plan. **47**(6), 379–389 (2014)
50. Agarwal, R., Helfat, C.E.: Strategic renewal of organizations. Organ. Sci. **20**(2), 281–293 (2009)
51. Velu, C.: A systems perspective on business model evolution: the case of an agricultural information service provider in India. Long Range Plan. **50**(5), 603–620 (2017)
52. Teece, D.J.: Dynamic capabilities as (workable) management systems theory. J. Manag. Organ. **24**(3), 359–368 (2018)
53. Alshurideh, M., Salloum, S.A., Al Kurdi, B., Al-Emran, M.: Factors affecting the social networks acceptance: an empirical study using PLS-SEM approach. In: 8th International Conference on Software and Computer Applications, 2019, pp. 1–5
54. Ashurideh, M.: Customer service retention—a behavioural perspective of the UK mobile market. Durham University (2010)
55. Kurdi: Healthy-Food Choice and Purchasing Behaviour Analysis: An Exploratory Study of Families in the UK. Durham University (2016)
56. Alshurideh, M.: A qualitative analysis of customer repeat purchase behaviour in the UK Mobile Phone Market. J. Manag. Res. **6**(1), 109–125 (2014)
57. Assad, N.F., Alshurideh, M.T.: Financial Reporting quality, audit quality, and investment efficiency: evidence from GCC economies. WAFFEN-UND Kostumkd. J. **11**(3), 194–208 (2020)
58. Al-Zu'bi, Z.M.F., Al-Lozi, M., Dahiyat, S.E., Alshurideh, M., Al Majali, A.: Examining the effects of quality management practices on product variety. Eur. J. Econ. Financ. Adm. Sci. **51**, 10–19 (2012)
59. Altamony, H., Masa'deh, R.M.T., Alshurideh, M., Obeidat, B.Y.: Information systems for competitive advantage: Implementation of an organisational strategic management process. In: Innovation and Sustainable Competitive Advantage: From Regional Development to

World Economies—Proceedings of the 18th International Business Information Management Association Conference, 2012, vol. 1, pp. 583–592

60. Shannak, R.O., Masa'deh, R.M.T., Al-Zu'bi, Z.M.F., Obeidat, B.Y., Alshurideh, M., Altamony, H.: A theoretical perspective on the relationship between knowledge management systems, customer knowledge management, and firm competitive advantage. Eur. J. Soc. Sci. **32**(4), 520–532 (2012)

61. Alshurideh, M., Masa'deh, R.M.T., Alkurdi, B.: The effect of customer satisfaction upon customer retention in the Jordanian mobile market: an empirical investigation. Eur. J. Econ. Financ. Adm. Sci. **47**, 69–78 (2012)

62. Al Dmour, H., Alshurideh, M., Shishan, F.: The influence of mobile application quality and attributes on the continuance intention of mobile shopping. Life Sci. J. **11**(10), 172–181 (2014)

63. Al-Dmour, H., Alshuraideh, M., Salehih, S.: A Study of Jordanians' television viewers habits. Life Sci. J. **11**(6), 161–171 (2014)

64. Nielsen-Englyst, L.: Operations strategy formation–a continuous process. Integr. Manuf. Sys. (2003)

65. Miller, J.G., Roth, A.V.: A taxonomy of manufacturing strategies. Manage. Sci. **40**(3), 285–304 (1994)

66. Paiva, E.L., Roth, A.V., Fensterseifer, J.E.: Organizational knowledge and the manufacturing strategy process: a resource-based view analysis. J. Oper. Manage. **26**(1), 115–132 (2008)

67. Riis, J.O., Johansen, J., Waehrens, B.V., Englyst, L.: Strategic roles of manufacturing. J. Manuf. Technol. Manage. **18**(8), 933–948 (2007)

68. Slack, N., Lewis, M.: Operations strategy. Pearson Education (2002)

69. Rytter, N.G., Boer, H., Koch, C.: Conceptualizing operations strategy processes. Int. J. Oper. Prod. Manage. (2007)

70. Ward, P.T., Bickford, D.J., Leong, G.K.: Configurations of manufacturing strategy, business strategy, environment and structure. J. Manage. **22**(4), 597–626 (1996)

Artificial Intelligence and Healthcare: The Effects of COVID-19 on Nigerians

T. Falana, J. Renner, A. Adekoya, and O. Abolurin

Abstract The purpose of this study is to determine the psychological and social impact of artificial intelligence and COVID-19 on Nigerians. A descriptive design was adopted for the study and four hundred and twenty-two Nigerians participated in the study. A self-administered questionnaire was used to obtain information on relevant issues and the gathered information was presented on frequency-percentage tables and charts while the hypotheses were tested using chi-square at 5% level of statistical error. Results from the study show that majority of the respondents 214 (50.7%) were not highly impacted on psychologically by Coronavirus and Artificial Intelligence while 208 (49.3%) were highly impacted on by artificial intelligence and Coronavirus with a mean score of 21.5 ± 1.9. Majority of the respondents 232 (55.0%) were less impacted on socially by Coronavirus and artificial intelligence while 190 (45.0%) were highly impacted on by Coronavirus and artificial intelligence with a mean score of 18.6 ± 2.3. The study discovered that majority of the respondents were not highly impacted psychologically by Coronavirus and Artificial Intelligence, majority of the respondents were less impacted socially by Coronavirus and artificial intelligence. The study revealed that there was a significant impact on the psychological welfare of Nigerians. Moreover, there was a significant impact on the social welfare of Nigerians. More studies should be carried out to identify the level of adoption of artificial intelligence in the management of pandemic in Nigeria.

Keywords Artificial intelligence · Healthcare · CoVid-19 · Psychological health · Social health

1 Introduction

Coronavirus belongs to a large family of infectious viruses, which are enveloped with a single positive-stranded huge RNA and cause illnesses ranging from common

T. Falana · J. Renner (✉) · A. Adekoya · O. Abolurin
Department of Pediatrics, Babcock University Teaching Hospital, Ilishan-Remo, Ogun State, Nigeria
e-mail: rennerjames@babcock.edu.ng

M. T. Alshurideh et al. (eds.), *The Effect of Coronavirus Disease (COVID-19) on Business Intelligence*, Studies in Systems, Decision and Control 334,
https://doi.org/10.1007/978-3-030-67151-8_7

cold to a more serious illness such as Severe Acute Respiratory Syndrome (SARS) and Middle East Respiratory Syndrome (MERS). Historically, in 2002, SARS CoV was transmitted to humans from Soviet cats in China while MERS-Cov was transmitted to humans from camels in 2012 in Saudi Arabia. Coronavirus entails four major families which are the alpha coronavirus, beta coronavirus, gamma coronavirus and lastly, delta coronavirus with a genome size that varies between 26 and 32 kb. The alpha coronavirus causes mild disease and fatalities while the beta coronavirus causes severe symptomatic infections. The novel coronavirus (COVID 19) formerly called nCoV is a family of beta coronavirus with a structural protein like membrane glycoprotein, spike protein, small membrane protein, nucleocapsid which trigger respiratory tract infection [43].

The active growth period of COVID-19 or SARS CoV 2 is about 7–14 days according to Anjorin [2] after being transmitted from humans to humans through droplets from coughing and sneezing. The early clinical sign of SARS CoV 2 according to Velavan et al. [43] was detected as pneumonia and symptoms of gastrointestinal and mild infections, especially in young children. The clinical manifestation of symptomatic patients includes cough, tiredness, fever, nasal congestion, dyspnoea, decrease in oxygen saturation and blood gas deviation. Globally Roser et al. [38] recorded COVID 19 deaths of 340 thousand and confirmed cases of over 5 million people, while, Craven et al. [7] stressed that the outbreak of the virus and the growing effects on the global economy especially in the erosion in consumer's confidence that the community spread of COVID-19 are not properly contained has led to low demand of some sectors like hotels, restaurant, tourism and aviation, thus, resulting in the global growth rate of 1.8% instead of 2.5% projected in early 2020.

As Martineau [25] opined, artificial intelligence (AI) are heavily being deployed in the fight against the highly infectious and deadly virus in the aspect of initial detection of symptoms to ensure healthier chances of survival of symptomatic patients, blocking the virus from binding to human cells through silk proteins, building machine learning models that forecast how susceptible an individual is to the virus and what tailored treatment might be most suitable. Martineau [25] further stated that AI is globally used to check through health records and medical claims that drugs use to fight severe health conditions might fight the deadly virus and also flattening the curve of the spread of the infectious virus through automated contact tracing. Apart from the advanced countries, in Africa, Rwanda deploys artificial intelligence machines or robots to prevent health care workers from getting infected with the virus by screening people for early detection of the virus, conveyance of drugs and food to symptomatic patients or isolation centres [27]. In Nigeria, 6 PCR laboratories with machines utilizing AI have been deployed to increase the testing capacity of COVID 19, besides, SORMAS digital tools have been rolled out for surveillance and to implement real-time reporting of data [26].

In Nigeria, 264 healthcare workers have been exposed to the virus, according to Onyedinefu [30], while, 113 health caregivers have tested positive to the virus according to Emi [12]. Also, NCDC [26] recorded 7526 confirmed cases of the virus with 2174 people have recovered from the virus but sadly 221 people have died from the complications of the virus. As Kalu [19] and Okwumbu [33] stated, the rising cases of the people testing positive and the number of causalities from the virus made the president declare a lockdown on March 30, 2020, for two weeks in Lagos, Abuja and Ogun State and extended the lockdown on April 13 for additional two weeks then later, declaration of curfew and extension of nationwide curfew from 8 pm to 6 am after the subsequent lockdowns to flatten the curve of the spread of the virus, in a country that is referred to as the headquarters of poverty. Ted [42], recorded an increase in the redundancy rate from 18.1% in 2018 to 23.1% among 15–64 years of age in the third part of 2019 Ojekunle [29] and 61% rate of redundancy, mostly in children and adults [28]. Furthermore, Onehi [31] stated that nCov pandemic has increased the redundancy rate of youths in Nigeria.

2 Statement of the Problem

There are rising cases of health caregivers contracting the virus COVID-19. There is no employment of artificial intelligence machines or robots in the fight against the novel coronavirus in Nigeria, unlike Rwanda which employs robots to reduce the risk of health caregivers from contracting the infectious virus and to boost testing of people for early detection of the virus. There have been cuts in salaries of workers and massive redundancies of labourers due to the effect of curfew and previous lockdown to curtail the spread of COVID 19. This research, therefore, studies the impact of artificial intelligence and COVID 19 on psychosocial welfare of Nigerians.

3 Objective of the Study

The focal purpose of this study was to investigate the impact of artificial intelligence and COVID-19. The specific objectives were to

(i) Evaluate the effects of artificial intelligence and COVID-19 on the psychological welfare of Nigerians.
(ii) Explore the influence of artificial intelligence and COVID-19 on the social welfare of Nigerians.

4 Research Questions

The study answered the following research inquiries

(i) How do artificial intelligence and COVID 19 impact the psychological welfare of Nigerians?
(ii) To what effect do artificial intelligence and COVID 19 have on the social welfare of Nigerians.

5 Research Hypotheses

This research study tested the following hypothesis at 5% level of significance

$H_0 1$ Artificial intelligence and COVID 19 does not affect the psychological welfare
          of Nigerians
$H_0 2$ Artificial intelligence and COVID 19 have no impact on the social welfare of
          Nigerians.

6 Justification of the Study

Nigeria is the headquarters of poverty, according to World Poverty Clock [45], which
revealed 91.8 million Nigerians are living below N693 per day. Lancer and Parolin
[23] opined that lockdown especially the closure of schools as part of physical
distancing for a long time has damaging health and social effects on vulnerable children, thus aggravating current inequalities The declaration of lockdown, working
hours from 9 am to 3 pm and curfew from 8 pm to 6 am by the federal government
to flatten the curve of the spread of the deadly virus from person to person posed big
issues to Nigerians' productivity because this period of curfew is the working hours
of Nigerians [3]. Also Anazia et al.[3] stated that social distancing in the workplace
disrupts the interaction of workers and also limit a large number of workers who are
working around this period. The CNBC Africa [10] asserted that the total or partial
lockdown led to vulnerable Nigerians losing their daily livelihood while Orjinmo
[32] opined that lockdowns led to quieter roads and reduction of traffic congestions.

7 Significance of the Study

This result of this study is beneficial not only on how artificial intelligence aids in
the fight against COVID-19 but also the psychological and social impact of artificial
intelligence and COVID-19 on Nigerians The significance of this research study is
that Nigeria government would know the implications of the various lockdown (total
or partial in form of curfew) on the psychological and social wellbeing of Nigerians.

8 Scope of the Study

The scope of this study was focused on Nigerians living in Nigeria. The sampling
size was derived by computing the minimum sample size required for accuracy
in estimating proportions by considering the standard normal deviation set at 95%
confidence level (1.96), percentage picking a choice or response (50% = 0.5) and

the confidence interval ($0.05 = \pm 5$). The population comprises all age groups from 15 to 80 currently living in Nigeria while the sampling methodology that was used as a convenience sampling method. This was done by sharing a Google form for individuals to fill. The research instrument for data collection was a self-structured questionnaire with a close-ended question to be administered to Nigerians through Google form. It was divided into 2 sections (SECTION A: Sociodemographic data of the respondents and SECTION B: Artificial Intelligence and CoVid-19 impact on the psychological welfare and the Social welfare of Nigerians). Lastly, this study was carried out within 4 weeks.

9 Conceptual Review

9.1 Artificial Intelligence

Artificial intelligence (AI) is defined by Habeeb [18] as a part of computer science that stresses the manufacture of intelligent machines that mimic humans. The scholar stressed that some activities with artificial intelligence include recognition of speech, deep learning, planning and solving the problem. Similarly, Frankenfield [13] defined AI as simulating the intelligence of human beings into a machine with the aid of programming languages that enable the machine to be able to reason like humans while Schroer [41] described AI as a comprehensive aspect of computer science concerned with developing smart machines capable of performing jobs that typically required human intelligence. deployed in the fight against the highly infectious and deadly virus in the aspect of initial detection of symptoms to ensure healthier chances of survival of symptomatic patients, blocking the virus from binding to human cells through silk proteins, building machine learning models that forecast how susceptible an individual is to the virus and what tailored treatment might be most suitable.

Historically, scholars have dated back the development of artificial intelligence fortified machines from the eighteenth century of duck with the ability to flap its wings, eat and digest grains created by Jacques de Vaucanon to the nineteenth century where the term robotics was defined by Asimov and the invention of human-like robot that can walk, talk and smokes called Elektro by Westinghouse according to Franck [14] to a programmable paint sprayer device created by Willard Pollard and Harold Roselund in 1942 which was later improved upon by George Devol to design the first truly programmable robot in 1954 called Unimate Although, Unimate was firstly used by Trallfa to spray their produce wheelbarrows because of the bad labour environment in Norway which later became a global success. Robotic technologies transformed from electrohydraulic robot arm of Unimate to self-replicating robots [24], now to living robots, having the capacity to heal themselves when assigned to dangerous tasks [17] and Velox-a generator that have fins capable of moving land and water [4].

Martineau [25] further stated that AI is globally used to check through health records and medical claims that drugs use to fight severe health conditions might fight the deadly virus and also flattening the curve of the spread of the infectious virus through automated contact tracing. Apart from the advanced countries, in Africa, Rwanda deploys artificial intelligence machines or robots to prevent health care workers from getting infected with the virus by screening people for early detection of the virus, conveyance of drugs and food to symptomatic patients or isolation centres [27]. In Nigeria, 6 PCR laboratories with machines utilizing AI have been deployed to increase the testing capacity of COVID 19, besides, SORMAS digital tools have been rolled out for surveillance and to implement real-time reporting of data [26].

Park and Han [35] examined the effect of employing AI and AI-equipped machines to make a high-dimensional or overparameterized predictive and diagnostic analysis of clinical images, epidemiology and biostatistics by summarizing the statistical systems for measuring the discriminating and standardizing efficiency of a diagnostic or extrapolative model. The scholars discovered that huge data of medical images are used to develop a deep learning algorithm through the use of deep neural networks but stress the need to employ external catalogue instead of internal record obtained from an exact medical cohort to circumvent overvaluing the medical performance as a consequence of overfitting in high-dimensional or over parametrized sorting model and gamut bias, and the fundamentals for attaining a more comprehensive quantifiable appraisal.

9.2 COVID-19

The novel coronavirus (COVID 19) according to WHO [44] is defined as a contagious illness caused by a newly revealed coronavirus. WHO [44] further expatiated that individuals who have contacted the virus initially experience minor to severe respiratory diseases, David et al. [11] defined COVID-19 as a sickness caused by severe acute respiratory syndrome coronavirus 2 (SARS-COV-2). David et al. [11] stress the fact that the novel coronavirus name which was firstly identified in Wuhan City of China, Hubel Province, the initial name of the virus was changed from 2019-n COV to SARS-COV-2 to sidestep stigmatization of Wuhan. Coronavirus entails of four major families which are alpha coronavirus, beta coronavirus, gamma coronavirus and lastly, delta coronavirus with a genome size that varies between 26 and 32 kb. The alpha coronavirus according to causes mild disease and while the beta coronavirus causes severe symptomatic infections.

The novel coronavirus (COVID 19) formerly called nCoV is a family of beta coronavirus with structural proteins like membrane glycoprotein, spike protein, membrane protein, nucleocapsid which trigger respiratory tract infection [43]. The active growth period of COVID-19 or SARS CoV 2 is about 7–14 days according to Anjorin [2] after being transmitted from humans to humans through droplets from coughing and sneezing. The early clinical sign of SARS CoV 2 according to Velavan et al. [43] was detected as pneumonia and symptoms of gastro-intestinal and mild infections, especially in young children. The clinical manifestation of symptomatic patients includes cough, tiredness, fever, nasal congestion, dyspnoea, decrease in oxygen saturation and blood gas deviation. Globally Roser et al. [38] recorded COVID 19 deaths of 340 thousand and confirmed cases of over 5 million people, while, Craven et al. [7] stressed that the outbreak of the virus and the growing effects on the global economy especially in the erosion in consumer's confidence that the community spread of COVID-19 are not properly contained has led to low demand of some sectors like hotels, restaurant, tourism and aviation, thus, resulting in the global growth rate of 1.8% instead of 2.5% projected in early 2020.

9.3   Psychological Welfare

Robertson [39] defined psychological welfare as having positive mind states such as joyfulness cheerfulness or fulfilment. The scholars further, asserted that psychological welfare composes of two major facets which are subjective welfare and purposeful living. Subjective welfare refers to the degree to which individuals experience constructive emotions or cheerfulness while a purposeful living entails the state of the mind that feels what an individual is doing with his or her life is filled with purpose and meaning. Also, Snyder et al. [40] defined psychological welfare as a positive association with other people, self-mastery coupled with purposeful mind and feeling of a life with meaning. The scholar further defined psychological wellbeing as a state of reaching a balance influenced by rewards and challenges from life's events. Snyder et al. [40] further stated that psychological wellbeing of a person is perceived self-competency, autonomy, self-confidence and personal achievement.

9.4   Social Welfare

Coleman [8] defined social welfare as a state of having social peace, social stability and having a good social association with people. Social welfare is further defined by the scholar as having positive social interactions with people. Social welfare is developed from the freedom of a person, and equal rights of the citizenry. Social welfare is cultivated by ethics, emotional intelligence, ability to adapt, sympathy and lastly, selflessness [8]. Conversely, Social welfare is defined by Cicognani [9] as the final condition in which basic human wants are met and individuals can live together tranquilly in societies with opportunities for development. The final state is pigeonholed by parity in the access to and delivery of essential amenities. Koutronas [22] also expanded social welfare as social security of the community or society against both social and economic violatility as a result of stoppage of livihood of the citizenry, thus aggravating unemployment, sickness, poverty and death. It is imperative for government to provide adequate medical care and subsidies through well-established nexus of social institutions with the motive of playing a vital role in the promotion and protection of the socio-economic welfare of the citizenries.

10   Empirical Review

10.1   *Artificial Intelligence and Healthcare*

Empirically, the study of Bennett and Hauser [5], investigated the effect of developing AI framework to address healthcare complexities like swift escalating cost, rapid expanding streams of information the growing myriad of treatment options in the modern healthcare system hampering healthcare givers in selecting optimal treatment decisions. The study employed Markov decision process and dynamic framework to study from patient data gotten from health record and make a complex plan by simulation of substitutes chronological decision paths while addressing the sometimes contradictory synergistic contacts of various mechanisms in the healthcare system. The study discovered that Artificial intelligence framework effortlessly outclasses the current treatment-as-usual (TAU), case-rate or fee-for-service models of hospitality. The scholar concluded that simulating AI framework can approximate the best decision even in difficult and indeterminate circumstances. Similarly, the study of Park and Han [35] examined the effect of employing AI and AI-equipped machines to make a high-dimensional or overparameterized predictive and diagnostic analysis of clinical images, epidemiology and biostatistics by summarizing the statistical systems for measuring the discriminating and standardizing efficiency of a diagnostic or extrapolative model. The scholars discovered that huge data of medical images are used to develop a deep learning algorithm through the use of deep neural networks but stress the need to employ external catalogue instead of internal record obtained from an exact medical cohort to circumvent overvaluing the medical performance as a consequence of overfitting in high-dimensional or over parametrized sorting model and gamut bias, and the fundamentals for attaining a more comprehensive quantifiable appraisal

However, the study of Fennema-Notestine et al. [15] oppose the involvement of robotic technologies in the healthcare system especially skull stripping or isolating brain tissues from non-brain tissues, the study used structural gradient (SPGR) T_1 weighted pulse sequence and GE 1.5T to gather data at the facility of VASan Diego Healthcare system MRI and also represented four (4) diverse diagnostic group\s which include depressed (DEPR), Alzheimer (AD), young (YNC) and elderly or advanced age normal controls (ENC) to opine that even though manual alteration to create an appropriate skull stripping volume which upsurges processing time and the rate of essential professionals and eventually leads to some errors, however, skull stripping method should be relatively robotized.

10.2 Artificial Intelligence: The Effect of COVID 19 on Psychological Welfare

The study of Rajkumar [37] investigated the mental health consequences as a result of COVID-19 outbreak. The study used prior studies of published literature on COVID-19 and mental health to discovered that symptoms of anxiety, depression of about 16–28%, while 8% of self-reported stress due to the mental reaction of the crown virus epidemic. The scholar, also asserted that COVID pandemic leads to subsyndromal psychological issues and thus recommended a precautionary measure must be put in place to avert mental illness. Subsyndromal mental health issues are a common response to the deadly virus contagion. Moreover, the study of Flesia et al. [16] empirically looked at the mental stress caused by the unexpected COVID-19 and subsequent lockdowns and how machine learning (ML) which is an aspect of artificial intelligence is employed to foresee human behaviour and are proficient on mental health data in recognizing people who are more likely to demonstrate a high level of perceived trauma during the COVID 19 the author stress the benefit of machine learning firstly, in making of complex models considering the explanatory variable, Secondly, ML makes it conceivable to make implications at the individual level while old-style statistical methods emphasized on the group level. The study employed 1628 participants of which 314 participants are suffering from a high level of anxiety to discover that COVID 19 increases the stress of Italians.

10.3 Artificial Intelligence: The Effect of COVID 19 on the Social Welfare of Nigerians

The study of Pirouz et al. [36] analyzed COVID-19 cases of Hubei province in China. The database was set for 30 days which consist of ranges of daily temperature, the population density of the municipal and wind spread and also the number of confirmed cases The study used binary cataloguing modelling was employed by the Group Method of Data Handling (GMDH) algorithm to predict confirmed cases. The study discovered that relative humidity and the maximum daily temperature had the highest impact on confirmed cases. The study also opined that isolation has a positive influence on reducing the number of confirmed cases.

10.4 Gaps in Literature

From all the works of literature reviewed none were on the effect of COVID-19 on Nigerians. Some authors reviewed how artificial intelligence has influenced the healthcare system in terms of prediction and diagnostic analysis of clinical images, epidemiology and biostatistics by summarizing the statistical systems for measuring the discriminating and standardizing efficiency of a diagnostic or extrapolative model, addressing healthcare complexities while some authors have investigated on the mental health consequences of the COVID-19 outbreak by empirically looking at prior studies of published literature on COVID-19 and mental health. This study assessed the effect of COVID-19 on the psychological and social wellbeing of Nigerians and how artificial intelligence aids in the fight against COVID-19.

11 Methodology

11.1 Study Design

This study was a non-experimental cross-sectional descriptive study investigating the impact of artificial intelligence and COVID-19 on Nigerians.

11.2 Study Area/Setting

This study was conducted in Nigeria. Officially the federal republic of Nigeria, it is a sovereign country located in West Africa bordering Niger in the North, Chad in the Northeast, Cameroon in the East, and Benin in the West. Its southern coast is on the Gulf of Guinea in the Atlantic Ocean. A federal republic comprising 36 states and the Federal Capital Territory, where the capital, Abuja is located. It has a population estimate of 204 million and inhabited by 250 ethnic groups with the three largest ethnic group being Hausa, Igbo and Yoruba. Nigeria announced its first CoVid-19 case on 28th of February, 2020. As at the 25th of May, Nigeria confirmed 229 new cases totalling 8068 confirmed cases out of 45,685 tests with 233 deaths, Lagos State had the highest number of cases with 3595.

11.3 Study Population

The study population comprised all age groups from 15 to 80 currently living in Nigeria.

11.4 Sample Size

Where the expected response population was unknown, the sample size was derived by computing the minimum sample size required for accuracy in estimating proportions by considering the standard normal deviation set at 95% confidence level (1.96), percentage picking a choice or response (50% = 0.5) and the confidence interval (0.05 = ± 5). The formula is:

$$n = Z^2 * (p) * (1 - p)/c$$

where
  z = standard normal deviation set at 95% confidence interval
  p = percentage picking a choice or response
  c = confidence interval
  $n = (1.96)^2 * 0.5 * (1 - 0.5)/0.05^2$
  $n = 3.8416 * 0.25/0.0025$
  $n = 384.16$
  Therefore, $n = 384$
  Attrition rate is 10% of the total sample size
  Attrition rate $= 384 * 10/100$
  Attrition rate $= 38.4$
  Attrition rate $= 38$
  Total sample size $= n +$ Attrition rate $(384 + 38)$
  Total sample size $= 422$.

11.5 Eligibility Criteria

Inclusion Criteria

- All age range in Nigeria.
- The individual must be living in Nigeria currently.

Exclusion Criteria

- A Nigeria who is not present within the shores of Nigeria.

11.6 Sampling Method

The sampling methodology was used as a convenience sampling method. This was done through sharing a Google form for individuals to fill, anyone could fill it at their convenience and as long as they had access to their mobile device.

11.7   Study Instrument

The research instrument for data collection was a self-structured questionnaire with closed-ended question administered to Nigerians through Google form. It was divided into 2 sections

SECTION A: Sociodemographic data of the respondents
SECTION B: Artificial Intelligence and CoVid-19 impact on the psychological welfare and the Social welfare of Nigerians.

11.8   Method of Data Collection

For data collection, a structured standard questionnaire was used. Structured English version questionnaire was developed from relevant literature and due consultation with the researchers' supervisor, which was an online-administered questionnaire. This is because the respondents are literate and this allowed for individual privacy as no name was required when filling the form. It was identified that consent has been accepted when filling the form. Data filled from the Google form was automatically recorded in an excel spreadsheet.

11.9   Validity and Reliability of Instrument

11.9.1   Validity of Research Instrument

A comprehensive review of relevant literature was carried out. This guided the formulation of research questions and the objectives were also be used in the development of the questionnaire for the study. A draft of the questionnaire was constructed, which has undergone an independent review from peers and experts in the field of medicine. Supervisor's review was then used to fine-tune the instrument. Special care was ensured to monitor the quality of data collected through supervision during the collection of data.

11.9.2   Reliability of Research Instrument

Reliability is the accuracy of a research-measuring instrument. The questionnaire was pre-tested to ascertain its consistency and reliability and to determine whether the questions were clear and simple enough for participants' comprehension. 10% of the sample populations [37] were used for the pre-test and this was done among randomly selected Nigerians. At the end of the exercise, questions that were not easily understood were reframed, those that were found to be irrelevant were removed and adequate spaces were provided for responses.

The reliability of the questionnaire was determined from the pre-test using Alpha Cronbach test. The result was interpreted as reliable if the result is greater than 0.7.

11.10   Ethical Consideration

Ethical approval to conduct this study and administer questionnaire was obtained from the Babcock University Health Research Ethics Committee (BUHREC) Ilishan. The respondents' consent was obtained after the provision of adequate, clear and complete information about what the study entails. Written informed consent was obtained from each participant. Ethical standard principles were strictly adhered to to ensure confidentiality. Participants were informed that participation was voluntary and that data collected would be used mainly for research purposes. Anonymity and confidentiality of responses were ensured by not requesting for names or any other personal identifiers from the respondents.

11.11   Study Duration

The study was carried out within 4 weeks.

11.12   Method of Data Analysis

Descriptive and inferential statistics were used for the analysis of data. The data collected from the questionnaire was manually sorted out, edited and coded in Microsoft Excel spreadsheet. It was thereafter analyzed using Statistical package for social services (SPSS) software version 21.

Frequency and percentage of tables were generated for the demographic characteristics of the respondents. Statistical significance for the association was tested using chi-square for discrete data with a p-value less than 0.05 considered statistically significant.

12   Data Analysis Result and Discussions, Findings

This chapter presents the analysis of information obtained from a study investigating the psychological and social impact of artificial intelligence and COVID-19 on Nigerians. Four hundred and twenty-two (422) questionnaires were distributed; all were

Table 1 Socio-demographic characteristics (N = 422)

Variables		Frequency	Percentage (%)
Gender	Female	282	66.8
	Male	140	33.2
Age	18–23	189	44.8
	24–29	87	20.6
	30–35	105	24.9
	36–41	37	8.8
	42 and above	4	0.9
Education	WASSCE or equivalents	20	4.7
	OND/NCE	98	23.2
	BSc/BA/HND	255	60.4
	Master's degree	43	10.2
	Ph.D.	6	1.4
Employment	Government employment	235	55.7
	Private employment	48	11.4
	Self-employed	59	14.0
	Unemployed	19	4.5
	Student	61	14.5

correctly and filled giving a response rate of 100.0%. The results of the study were presented on frequency-percentage tables, charts and the hypotheses were tested at 5% significance level using Chi-square analysis.

12.1 Socio-Demographic Characteristics

Table 1 reveals that majority of the respondents 282 (66.8%) are females while 140 (33.2%) are males, 189 (44.8%) are aged within 18-23 while minority 4 (0.9%) are aged 42 and above. 225 (60.4%) have BSC/BA/HND certificates of education while 6 (1.4%) have a PhD. 235 (55.7%) are government employed while 19 (4.5%) are unemployed.

12.2 Artificial Intelligence and COVID 19 Impact on Nigerians

Table 2 shows that majority of the respondents 268 (63.5%) are not sure if COVID-19 is highly infectious, 255 (60.4%) are not sure if total lockdown would flatten the

Table 2 Artificial intelligence and COVID 19 impact on the psychological welfare of Nigerians (N = 422)

	SA (%)	A (%)	D (%)	SD (%)	NS (%)
I believe COVID-19 is a highly infectious disease	24 (5.7)	31 (7.3)	50 (11.8)	49 (11.6)	268 (63.5)
I believe total lockdown or curfew to flatten the curve of the spread of the virus makes me feel safer	33 (7.8)	49 (11.6)	56 (13.3)	29 (6.9)	255 (60.4)
Extending curfew to flatten the curve of the spread of the virus makes me feel safer	28 (6.6)	71 (16.8)	251 (59.5)	47 (11.1)	25 (5.9)
Curfew to flatten the curve of COVID-19 infection increased my salary	0 (0.0)	2 (0.5)	90 (21.3)	330 (78.2)	0 (0.0)
Extension of curfew to flatten the curve of the spread of the virus makes me feel safer	11 (2.6)	296 (70.1)	30 (7.1)	39 (9.2)	46 (10.9)
Washing of hands regularly minimize the risk of contracting the virus	53 (12.6)	363 (86.0)	0 (0.0)	0 (0.0)	6 (1.4)
Extension of curfew increases my stress	290 (68.7)	80 (19.0)	10 (2.4)	23 (5.5)	19 (4.5)
Extension of curfew makes me poorer	1 (0.2)	370 (87.7)	31 (7.3)	1 (0.2)	19 (4.5)
I believe that artificial intelligence is used for early detection of sepsis of COVID 19	11 (2.6)	96 (22.7)	30 (7.1)	39 (9.2)	246 (58.3)
I believe that machine learning can expedite drugs that can cure the COVID-19	7 (1.7)	16 (3.8)	329 (78.0)	12 (2.8)	58 (13.7)
Artificial intelligence provides data that will assist healthcare givers in finding a better ventilator setting for COVID 19 patients	323 (76.5)	32 (7.6)	15 (3.6)	19 (4.5)	33 (7.8)
Artificial intelligence machine (Robot) will prevent healthcare givers from having the virus when treating COVID-19 patients	1 (0.2)	390 (92.4)	3 (0.7)	6 (1.4)	22 (5.2)

curve of the coronavirus making them feel safer, 251 (59.5%) disagree that extending the curfew to flatten the curve of COVID-19 infection, 330 (78.2%) strongly disagree that the curfew has led to increase in their salary, 296 (70.1%) agree that extension of the curfew makes them feel safer. 363 (86.0%) agree that washing of hands regularly minimizes the risk of contracting the virus, 290 (68.7%) strongly agree that the extension of the curfew increases their stress. 370 (87.7%) agree that the extension of the curfew makes them poorer, 246 (58.3%) are not sure if artificial intelligence

Table 3 Level of impact on respondents' psychological welfare *Source* Author's Computation, 2020

Value	Score	Frequency	Per cent (%)
Mean score = 21.5 ± 1.9	<22	214	50.7
	≥22	208	49.3
Total		422	100.0

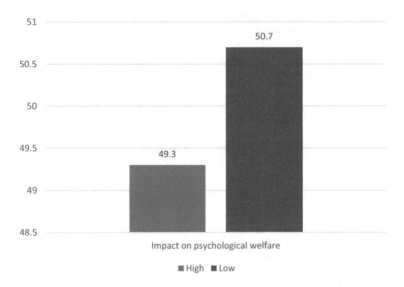

Fig. 1 Impact on the psychological welfare

is used for early detection of sepsis. 329 (78.0%) disagree that machine learning can expedite drugs that can cure the COVID-19. 323 (76.5%) strongly agree that artificial intelligence provides data that will assist healthcare givers in finding a better ventilator setting for CoVid-19 patients. 390 (92.4%) agree that artificial intelligence (machine) will prevent healthcare givers from having the virus when treating COVID-19 patients (Table 3).

Figure 1 shows that majority of the respondents 214 (50.7%) were not highly impacted on psychologically by Coronavirus and Artificial Intelligence while 208 (49.3%) were highly impacted on by artificial intelligence and Coronavirus.

12.3 Artificial Intelligence and COVID 19 Impact on Nigerians

Table 4 shows that regarding the social impact of COVID-19 on Nigerians, majority of the respondents 383 (90.8%) strongly agree that avoiding shaking makes them lose

Table 4 Artificial intelligence and COVID 19 impact on the social welfare of Nigerians (N = 422)

	SA (%)	A (%)	D (%)	SD (%)	NS (%)
Avoiding handshakes makes to flatten the curve of the spread of the virus makes me lose more friends	383 (90.8)	35 (8.3)	0 (0.0)	0 (0.0)	4 (0.9)
Social distancing in the office makes me bored	6 (1.4)	28 (6.6)	23 (5.5)	10 (2.4)	355 (84.1)
Lockdown makes me increases togetherness with my family	339 (80.3)	14 (3.3)	36 (8.5)	26 (6.2)	7 (1.7)
Curfew makes increases family togetherness	324 (76.8)	20 (4.7)	43 (10.2)	28 (6.6)	7 (1.7)
I believe lockdown increases domestic violence	12 (2.8)	346 (82.0)	23 (5.5)	22 (5.2)	19 (4.5)
I believe curfew increases domestic violence	16 (3.8)	350 (82.9)	23 (5.5)	17 (4.0)	16 (3.8)
Lockdown increases societal unrest	56 (13.3)	300 (71.1)	16 (3.8)	22 (5.2)	28 (6.6)
I believe talking with face mask reduces my friendship	11 (2.6)	3 (0.7)	396 (93.8)	8 (1.9)	4 (0.9)
Automated contact tracing will help limit the spread of COVID-19	383 (90.8)	10 (2.4)	6 (1.4)	9 (2.1)	14 (3.3)
Machine learning can provide data on the effect of curfew on the mental health of people	42 (10.0)	299 (70.9)	10 (2.4)	12 (2.8)	59 (14.0)
Machine learning can predict how COVID-19 cases and death across cities	308 (73.0)	19 (4.5)	19 (4.5)	32 (7.6)	44 (10.4)
Artificial intelligence can solve boredom	320 (75.8)	13 (3.1)	29 (6.9)	39 (9.2)	21 (5.0)

more friends, 355 (84.1%) is not sure if social distancing in the office makes them bored, 339 (80.3%) strongly agree that lockdown makes them increase togetherness with their family, 324 (76.8%) strongly agree that curfew increases family togetherness, 346 (82.0%) agree that curfew increases domestic violence, 300 (71.1%) agree that lockdown increases social unrest, 396 (93.8%) disagree that talking with facemask reduces their friendship, 383 (90.8%) strongly agree that automated contact tracing would help limit COVID-19 spread, 299 (70.9%) agree that machine learning can provide data on the effect of curfew on mental health. 308 (73.0%) strongly agree that machine learning can predict how COVID-19 cases and deaths across cities. 320 (75.8%) strongly agree that artificial intelligence can solve boredom (Table 5).

Figure 2 shows that majority of the respondents 232 (55.0%) were less impacted on socially by Coronavirus and artificial intelligence while 190 (45.0%) were highly impacted on by Coronavirus and artificial intelligence.

Table 5 Level of impact on respondents' social welfare *Source* Author's Computation, 2020

Value	Score	Frequency	Per cent (%)
Mean score = 18.6 ± 2.3	<19	232	55.0
	≥19	190	45.0
Total		422	100.0

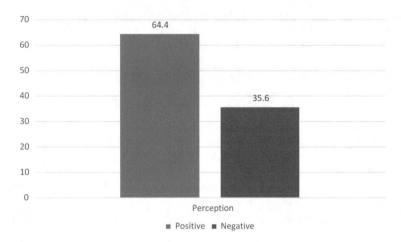

Fig. 2 Impact on respondents' social welfare

13 Test of Hypotheses

Hypothesis One Artificial intelligence and COVID 19 does not affect the psychological welfare of Nigerians.

Table 6 above reveals that there was a significant impact on the psychological welfare of Nigerians with a *p*-value <0.05 (0.000). Therefore, we rejected the null hypothesis and also a significant impact on the social welfare of Nigerians with a *p*-value <0.05 (0.000). Therefore, we rejected the null hypothesis.

Table 6 Cross-tabulation of artificial intelligence and COVID 19 impact on the psychological and Social welfare

	Respondents		N	Mean	Standard deviation	Mean difference	t-value	*p*-value for $H_0 1$ & $H_0 2$
Social welfare	Low	$H_0 1$	214	20.056	1.060	−2.948	−26.850	0.000
		$H_0 2$	232	16.918	0.952	−3.75	−27.372	
	High	$H_0 1$	208	23.004	1.194	−2.948	−26.850	
		$H_0 2$	190	20.668	1.802	−3.75	−25.874	

13.1   Answering Research Questions

Research question 1: How do artificial intelligence and COVID 19 impact the psychological welfare of Nigerians?
This can be identified from Table 3 above which shows that majority of the respondents 214 (50.7%) were not highly impacted on psychologically by Coronavirus and Artificial Intelligence while 208 (49.3%) were highly impacted on by artificial intelligence and Coronavirus with a mean score of 21.5 ± 1.9.

Hypothesis One Artificial intelligence and COVID 19 does not affect the psychological welfare of Nigerians.

The study revealed that there was a significant impact on the psychological welfare of Nigerians with a p-value <0.05 (0.000). Therefore, we rejected the null hypothesis.

Research question 2: To what effect do artificial intelligence and COVID 19 have on the social welfare of Nigerians?
From the analysis of findings, Table 5 shows that majority of the respondents 232 (55.0%) were less impacted on socially by Coronavirus and artificial intelligence while 190 (45.0%) were highly impacted on by Coronavirus and artificial intelligence with a mean score of 18.6 ± 2.3.

Hypothesis Two Artificial intelligence and COVID 19 have no impact on the social welfare of Nigerians.

The study reveals that there was a significant impact on the social welfare of Nigerians with a p-value <0.05 (0.000). Therefore, we rejected the null hypothesis.

13.2   Discussion of Findings

13.2.1   Sociodemographic Findings

The result of the study has revealed that majority of the respondents 282 (66.8%) are females while 140 (33.2%) are males, 189 (44.8%) are aged within 18-23 while minority 4 (0.9%) are aged 42 and above. 225 (60.4%) have BSC/BA/HND certificates of education while 6 (1.4%) have a PhD. 235 (55.7%) are government employed while 19 (4.5%) are unemployed.

13.2.2   Artificial Intelligence and COVID 19 Impact the Psychological Welfare of Nigerians

The findings revealed that majority of the respondents 214 (50.7%) were not highly impacted on psychologically by Coronavirus and Artificial Intelligence while 208

(49.3%) were highly impacted on by artificial intelligence and Coronavirus with a mean score of 21.5 ± 1.9. This is not in close relation to a study by Panchal et al. [34], where nearly half the adults in the United States (45%) reported that their mental health has been negatively impacted due to worry and stress over the virus.

The study under review also found out that 330 (78.2%) strongly disagree that the curfew has led to increase in their salary, this goes in tandem with existing researches as noted in Panchal et al. [34] where it was identified that job loss is associated with increased depression, anxiety, distress, and low self-esteem and may lead to higher rates of substance use disorder and suicide. Recent polling data shows that more than half of the people who lost income or employment reported negative mental health impacts from worry or stress over coronavirus, and lower-income people report higher rates of major negative mental health impacts compared to higher-income people. The recent KFF Tracking Poll found that older adults were less likely than adults ages 18–64 to report that worry or stress related to the coronavirus has hurt their mental health (31% vs. 49%, respectively) [21]. However, research also shows that older adults are already at risk of poor mental health due to experiences such as loneliness and bereavement. The American Psychiatric Association [1] also emphasized the negative psychological crisis causing changes in our daily lives, job loss, financial hardship and grief over the death of loved ones has the potential to affect the mental health and well-being of many.

Findings from the study also showed that 323 (76.5%) strongly agree that artificial intelligence provides data that will assist healthcare givers in finding a better venti-lator setting for CoVid-19 patients. Panchal et al. [34] also identified the importance of data to improve healthcare givers work and to improve the psychological health of people as they stay at home.

13.2.3 Artificial Intelligence and COVID 19 Have on the Social Welfare of Nigerians

The study has shown that majority of the respondents 232 (55.0%) were less impacted on socially by Coronavirus and artificial intelligence while 190 (45.0%) were highly impacted on by Coronavirus and artificial intelligence with a mean score of 18.6 ± 2.3.

Findings from the study under review have also shown that 300 (71.1%) agree that lockdown increases social unrest. According to Kazeem [20], the Covid-19 outbreak, its abruptnes, has left millions of low-income households battling hunger in Nigeria causing social unrest depicting how much the impact has been on the social life of Nigerians.

324 (76.8%) agree that the lockdown would increase family togetherness, this is in tandem with a study by Coulson [6], who identified that likely impact on families and even though there is likely danger on families, many families will use this time to "get back to basics." They will find ways to connect with frequent touchpoints via technology or with others who have also successfully isolated themselves, allowing for strengthened bonds. They may reconnect with tradition, faith, or other activities

that build hope and community. Parents are likely to thrive as they give of themselves to help their children with reading and other mutually enjoyable learning activities. Families who flourish through this challenging time will learn to listen, understand, play, learn, laugh, and grieve together.

14   The Implication of Findings to Healthcare Givers

This study has shown that there is an impact on the psychological and social life of Nigerians. It also shows that there is a need to implement the use of artificial intelligence in managing social and psychological problems. Healthcare givers also need to identify the specific use of artificial intelligence and how it would be implemented in the management of patients in the ward settings, community and after the Covid-19 period. Healthcare givers need more skills including technological skills to aid faster prediction and early determination of pandemics; this would inadvertently aid the management of individual social lives and psychology of Nigerians.

15   Recommendations

Governments and private workers are encouraged to:

- Ensure an active database of the population to aid contact tracing, detect signs of depression and abnormal social lifestyle, as the backbone of Artificial Intelligence is data.
- Evaluate pre-existing policies on the use of Artificial Intelligence, artificial intelligence fortified machine for the fight against COVID-19 and other pandemics and its effects on psychological and social welfare.
- Strengthen policies on the affordability and accountability of the Artificial Intelligence system to intensify the accuracy of data to reduce burnout of health workers and improve the psychological and social welfare of Nigerians by increasing detection and survival rates.
- Boost multi-disciplinary collaboration between the artificial intelligence community, healthcare workers and the community at large to fill gaps in health data and to discuss possible solutions.
- Ensure the use of various applications, such as mobile games which have been linked to reduced boredom and anxiety of Nigerians.
- Encourage the use of various mental health apps such as Bark, which monitors mental health and profers suitable solutions thereby improving the psychological welfare of Nigerians.
- Generate revenue and employment through the creation of apps to cushion the financial effects of the pandemic such as Google COVID 19 Tracker and FCT COVID Response App.

- Integrate AI into healthcare to improve quality of care to improve the psychology and social life of the masses. This would also create the opportunity for new tools for diagnosis, monitoring and treatment of patient management in future outbreaks.
- Develop an early warning system to serve as a national epidemiologic tool for recognition diseases and psychological status of Nigerians.

Suggestion for Further Study

- Studies should be carried out to evaluate the effect of COVID 19 and Malaria on Nigerians since both illnesses shut down the kidney and the body systems.
- More studies should be carried out to identify the level of adoption of artificial intelligence in the management of pandemic in Nigeria.
- There is a need to assess the level of health data Nigeria has to determine the applicability of AI in the social and psychological welfare of Nigerians.

16 Limitations

Some limitations were encountered when carrying this study, firstly, the nationwide lockdown which hindered us from physically administering questionnaires to the respondents. Secondly, socio-cultural issues such as superstitions in the provision of personal details in African countries made some respondents ignore the questionnaire administered. We overcame these limitations by electronically administering the questionnaire via Google Form.

References

1. American Psychiatric Association: The psychological impact of COVID-19. Know the signs of anxiety, panic attacks, depression and suicide (2020). Retrieved from https://www.apa.org/topics/covid-19/psychological-impact
2. Anjorin, A.A.: The coronavirus disease 2019 (COVID-19) pandemic: a review and an update on cases in Africa. Asian Pac. J. Trop. Med. **13**, 1–5 (2020). https://doi.org/10.4103/1995-7645.281612
3. Anazia, D., Awodipe, T., Adelowo, A.: Lagosians worry over the impact of curfew on livelihood (2020). Retrieved from https://t.guardian.ng/news/lagosians-worry-over-impact-of-curfew-on-livelihood/
4. Barker, P.: The accidental robot that can crawl, swim and even charge itself (2019). Retrieved from https://www.redbull.com/ng-en/plant-energy-systems-velox-drone on March, 18, 2020
5. Bennett, C.C., Hauser, K.: Artificial Intelligence framework for simulating clinical decision-making: a Markov decision process approach. Artif. Intell. Med. **57**, 9–19 (2013)
6. Coulson, J.: The Potential Social and Psychological Effects of Coronavirus Isolation on Families, (2020). Retrieved from https://ifstudies.org/blog/the-potential-social-and-psychological-effects-of-coronavirus-isolation-on-families
7. Craven, M., Liu, L., Mysore, M., Wilson: COVID-19: implications for business McKinsey & Company **3**, 1–8 (2020)

8. Coleman, D.: Social well-being (2020). Retrieved from https://econation.co.nz/social-well-being/. 24 May 2020
9. Cicognani, E.: Social well-being. In: Michalos, A.C. (ed.) Encyclopedia of Quality of Life and Well-Being Research. Springer, Dordrecht (2014). Retrieved from https://doi.orrg/10.1007/978-94-007-0753-5_2797
10. CNBC Africa: Nigeria to phase out lockdown in Abuja, Lagos and Ogun States, these are the details (2020). Retrieved from https://www.cnbcafrica.com/coronavirus/2020/04/27/nigeria-to-phase-out-lockdown-in-abuja-Lagos-ogun-states-these-are-the-details/
11. David, J.C., Scott, J.B., Keith, M.O.: What is COVID-19? (2020) Retrieved from https://www.medscape.com/answers/2500114-197401/what-is-covid-19/. 24 May 2020
12. Emi, I.: Nigeria records a rise in COVID-19 infections among doctors (2020). Retrieved from http://venturesafrica.com/nigeria-records-a-rise-in-covid-19-infections-among-doctors/. 24 May 2020
13. Frankenfield, J.: Artificial intelligence (AI) (2020). Retrieved from https://www.investopedia.com/terms/a/artificial-intelligence-ai-asp. 24 May 2020
14. Franck, L.: A brief history of robots (2013). Retrieved from parisinnovationreview.com/articles-en/a-brief-history-of-robots. 6 Feb 2020
15. Fennema-Notestine, Ozyurt, Clark, … Brown: Quantitative evaluation of automated skull-stripping methods applied to contemporary and legacy images: effects of diagnosis, bias correction and slice location. Hum. Brain Mapp. **27**, 99–113 (2006)
16. Flesia, L., Fietta, V., Colicino, E., Segatto, B., Monaro: Stable psychological traits predict perceived stress related to the COVID-19 (2020). Retrieved from https://doi.org/10.31234/osf.io/yb2h8
17. Griffin, A.: Scientists create first 'living robots' in major breakthrough (2020). Retrieved from https://www.independent.co.uk/life-style/gadgets-and-tech/news/living-robots-xenobots-living-cells-frog-embryos-a9282251.html. 2 Feb 2020
18. Habeeb, A.: Introduction to Artificial Intelligence. University of Mansoura (2017)
19. Kalu, B.: COVID-19 in Nigeria: a disease of hunger (2020). Retrieved from https://www.thelancet.com/journals/lanres/articles/PIIS2213-2600(20)30220-4/fulltext 24 May 2020
20. Kazeem, Y.: Ordinary Nigerians are filling the country's major social welfare gaps amid coronavirus (2020). Retrieved from https://www.google.com/amp/s/qz.com/africa/1843839/nigerias-coronavirus-lockdown-is-hitting-poor-families-hard/amp/
21. KFF Health Tracking Poll: Conducted March 25–30, 2020 (2020). Retrieved from https://www.kff.org/report-section/the-implications-of-covid-19-for-mental-health-and-substance-use-issue-brief/#endnote_link_460419-4
22. Koutronas, E.: Understanding social welfare: basic concepts and definitions (2017). Retrieved from https://www.researchgate.net/publication/315454313_Understanding_Social_Welfare_Basic_Concepts_and_Definitions 30 June 2020
23. Lancer, V.Wim, Parolin, Z.: COVID-19, school closures and child poverty: a social crisis in the making. Lancet Public Health **5**(5), 243–244 (2020)
24. Langford, W.K., Ghassael, A., Gershenfeld, N.: Hierarchical Assembly of Self-replicating Spacecraft Using Distributed Mechanisms and Actuation in Digital Materials. Massachusetts Institute of Technology, Cambridge, MA. US20190077030A1 (2019)
25. Martineau, K.: Marshalling artificial intelligence in the fight against COVID-19 (2020). Retrieved from http://news.mit.edu/2020/mit-marshaling-artificial-intelligence-fight-against-covid-19-0519 24 May 2020
26. NCDC:The national strategy to scale up access to coronavirus disease testing in Nigeria (2020). Retrieved from https://covid19.ncdc.gov.ng 24 May 2020
27. Nnabiugwu, T.: Rwanda set to deploy robots in the fight against the COVID 19 pandemic (2020). Retrieved from http://venturesafrica.com/rwanda-set-to-deploy-robots-in-the-fight-against-the-COVID-19-pandemic/. 24 May 2020
28. OHCHR Office of the High Commission Human Right: OHCHR in Nigeria 2019: human right context. OHCHR (2019). Retrieved from https://www.ohchr.org/EN/Countries/AfricaRegion/Pages/NGSummary2019.aspx

29. Ojekunle, A.: Nigeria's rising public debt continue to be a cause for concern at $83.88 billion Pulse (2019). Retrieved from https://www.pulse.ng/bi/finance/nigerias-rising-public-debt-con tinue-to-be-a-cause-for-concern-at-dollar8388-billion/4ftbkrp

30. Onyedinefu, G.: Poorly equipped & unprotected more doctors risk contracting coron-avirus (2020). Retrieved from https://businessday.ng/coronavirus/article/poorly-equipped-unp rotected-more-doctors-risk-contracting-coronavirus/. 24 May 2020

31. Onehi, V.: Nigeria: COVID-19 will worsen Nigeria's unemployment crisis-expert (2020). Retrieved from https://allafrica.com/stories/202004020021.html. 24 May 2020

32. Orjinmo, N.: Coronavirus lockdown: Nigerians cautious as restrictions eased in Lagos and Abuja (2020). Retrieved from https://www.bbc.com/news/world-52526923

33. Okwumbu, R.: FG declares nationwide curfew 8 pm to 6 am (2020). Retrieved from https://nai rametrics.com/2020/04/27/just-in-fg-declares-nationwide-curfew-8pm-to-6am/. 24 May 2020

34. Panchal, N., Kamal, R., Orgera, K., Cox, C., Garfield, R., Hamel, L., Munana, Chidambaram, P.: The Implications of COVID-19 for Mental Health and Substance Use. (2020). Retrieved from https://www.kff.org/report-section/the-implications-of-covid-19-for-mental-health-and-substance-use-issue-brief/#endnote_link_460419-4

35. Park, S.Ho., Han, K.: Methodologic Guide for evaluating clinical performance and the effect of artificial intelligence technology for medical diagnosis and prediction. Radiology **000**, 1–10 (2018)

36. Pirouz, B., Haghshenas, S.Sina, Haghshenas, S.Sami, Piro, P.: Investigating a serious challenge in the sustainable development process: analysis of confirmed cases of COVID-19 (Newtype of coronavirus) through a Binary classification using artificial intelligence and regression analysis. Sustainability **12**, 2427 (2020)

37. Rajkumar, P.Ravi: COVID-19 and mental health: a review of the existing literature. Asian J. Psychiatry **52**, 102066 (2020)

38. Roser, M., Ritchie, H., Ortiz-Ospina, E., Hasell, J.: Statistics and Research: Coronavirus (COVID-19) Deaths (2020). Retrieved from https://ourworldindata.org/covid-deaths. 23 May 2020

39. Robertson, I.: What is psychological wellbeing? (2020) Retrieved from https://www.robertson cooper.com/blog/what-is-psychological-wellbeing/. 24 May 2020

40. Snyder, C.R., Lopez, S.J., Pedrott, J.T.: Positive Psychology: The Scientific and Practical Explorations of Human Strengths. Sage (2011). ISBN 978-1-4129-8195-8

41. Schroer, A.: 32 Artificial intelligence companies building a smarter tomorrow (2020). Retrieved from https://builtin.com/artificial-intelligence/ai-companies-roundup. 24 May 2020

42. Ted, O., Mathew, R.: Optimal sin taxes. Sciencedirect (2006). Retrieved from https://www.sci encedirect.com/science/article/pii/S0047272706000247. The European Union. Addiction **107**, 188–196 (2011)

43. Velavan, T.P., Meyer, C.G.: The COVID-19 epidemic. Tropical Med. Int. Health TM & IH **23**(3), 278–280 (2020). https://doi.org/10.1111/tmi.13383

44. W.H.O.: Coronavirus (2020). Retrieved from https://www.who.int/health-topics/coronavirus#tab=tab_1. 24 May 2020

45 World Poverty Clock.: Explore Poverty Data by Age, Gender and Region for Every Country in the World Until 2030. World Data Lab. Retrieved from https://worldpoverty.io/headline on February 10, (2020)

E-payment/M-payment Systems
and E-learning

Factors Influencing the Adoption of E-Payment During Pandemic Outbreak (COVID-19): Empirical Evidence

A. Al-Dmour, H. Al-Dmour, R. Al-Barghuthi, R. Al-Dmour, and M. T. Alshurideh ⓘ

Abstract This study aimed at identifying the main factors influencing the adoption of the electronic payment system (EPS) by Jordanian consumers based on the technology acceptance model (TAM) during the emergence of the pandemic crisis (COVID-19). A quantitative approach and a convenience sample of (567) Jordanian banking customers were employed to collect the required data via an online survey questionnaire. The study revealed that the most important predictors of behavioural intention of e-payment adoption by Jordanian consumers during the pandemic crisis (COVID-19) are: perceived usefulness, trust and ease-of-use Furthermore, education level, age and income as demographic variables have been found to affect the intention of e-payment service adoption significantly, while the effect of gender was found insignificant during this time. Based on these findings, several practical recommendations were suggested to enhance the electronic payment adoption process in Jordan. This study has provided its contribution concerning e-payment adoption, particularly during the breakout of Covid-19 pandemic in the world for the first time.

A. Al-Dmour (✉)
School of Business, Al Ahliyya Amman University, Amman, Jordan
e-mail: Ahmadmourh@gmail.com

H. Al-Dmour · R. Al-Barghuthi · M. T. Alshurideh
Department of Marketing, School of Business, The University of Jordan, Amman, Jordan
e-mail: dmourh@ju.edu.jo

R. Al-Barghuthi
e-mail: rawanwajed@hotmail.com

M. T. Alshurideh
e-mail: m.alshurideh@ju.edu.jo; malshurideh@sharjah.ac.ae

R. Al-Dmour
Department of Management Information Systems, School of Business, The University of Jordan, Amman, Jordan
e-mail: rand.aldmour@ju.edu.jo

M. T. Alshurideh
Department of Management, College of Business Administration, University of Sharjah, Sharjah, United Arab Emirates

Keywords E-payment · Adoption · Intention · Trust · Security banking service

1 Introduction

Infection with Coronavirus (COVID-19) has become a severe public health issue all over the world causing the severe acute respiratory syndrome known as Coronavirus 2 (SARS-CoV-2), which is a novel coronavirus that has recently emerged from China. In March 2020, the WHO made the declaration that COVID-19 can be characterized as a pandemic. It is therefore of utmost importance to prevent any further spread of the pandemic in public and healthcare settings. Among the strategies, measures and guidelines that many countries have enforced are social distancing, testing every suspected case, staying home, avoiding social gatherings, treating patients and contact tracing [1, 2]. Some countries are, however, taking stricter measures to contain the pandemic outbreak, such as lockdown and mass testing. Furthermore, the World Health Organization has warned that banknotes may spread the Coronavirus. The WHO recommended using contactless payments that are likely to help control the virus. COVID-19 further undercut the use of cash by forcing many retailers to close the doors of their shops and sell exclusively through online orders for delivered goods.

In order to effectively implement social distancing measures aimed at containing the additional spread of COVID-19, many governments and health official bodies around the world are encouraging people to adopt online shopping as an alternative to actual shopping and consumers have adapted their shopping patterns and behaviors to minimize the risk of getting infected. The enforcement of social distancing, lockdowns and other measures in response to the COVID-19 pandemic has increased online shopping, use of social media, online telephones, teleconferencing and video and movie broadcasts as well. Digital payments allow people to access goods and services while in quarantine or social distancing and can reduce the possibility of the virus spreading to others through cash exchanges. Digital payments limit personal transactions and ensure consumer purchase essentials from their homes in comfort. Digital payments are even helping to put stimulus money in the hands of consumers faster. For example, local governments in China have distributed vouchers through WeChat Pay to encourage instant spending. Certainly, digital and contactless payment methods require less physical interaction and are therefore safer. Social distancing and other pandemic-mitigation measures are likely to boost demand for digital banking services. However, according to one study, credit cards carry more bacteria than coins or cash, while smartphones can be a "haven for germs". This is why some experts still recommend keeping additional cash on hand because banks are already taking steps to make sure that any cash they handle is sanitized.

As the COVID-19 public health emergency continues to emerge, one thing is obvious: the economic and human impacts of this virus are high. Primarily, the coronavirus radically changes how consumers shop. They have become more cautious about shopping in public places. Furthermore, since many people work from home, chances are occurring in how they shop and how they pay. [3] indicated that US consumers who do not trust digital transaction had become more interested in contactless payment, while more than a half of the payments currently made by card in Germany are untouched, compared to 35% before the coronavirus crisis occurred.

Similarly, cash use in the UK has also halved after the imposition of a nationwide lockdown, where people tended increasingly to avoid cash transactions. For these, as well as other reasons, the pandemic is driving the adoption of contactless payment in a major way. In Jordan, like in other countries, major banks are encouraging their customers to use online banking applications services to avoid crowding at bank branches and ensure social distancing. The aim is to deliver essential services with reduced staff while ensuring the safety and well-being of customers and employees.

According to [4], the availability of digital payment technologies, such as online banking, mobile and credit/debit cards has increased rapidly in the developing world. These technologies form the cornerstone of financial inclusion initiatives in developing countries. Despite significant efforts to boost digital payments, the adoption rates remain modest, particularly in low-income countries. However, the pandemic of COVID-19 significantly increased the adoption of digital payment solutions. Although different economies are still in different stages of development around the world, the outbreak has forced people to change their shopping habits in favor of e-business. It paved the way for digital payments. [5] conducted a research study on shopper's behaviour that started during the first phase of the epidemic that started in China and spread to other affected countries. His study also aimed to watch consumer trends. It was found that as COVID-19 news reach the audience; consumers go through six behavioral stages based on their awareness of the spread of COVID-19 in their communities. The study found that consumers typically move from one stage to another in two weeks in areas close to the initial outbreak. However, this happened faster in other countries where the outbreak began later, such as Italy and the United States. The task of continuing to attract consumers to adopt digital payment is grounded on trust to boost lasting loyalty.

A remarkable number of studies have investigated e-payment from technical and user acceptance perspectives [6–8]. In conjunction with these studies, several key factors influencing the perception of e-payment have been proposed. For example, [9] stated that cash and cheques remain as popular payment tools which results from that consumers is not convinced with the usefulness of e-payment. The most critical question facing e-payment service providers is associated with the reason why there is slow adoption of e-payment. So, it is required to explore the factors influencing e-payment adoption intention in Jordan. An examination by the Central Bank of Jordan, [10] refers to the absence of awareness is one reason why consumers are not using e-payment. However, the Central Bank concludes that Jordan is moving toward better e-payment acceptance in the coming years [11]. 1 attribute e-payment inability to the system design and deployment that do not meet the clients' needs and desires, while numerous studies view security and trust to be among the vital concerns [12]. Interestingly, these studies evoke that security, trust, usefulness, self-efficacy and ease-of-use are significant factors influencing the perception of e-payment. Very few studies to date have attempted to study these factors in a single setting [4], and these relevant studies were conducted outside Jordan. It is interesting to inspect these factors, particularly in the Jordanian setting, due to the fascinating developments that are currently occurring. Thus, it is predicted that e-payment use will expand at a significant rate within the next few years.

Understanding and meeting customers' requirements and expectations are key factors to e-banking success. To increase the adoption rate of e-payments, the factors that affect consumer adoption should be better managed [13]. Most likely, the selection of e-payment services is closely related to the design characteristics of the e-payment system. For any e-payment type to be adopted, its design should meet the users' needs and desires [13]. Furthermore, studied the effect of customers' demographic characteristics is essential for banks in developing appropriate marketing strategies and financial resources. For example, the banks' understanding of the customers' demographic influences on the adoption of e-payment services may require the deployment of target marketing strategies. Also, challenges arising from the spread of (COVID-19) are likely to accelerate the use of both existing and new technologies.

Moreover, as consumers go into lockdowns, large numbers of customers over the world are forced to work from their homes and digital connectivity takes even more of a hold on everyday habits. Accordingly, identifying the main predicting factors that influence consumers' acceptance to adopt digital payment during pandemic emergence would be useful for bankers as well as for practitioners during this time. Given the limited number of relevant studies, attempts to assess consumers' intention to adopt digital payment during the pandemic outbreak are not widely available to academics and practitioners.

2   The Study's Model and Hypotheses Development

This study uses the technology acceptance model (TAM) as a theoretical base to hypothesize and examine an integrated model to explore factors affecting e-payment adoption intention during the breakout of Covid-19 pandemic in Jordan. TAM is a theory, which is utilized to model how users accept and use technology [14]. In TAM, two main factors influence users' decision to use the technology; namely, perceived usefulness and perceived ease-of-use. Perceived usefulness (usefulness) is defined as "the degree to which users believe that using a particular system would enhance their job performance", while perceived ease-of-use is defined as "the degree to which users believe that a particular system is easy to use" [14]. In addition to TAM, constructs, such as trust, security and self-efficacy, have been added as the antecedents of adoption according to previous studies [15–17].

Consequently, this study groups all these factors into one integrated model in order to examine which of these factors appear to be the most influential on Jordanian intention to adopt e-payment during the breakout of Covid-19 pandemic. The proposed research theoretical model is illustrated in Fig. 1. It shows the influence

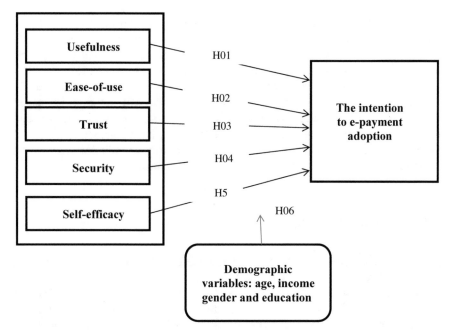

Fig. 1 Study model

of the five independent variables (security, trust, usefulness, self-efficacy and ease-of-use) and demographic variables (age, gender and education) on the dependent variable (consumers' intention to adopt e-payment). Behavioral intention refers to the desire of individuals to perform or not to perform some specified future behavior [18]. In this study, behavioral intention refers to the intention of consumers to adopt e-payment services. E-payment refers to any non-cash payment that does not include paper cheques [3]. Vinitha and Vasantha [19] defined e-payment as the transfer of an electronic value of payment from a payer to the payee through an e-payment tool which allows customers to remotely enter and manage their bank accounts and transactions, performed through an electronic network. This study adopts these definitions and mentions the Central Bank of Jordan definition of e-payment systems that was stated in the Bylaw of Electronic Payment and Transfer of Funds 2017 [10], which is, "the group of programs or instruments prepared for payment, transfer, clearance or settlement of funds electronically and approved by the Central Bank." Under the bylaw mentioned above, the business activities of electronic payment and fund transfer companies shall be subject to the supervision and control of the Central Bank of Jordan. This will make payment and fund transfer activities, such as payment, transfer, financial settlement or electronic clearing or issuing and managing electronic payment tools and systems, subject to the Central Bank of Jordan's supervision and control, within a new system of legislation and regulations governing this framework, requiring any entity that exercises such activities to obtain a license from the Central Bank of Jordan. All of this aims to enhance the role of the Central Bank of Jordan in supervising and monitoring the activities of payment and electronic transfer of funds, as well as to improve the efficiency, soundness and effectiveness of payment, clearing and settlement systems and achieve transparency, efficiency, competitiveness and protection of financial consumers in the payment sector [19].

On the other hand, this will lead to evaluate the feasibility of the technology under investigation in terms of security, trust and efficiency that will, in turn, affect users' decision to use e-payment. Based upon the above study's conceptual framework and the content analysis of previous studies, the following most common and relevant factors were selected to be examined in the relationship with the intention of the adoption of e-payment services in Jordan's cultural context.

2.1 Perceived Usefulness

[20] argued that perceived usefulness includes fixed and transaction costs in adopting e-payment. Fixed costs refer to the costs of getting payment tools, such as card readers and payment software, while transaction costs are those incurred by consumers and merchants every time they carry out a financial transaction [21]. Accordingly, customers can enjoy the usefulness of low cost when they involve in electronic trans-actions, as they only need to pay a nominal fee to their respective banks, payment service providers and card issuers for the services used [16]. Therefore, the useful-ness of the service is a key factor that explains the intention to adopt the service [14, 22], and it has been empirically authenticated as a vital forerunner of new technology adoption [23, 24]. The statistics of e-payment use in Jordan show that the percep-tion of Jordanians is slowly changing from using cash to e-payment due to many reasons; for instance, the attractiveness of cards and mobile wallets as transactional e-payment tools. This attractiveness stems from two important sources. First, they allow individuals to minimize their handled cash. Second, there has been a gradual increase in consumer demand for the use of cards and mobile wallets driven by the convenience of using them for conducting transactions [10]. Thus, they offer a transactional advantage over cash and cheques through providing consumers with a convenient means of payment which includes users' ability to spend, store and trans-port a currency value through the payment systems [24], in addition to the primary advantages of e-payment, which include time, speed, cost and effort savings. There-fore, the perceived usefulness for customers represents a valid dimension of the adoption process.

Before accepting any new technology, customers critically assess the usefulness they will get after using it. Therefore, perceived usefulness is referred to as the degree of consumers' belief that adopting the new technology will enhance their performance [14]. To start with, the relative advantage of a new product or service compared to those of existing ones is viewed as one of the most important factors affecting consumer adoption. [25] viewed usefulness as a significant driver for e-payment systems. Similarly, [26] who studied four e-commerce activities (online shopping, banking, investing and e-payment systems) revealed that perceived suitability and financial usefulness predict the adoption decision.

In order to encourage customers to adopt digital payment during the outbreak of Covid-19, banks could show them the benefits of using these channels. These benefits include, but are not limited to the ease of carrying out any transaction 24/7 and enjoying permanent access to all financial information in real-time. Banks could also try to promote online banking by sharing tutorials and expanding the types of transactions customers can carry out remotely. For users themselves, there are also other factors driving the increased use of online banking, among which is the fear that cash could spread coronavirus. Based on the above discussion, this study relates positively to consumers' attitudes to adopt e-payment in Jordan during this time; therefore, the following hypothesis can be formulated:

H1.1: Perceived usefulness factor will have significant effects on consumers' intention to adopt e-payment during the breakout of Covid-19 in Jordan.

2.2 Ease-of-Use

This factor has been defined as the level to which an innovation is perceived as easy to understand and use [23]. Similarly, [14] defined ease-of-use as the degree of belief that using a particular technology will be effortless. Besides, [4] stated that a fruitful design of e-payment systems from the user position is important to attract the user's acceptance toward e-payment. Because of this, problems with usability have been found to justify low adoption of a variety of payment systems and tools, including cards, mobile wallets and internet banking [15, 23]. As such, perceived ease-of-use is regarded as a decisive factor in e-payment adoption.

Furthermore, this variable may explain consumers' view on new technologies, which relates to the amount of complexity in the technology [27]. In short, content, design and speed are essential features leading to perceived ease-of-use and later influence consumers' adoption of e-payment [28]. [29] proposed a model based on TAM and innovation diffusion theory (IDT) and found that ease-of-use is a crucial determinant of e-payment acceptance. [23] concluded that ease-of-use of a new payment service or lack of complexity would lead to a positive attitude by individuals. Suppose it is challenging to start using new technology or service when there are usage barriers [30], which hinder its adoption [30]. In addition to whether customers can perform banking functions through alternative channels or not during the breakout of Covid-19, greater use will test how easy these channels are to use, especially for first-time users whose use is a direct result of coronavirus concerns. For this reason, banks should strive to simplify the interfaces of a call centre over the Internet or to publish information and educational materials that guide customers on how to make the most of the non-affiliate channels to manage their money. As such, the following hypothesis was developed to relate ease-of-use to consumers' attitudes to adopt e-payment in Jordan.

H1.2: Ease-of-use factor will have significant effects on consumers' intention to adopt e-payment during the breakout of Covid-19 in Jordan.

2.3 Security

Safety and security concerns of users about electronic payment transactions are commonly connected to authentication and confidentiality issues, as well as to anxieties about secondary use and unauthorized entrance to payments and consumer data [31]. Commonly, security is associated with a set of procedures and programs to confirm the data source and guarantee the integrity and privacy of the information [19]. Further, security refers to the perception regarding payment instruments and mechanisms for saving and transmission of information [32, 33]. Therefore, it gets back to the technical aspects that guarantee integrity, authentication, confidentiality and non-recognition of relationships in e-payment.

Regarding e-payment, [15] classified security into three areas; system, transaction and legal security. This is because e-payment can only be considered as confidential when the transaction process stages are qualified to satisfy users' needs and security expectations [34]. [28] stated that security is an essential obstacle to online banking usage, which affects the use of e-payment systems because though consumers' confidence regarding their selected bank is strong, their confidence in new technologies remains weak. It is for these reasons that security could be a determinant of users' decision to adopt e-payment [34].

Security is always a hot topic when discussing digital payment applications. A study done by [35], shows that 67% of respondents are concerned about the security of mobile payment, making this factor the second most cited one after use. Several studies have revealed that security is a major concern for most customers who do not support online banking. According to [36], most attacks directed to online banking systems target the user, with a focus on obtaining authentication and identification information through the use of social engineering and bargaining for the user's online banking services device access to install malware that automates banking transactions, regardless of obtaining authentication data. Although this is a justified concern, it is interesting to note that no one is ever talking about the security of plastic cards anymore. While security is an important issue during the breakout of Covid-19 pandemic, it might not be of any daily concern for consumers. Therefore, the following hypothesis was proposed to confirm the effects of security on e-payment and relate security positively to consumers' attitudes to adopt e-payment in Jordan.

H1.3: Security factor will have significant effects on consumers' intention to adopt e-payment during the breakout of Covid-19 in Jordan.

2.4 Self-efficacy

The degree to which an individual believes in executing any activity with his/her skills is known as self-efficacy [36, 37]. [28] stated that self-efficacy means that a person understands and believes in his/her skills and capability to perform a task given. In this regard, self-efficacy is derived from the experience of one's mastery. Such Self-efficacy beliefs are developed as a response to four sources of data; experience (success and failure), vicarious experience (watching others' successes and failures), verbal influence (from peers, partners and relatives) and affective state (emotional states, such as tension) [28]. So, it has been an important determinant of users' perception toward adopting electronic payment.

Many studies revealed that self-efficacy has a significant positive influence on perception and behavioral intention to use information systems [28, 38]. Users with higher self-efficacy tend to experience more types of communication media and functions, while users with lower self-efficacy may be restricted to fewer operations [39]. In the context of e-payment, self-efficacy refers to the judgment of one's ability to use e-payment systems [15]. During the emergence of Covid-19 pandemic, [5] expects that consumers may have higher motivations and fewer perceived barriers to more actively seeking technology-enabled solutions to assist in everyday tasks, like shopping. For some consumers, this may be a new behavior (such as shopping from groceries online for the first time). For others, this may mean increased online usage or the addition of new technologies, tools and software. The following hypothesis was formulated to relate self-efficacy to consumers' attitudes to adopt e-payment in Jordan.

H1.4: Self-efficacy factor will have significant effects on consumers' intention to adopt e-payment during the breakout of Covid-19 in Jordan.

2.5 Trust

Numerous studies argue that trust is a vital factor for understanding interpersonal behaviour and economic exchanges influencing consumers' perception toward e-payment systems [40] and the success of usage of these systems [41]. Trust can be deemed as a positive expectation of the consumer towards the service provider [40, 42]. [43] concluded that trust consists of three beliefs; ability, integrity and benevolence. Ability means that parties possess enough specialized knowledge to fulfil their promises. Integrity is seen as the capability of e-payment service entities to keep their obligations. Consumer acceptance of electronic payment needs a belief that concerns will be addressed [44]. Benevolence is defined as the service provider's concern to defend consumer interest. In electronic services, trust is the most important factor affecting consumer perception [41].

Trust is characterized as a component of the level of risk involved in financial transactions, where trust results in reduced perceived risk, leading to positive intention toward e-payment adoption [45]. Prior studies found trust to be a critical determinant impacting consumers' willingness to make e-commerce deals and engage in online money exchanges [15, 16]. During the emergence of Covid-19 pandemic, [3] indicated that US consumers with no trust had become more interested in contactless payments, while more than a half of the payments currently made by card in Germany are untouched, compared to 35% before the coronavirus crisis occurred. Similarly, cash use in the UK has also halved after the imposition of a nationwide lockdown, where people avoid cash transactions. Accordingly, the following hypothesis was developed to relate trust to consumers' attitudes to adopt e-payment in Jordan.

H1.5: Trust factor will have significant effects on consumers' intention to adopt e-payment during the breakout of Covid-19 in Jordan.

2.6 Demographic Characteristics

According to [46], new technology adopters are typically younger, people having a good income as well as an appropriate level of education and are more reactive to innovation compared to non-adopters; hence, the attitude towards using new technology was found to be more salient in the younger generation [47]. This has been confirmed by many researchers, such as [20, 47, 48], who concluded that adopters are younger, wealthier and usually have a better level of education, there is a general agreement among researchers that demographic characteristics play an important role in determining Internet banking behavior. High income, young age and good education have been found out to explain the acceptance of Internet banking. Gender has sometimes been suggested to be a factor influencing Internet banking adoption, while some studies argued that the Internet is male-dominated [49, 50]. [51] conducted research aimed at understanding and explaining customers' intention to adopt Internet banking in Egypt. Using a sample of users of Internet banking services, it was shown that demographic variables had no significant effect on continued usage of Internet banking services. Also, [52, 53] showed that age, gender, education, employment and marital status are not significantly related to electronic payment card acceptance. [54] concluded that age, income and education level have a direct impact and gender has no direct impact on technology adoption. Besides, results obtained from other studies revealed no significant influence of demographics on consumer adoption of Internet banking [55]. On the other hand, [56] found that age has a significant influence on user acceptance of Internet banking.

Moreover, [57] asserted that young individuals are more likely to adopt Internet banking. [58] revealed that while the influences of marital status, age and education level on the adoption of e-banking are significant, reversed is the case with other demographic variables, such as gender, religion and income. [59] found that age, education and income are particularly strongly correlated with adoption and use of most payment instruments in the United States. [60, 61] used a single year's SCPC survey data and showed that demographics and income have significant effects on the adoption and use of payment instruments. [19] revealed no significant differences between male and female customers in the use of e-payment services. However, they indicated that younger and higher educated customers had more ICT skills and use e-payment services much more than their less-educated counterparts do. A critical look at the previous studies clearly shows that the influences of demographic characteristics of customers on the adoption of e-banking services were mixed. Therefore, the following hypothesis was formulated.

H1.6: Demographic characteristics of customers (age, education and gender) will have significant effects on their intention of the adoption of e-banking payment services during the breakout of Covid-19 in Jordan.

3   Research Methodology

Based on the study objectives, a web survey was considered as a valid approach. Thus, a questionnaire was developed in line with the study's conceptual framework and previous studies' background. In this survey, some variables are factual (for example, customers' demographic characteristics, such as gender, age, income and education), whereas others are perceptual (i.e., factors influencing the adoption of e-payment and customers' intention of adoption). A five-point Likert scale was used with the levels (1 = strongly disagree; 2 = disagree; 3 = neutral; 4 = agree; 5 = strongly agree). The questionnaire's content (constructs and items) was mainly selected from previous studies and adoption theories [10, 16, 17, 62]. Moreover, the measurement scales of e-payment adoption were adopted by [28]. To ensure the reliability of the questionnaire content, all selected measures were modified to suit the Jordanian environment context based on the results of a pilot study and feedback from five professional academic experts in this filed. Simple statistical tools, like means, standard deviations and multiple regression analysis, were applied.

The target population consists of bank customers of commercial banks in Jordan. To reach them, a web link for the online questionnaire was sent out to the potential respondents during the period between 15th of March to 15th May 2020. A **convenience** sample size of 567 respondents was reached. In this study, the respondents comprised 56.5% males and 43.5% females. 42% were in the age group between 18 and 35 years. In terms of education level, 62.6% of the respondents have obtained a bachelor degree, and in terms of *per capita* income, about 46% of the respondents have a monthly income between 500 and 1500 JD. Also, 26% of the respondents were first-time users of e-payment applications during the breakout of Covid-19, and about 18% will intend to adopt e-payment shortly. 89.2% of the respondents have tools which can be used in e-payment, and 76.2% of the respondents used point-of-sale terminals for shopping as the most frequently used e-payment service for their different activities, such as online shopping, payment of bills, money transfer, mobile credit recharge and cash withdraw.

3.1   Descriptive Statistics

All items related to independent and dependent variables included in the study were tested for their means, standard deviations, skewness, kurtosis and reliability. The descriptive statistics presented in Table 1 indicate a positive disposition towards the items. While standard deviation (SD) values ranged from 0.75431 to 0.91622, these values indicate a narrow spread around the means. Also, the mean values of all items were higher than the midpoint (3) and ranged from 3.3537 to 3.9952. However, after careful assessment by using skewness and kurtosis, the data was found to be normally distributed. Indeed, skewness and kurtosis were usually distributed, since most of the values were inside the adequate ranges for normality (i.e., −1.0 to +1.0) for skewness and less than 10 for kurtosis [63]. Cronbach's alpha coefficients for all the study variables were calculated and ranged between 0.682 and 0.883, which means that all the items used were reliable [64].

3.1.1   Multicollinearity Analysis Results

In statistics, multicollinearity (collinearity) is a phenomenon that occurs in multiple regression analysis. It arises if there is a high level of correlation (positive or negative) between two or more independent variables. The most widely used measures can assess multicollinearity; Tolerance and Variance Inflation Factor (VIF), which is the proportion of variance in the independent variable that is not explained by its relationships with the other independent variables. [65] stated that the minimum cut-off value for tolerance is typically (0.10); that is, a tolerance value less than 0.10 should be further investigated. On the other hand, the Variance Inflation Factor (VIF) shows how much multicollinearity problems inflate the variance of regression coefficients, nothing that (VIF) value is favorable when being less than five and in more relaxed criterion when being less than 10 [65]. Collinearity statistics results for the independent variables of the study are illustrated in Table 2.

From this perspective, the multicollinearity test for the five main factors (usefulness, ease-of-use, security, self-efficacy and trust) was accomplished, and all (VIF) values were less than (5 and 10), whereas tolerance values were higher than (0.10),

Table 1   Means, standard deviations and normality of scale items

Factors	Mean	S.D	Skewness	Kurtosis	Cronbach's alpha
Usefulness	3.9952	0.78022	−0.890	0.756	0.858
Ease-of-use	3.6254	0.75431	−1.131	1.942	0.773
Security	3.5926	0.84073	−0.645	0.034	0.682
Self-efficacy	3.3537	0.81313	−0.542	−0.643	0.703
Trust	3.7852	0.86841	−0.487	0.114	0.866
E-payment adoption intention	3.9331	0.91622	−0.315	−1.11	0.883

Table 2 Collinearity statistics for independent variables

Independent variables	Collinearity statistics	
	Tolerance	VIF
Usefulness	0.422	2.369
Ease-of-use	0.425	2.352
Security	0.936	1.068
Self-efficacy	0.853	1.172
Trust	0.493	2.028

as shown in Table 2. Consequently, there is no collinearity within the collected data, which reinforces the model by avoiding the problem of having interchangeable beta values between independent variables and confirms that there was no bias.

3.2 Testing Hypotheses Results

Multiple regression analysis was used to examine the main hypothesis and sub-hypotheses.

Table 3 summarizes the results of multiple regression analysis, with the F-ratio test, for the study's main hypothesis. The multiple correlation coefficient ($R = 0.802$) value indicates that the main hypothesis is accepted at $\alpha \leq 0.000$. Accordingly, it may be concluded that five predicting values (perceived usefulness, ease-of-use, security, self-efficacy and trust) have significant effects on the adoption of e-payment when taken together. The value of the coefficient of determination (R Square) also indicates that all these factors could explain 68.1% of the variance in the extent of the intention of adoption of e-payment. Adjusted R square value of this study was 0.674, as shown in Table 3. Since adjusted R values are always less than or equal to R square, if the model has been fitted when the whole population would have participated rather than those who responded to the study questionnaire, there would be 0.007 (0.681–0. 674) less variance in the model outcome.

By implementing the ($\alpha \leq 0.05$) level of significance either to reject or accept the main hypothesis, the F-value for the collected primary data was 91.245, which is significant at the level of $p < 0.05$ (Sig. = 0.000). Meaning that there is a statistically

Table 3 A summary of multiple regression RESULTS

Model	R	R square	Adjusted R square	Change statistics				
				R square change	F change	df1	df2	Sig. F change
1	0.802[a]	0.681	0.674	0.681	93.345	5	562	0.000

[a]Dependent Variable: E-payment Adoption Intention

Table 4 ANOVA results for independent and dependent variables

Model		Sum of squares	Df	Mean square	F	Sig
1	Regression	197.760	5	39.352	91.245	0.000[a]
	Residual	212.881	562	0.422		
	Total	409.641	562			

[a]Predictors: (Constant), Trust, Security, Self-efficacy, Ease-of-use, Usefulness

significant impact of the main five factors (usefulness, ease-of-use, security, self-efficacy and trust) on e-payment adoption intention, as shown in Table 4.

Table 5 shows an estimation of beta coefficients for all independent variables. It gives a measure of the contribution of each variable to the model. Clearly, a tremendous value shows that a small change in the independent variable has a substantial effect on the dependent variable. Perceived usefulness, trust and ease-of-use had the highest beta values, which were 0.331, 0.275 and 0.194, respectively, while security and self-efficacy factors had the lowest beta values, which were 0.161and 0.12, respectively. Therefore, usefulness, trust and ease-of-use factors have the highest contributions in the research model. The second part of Table 5 demonstrated t-values and Sig. Values, which give a rough indication of the impact of each independent variable on the dependent variable. For instance, a big absolute (t) value and a small (p)-value indicate that a predictor variable is having a large effect on the criterion variable.

ANOVA analysis was used to test H1.6 and to assess the differences among respondents in terms of their intention to adopt e-payment services based on their demographic characteristics, such as gender, income, age and education level. Table 6 summarizes the results of ANOVA for the above hypothesis and the results indicate that education level, income and age were significantly related to the participants' intention to adopt e-payment services, while gender was not. Therefore, it could be concluded that education level, income and age play essential roles in the intention of the adoption of e-payment services in Jordan, were higher educated, younger (25–35 years) and middle-income (1000–1500 per month) costumers are more likely to adopt e-payment services.

Table 5 Coefficients for variables

Model	Unstandardized coefficients		Standardized coefficients	T value	Sig
	B	Std. error	Beta		
(Constant)	−0.954	0.220		−5.416	0.000
Usefulness	0.361	0.069	0.331	6.622	0.000
Trust	0.332	0.068	0.275	5.716	0.000
Ease-of-use	0.186	0.039	0.194	2.344	0.000
Security	0.131	0.043	0.163	6.689	0.000
Self-efficacy	0.105	0.059	0.112	2.366	0.015

Table 6 ANOVA results for hypothesis H1.6

Demographic variables		Sum of squares	Df	Mean square	F	Sig
Age	Between groups	32.511	2	7.420	11.44	0.005
	Within groups	395.307	562	0.844		
	Total	389.841	562			
Income	Between groups	15.312	3	6.324	9.321	0.006
	Within groups	365.317	562	0.690		
	Total	389.841	562			
Gender	Between groups	2.922	1	2.876	2422	0.087
	Within groups	393.805	562	0.815		
	Total	389.731	562			
Education	Between groups	35.401	4	8.901	13.66	0.000
	Within groups	367.430	562	0.733		
	Total	399.721	562			

4 Discussion and Conclusion

In this research, the quantitative approach was applied to examine the relationship between the five factors (usefulness, ease-of-use, trust, security and self-efficacy) and the dependent variable (e-payment services' consumer intention of adoption) proposed in the conceptual framework. Multiple regression analysis was conducted to test the study hypotheses. The population of the study was the banking customers in Jordan (N = 600). Data was collected using a web questionnaire which was administered to a convenience sample of 567 respondents. The questionnaire consists of 33 items that represent the variables of the study. The following paragraphs review and discuss the results for each sub-hypothesis, worth mentioning that the results of the analysis and the hypotheses proposed were relatively consistent.

H1.1: Perceived usefulness factor will have significant effects on consumers' intention to adopt e-payment during the breakout of Covid-19 in Jordan.

The study findings indicate that the perceived usefulness factor has a significant effect on the adoption of e-payment during the breakout of Covid-19 in Jordan. The perceived usefulness factor scored the highest overall mean and the highest correlation. This finding received support from previous studies [16, 22, 28, 65–67] that showed perceived benefit to be a crucial driver of e-payment adoption. [28] indicated that individuals find it beneficial when they made their purchases at unconventional locations and at whichever time of the day. This is supported by the study results, whereby the respondents stated that they found it more comfortable and more convenient to make financial transactions electronically. The study found that Jordanian people perceive that e-payment adoption helps them save time, cost and effort when they conduct financial transactions. Also, the speed and accuracy of the process are considerably improved compared to traditional payment methods. [16] indicated that this factor has a significant impact on consumers' intention to adopt technology-enabled products, which is consistent with the findings of many previous e-payment adoption-related studies [67, 68]. The findings of this study emphasized that customers would adopt a new technology only when they found it useful to fulfil their specific needs and would switch to another payment method for a financial transaction if they perceived an extra benefit at a less or similar cost.

H1.2: Ease-of-use factor will have significant effects on consumers' intention to adopt e-payment during the breakout of Covid-19 in Jordan.

Ease-of-use scored the third-highest overall mean. This factor is correlated with consumers' intention to adopt e-payment during the breakout of Covid-19 in Jordan, and the correlation is highly significant. Some banks, card issuers and payment service providers (PSPs) in Jordan have also offered tutorials and advice to their customers on how to use the various e-payment tools and channels. To some extent, the ease-of-use factor allows individuals to think that they are in control of the transaction process. The conclusions mentioned above were consistent with previous literature [24, 67], where the respondents feel that e-payment instruments are user-friendly with easy-to-understand process and content. Accordingly, customers found it easy to learn the use of e-payment, where only minimal efforts are required. [16] emphasized that individuals can adopt e-payment only when they find it easy to use in comparison with other traditional payment methods for a financial transaction.

H1.3: Security factor will have significant effects on consumers' intention to adopt e-payment during the breakout of Covid-19 in Jordan.

While the security factor scored the lowest overall mean among the five factors under investigation and therefore, the lowest level of correlation with consumers' intention to adopt e-payment variable, multiple regression results showed other findings. This result is a bit inconsistent with prior studies [17, 38, 68], but it was supported by the result of [24]. The point can clarify the implication that Jordanian people do not perceive the security factor to be a crucial issue that customers more and more recognize the steps taken by the Central Bank of Jordan, local banks, payment service providers and card issuers to address the challenges associated with security. These moves have instilled customers' confidence to use e-payment tools and channels. However, the significant correlation coefficient reported implies that security is another important factor requiring attention because of the mean score, positive correlation and multiple regression results, which indicate that security is significantly responsible for the change in consumers' intention to adopt e-payment in Jordan during this time.

H1.4: Self-efficacy factor will have significant effects on consumers' intention to adopt e-payment during the breakout of Covid-19 in Jordan.

Self-efficacy factor scored a somewhat high overall mean, and this factor is significantly correlated with consumers' intention to adopt e-payment in Jordan.

The conclusions as mentioned above were consistent with previous literature [24, 38, 54, 69–71]. Since the majority of individuals have tried e-payment tools and channels, their positive experience allows them to continue adopting e-payment. Friends, family and other people who have used the e-payment method would have passed positive comments, which further influence the consumers' intention to adopt e-payment. In addition, if e-payment were easy to use, the consumers would have perceived that they own the skills and capabilities to complete their financial transactions, and this will increase their confidence [24].

H1.5: Trust factor will have significant effects on consumers' intention to adopt e-payment during the breakout of Covid-19 in Jordan.

The study findings indicate that the trust factor has ranked as the 2nd significant factor affecting customers' behavioural intention to adopt e-payment during this time. Consumer trust is an essential factor while adopting technology-enabled products, as consumer trust reduces perceived risk, which leads to adoption [27, 72]. The results also indicated that initial trust has a significant impact on adoption intention during the breakout of Covid-19 in Jordan. This result is consistent with the findings of many e-payment adoption studies [16, 23]. If consumers do not trust e-payment tools, channels and service providers, their intention to adopt e-payment will be negatively affected. Even though the Central Bank of Jordan, local banks, payment service providers and other related parties are taking initiatives by actively enhancing and developing privacy policy in their codes of practice. They are improving their security system to protect consumers' privacy and prevent the occurrence of any frauds; the trust issue has a significant impact on consumers' intention to adopt e-payment because of the few knowledge about digital financial services and electronic payment methods in Jordan. However, the Central Bank of Jordan launched the digital financial literacy campaign in 2019 to raise awareness on e-payment and digital financial services, and this will probably solve the trust issue.

H1.6: Demographic characteristics of customers (age, education and gender) will have significant effects on their intention of the adoption of e-banking payment services during the breakout of Covid-19 in Jordan.

The study findings indicate that the level of education, income and age factors as personal demographic characteristics have significant effects on the intention of adoption of e-payment services, while the effect of gender was found insignificant. This result is supported by previous studies [17, 65, 73]. Therefore, bank policy-makers should pay more attention to those who are highly educated, middle income and young customers for the adoption of e-payment services in Jordan, regardless of their gender.

5   Research Theoretical and Managerial Implications

The main aim of this study is to investigate the impact of (usefulness, ease-of-use, security, self-efficacy and trust) factors on consumers' intention to adopt e-payment during the breakout of Covid-19 in Jordan. The literature review provided in this study was to explain the relationships between study constructs as clarified in the conceptual framework. The proposed model of this research considers the customers of e-payment in Jordan as a case study to test the research hypotheses. Since there is a lack of studies conducted in Jordan to investigate this issue, it is interesting to inspect these factors, particularly in the Jordanian setting due to the breakout of Covid-19 in Jordan. Accordingly, this study is the first of its kind to consider the TAM model, which consists of five major factors (usefulness, ease-of-use, security, self-efficacy and trust) and demographic variables (age, gender and education) within the context of Jordan in this time. It contributed and supported the existing literature on the relationships between the five mentioned factors and e-payment adoption. This study has provided its contribution concerning e-payment adoption, particularly during the breakout of Covid-19 pandemic in the world. It is worth noting that similar studies shall use the validated instruments in this research to confirm whether the findings are similar or different.

This study has a significant managerial perspective; its findings hold several implications for the development of e-payment to increase the pace of adoption in Jordan. Accordingly, the following practical recommendations are suggested: Jordanian banks, payment service providers and the card issuers, as well as other related parties, should continually develop their e-payment services because of the promising growth rate of e-payment adoption during the breakout of Covid-19. Services provided by the mentioned parties should meet consumers' expectations. As stated by [28, 38]. According to [38], e-payment methods should be convenient and effective to win more significant market share from cash and raise consumers' awareness. Moreover, the parties, as mentioned above, must ensure that e-payment tools, channels and systems are always secure to maintain trust and confidence for their consumers during this period.

The findings of this study on usefulness trust and ease-of-use show that consumers need to be educated on how to use the available e-payment tools and channels. Employees in banks, payment service providers and the remaining parties can play a role to inform and educate consumers about the safety of e-payment method during this time to avoid coronavirus. Furthermore, to boost confidence and enhance trust within consumers during this pandemic period, which is a crucial issue that related parties should prioritize, they should provide error-free processes, reliable and responsive customer services and speedy transactions. Hence, they should ensure that users have good experiences with the technology, as these experiences could translate into a positive impact on their intention to adopt e-payment. Also, they should organize campaigns to increase the awareness of usefulness and convenience, particularly in highly educated, middle income and young customers, regardless of their gender.

The Central Bank and the Government of Jordan, in general, have to maintain a competitive environment in the country, where innovation can continue to foster new tools, channels and services, as well as to lower the transaction costs for customers, merchants and overall businesses. Furthermore, policymakers, overseers in the Central Bank of Jordan, local banks and payment service providers all have important roles to play to guarantee the security and trustworthiness of e-payment services. The regulator should continue to ensure stability and financial integrity by controlling e-payment services to protect consumers.

6 Research Limitations

However, this study is not without limitations, the applications of the study's conclusions may be limited by time and geographical location. This study is a cross-sectional survey; the underlying identified associations may contrast across divisions and countries or may even lose their meaning over time, most importantly these are self-reported data from self-selected participants, and the lockdown period was a constraint to gather more representative data.

References

1. Kampf, G.: Antiseptic stewardship: biocide resistance and clinical implications. Springer International Publishing, Cham (2018)
2. Chan, J.F., Yuan, S., Kok, K.H., To, K.K., Chu, H., Yang, J., et al.: A familial cluster of pneumonia associated with the 2019 novel coronavirus indicating person-to-person transmission: a study of a family cluster. Lancet **395**(10223), 514–523 (2020)
3. Ozili, P.K.: Financial inclusion and Fintech during COVID-19 crisis: Policy solutions. Available at SSRN 3585662 (2020)
4. Ligon, E., Malick, B., Sheth, K., Trachtman, C.: What explains the low adoption of digital payment technologies? Evidence from small-scale merchants in Jaipur, India. PloS one **14**(7) (2019)
5. Nielsen Retail Measurement Services: Total U.S. All Outlets Combined (AOC), the one week ended Feb 22 (2020)
6. Lwoga, E.T., Lwoga, N.B.: User acceptance of mobile payment: the effects of user-centric security, system characteristics and gender. Electron. J. Inf. Syst. Dev. Countries **81**(1), 1–24 (2017)
7. Lu, J., Wei, J., Yu, C.S., Liu, C.: How do post-usage factors and espoused cultural values impact mobile payment continuation? Behav. Inf. Technol. **36**, 140–164 (2017)
8. Park, J.K., Ahn, J., Thavisay, T., Ren, T.: Examining the role of anxiety and social influence in multi-usefulness of mobile payment service. J. Retail. Consum. Serv. **47**, 140–149 (2019)
9. Panhwer, P., Pitafi, A., Memon, M.S., Memon, A.: Awareness and reason towards slow adoption of e-payment system: study of Hyderabad. Ann. Contemp. Dev. Manage. HR (ACDMHR) **2**(1), 6–23 (2020)
10. Central Bank of Jordan: The Bylaw of Electronic Payment and Transfer of Funds No. 111 for 2017. Amman, Jordan (2017)
11. Lashitew, A.A., van Tulder, R., Liasse, Y.: Mobile phones for financial inclusion: what explains the diffusion of mobile money innovations? Res. Policy **48**, 1201–1215 (2019)

12. Fatonah, S., Yulandari, A., Wibowo, F.W.: A review of e-payment system in e-commerce. J. Phys. Conf. Ser. **1140**(1), 012033
13. Davis, F.D.: Perceived usefulness, perceived ease of use, and user acceptance of information technology. MIS Q. **13**(3), 319–340 (1989). https://doi.org/10.2307/249008,JSTOR249008
14. Qatawneh, A., Aldhmour, F., Alfugara, S.: The adoption of electronic payment system (EPS) in Jordan: case study of orange telecommunication company. Res. J. Fin. Account. **6**(22), 20–25 (2015)
15. Shankar, A., Datta, B.: Factors affecting mobile payment adoption intention: an Indian perspective. Global Bus. Rev. J. **19**(3), 72–89 (2018)
16. Isaac, O., Mutahar, A.M., Daud, N.M., Ramayah, T., Aldholay, A.H.: The effect of awareness and perceived risk on the technology acceptance model (TAM): mobile banking in Yemen. Int. J. Serv. Stand **12**, 180–204 (2018)
17. Alshurideh, M., Salloum, S.A., Al Kurdi, B., Al-Emran, M.: Factors affecting the acceptance of the social network: an empirical study using PLS-SEM approach. In: Proceedings of the 2019 8th International Conference on Software and Computer Applications, pp. 414–418 (2019)
18. Vinitha, K., Vasantha, S.: factors influencing consumer's intention to adopt digital payment-conceptual model. Indian J. Public Health Res. Dev. **8**(3), 170–175 (2017)
19. Goh, S.W.: Factors affecting the adoption of e-payment among private university students in Klang Valley. Doctoral dissertation, UTAR (2017)
20. Chogo, P.J., Sedoyeka, E.: Exploring factors affecting mobile money adoption in Tanzania. Int. J. Comput. ICT Res. **8**(2), 53–64; Comm. Res. **7**(1), 21–33 (2014)
21. Kim, C., Mirusmonov, M., Lee, I.: An empirical examination of factors influencing the intention to use mobile payment. Comput. Human Behav. J. **26**(3), 310–322 (2010)
22. Arvidsson, N.: consumer attitudes on mobile payment services—results from a proof of concept test. Int. J. Bank Market. **32**(2), 150–170 (2014)
23. Apanasevic, T., Markendahl, J., Arvidsson, N.: Stakeholders' expectations of mobile payment in retail: lessons from Sweden. Int. J. Bank Market. **34**(1), 37–61 (2016)
24. Chou, Y., Lee, C., Chung, J.: Understanding M-commerce payment systems through the analytic hierarchy process. J. Bus. Res. **57**, 1423–1430 (2004)
25. Eastin, M.S.: Diffusion of e-commerce: an analysis of the adoption of four e-commerce activities. Telematics Inf. **19**(3), 251–267 (2002)
26. Mallat, N.: Exploring consumer adoption of mobile payments—a qualitative study. J. Strateg. Inf. Syst. **16**(4), 413–432 (2007)
27. Wendy, M.T., Siong, C.C., Binshan, L., Jiat, W.C.: Factors affecting consumers' perception of electronic payment: an empirical analysis. Internet Res. J. **23**(4), 465–485 (2013)
28. Chen, L.D.: A model of consumer acceptance of mobile payment. Int. J. Mobile Commun. **6**(1), 32–52 (2008)
29. Wang, W.T., Li, H.M.: Factors influencing mobile services adoption: a brand-equity perspective. Internet Res. J. **22**(2), 142–179 (2011)
30. Alyabes, A.F., Alsalloum, O.: Factors affecting consumers' perception of electronic payment in Saudi Arabia. Eur. J. Bus. Manage. **10**(10), 21–34
31. Laukkanen, T., Kiviniemi, V.: The role of information in mobile banking resistance. Int. J. Bank Market. **28**(5), 372–388 (2010)
32. Tsiakis, T., Sthephanides, G.: The concept of security and trust in electronic payments. Comput. Secur. J. **24**(1), 10–15 (2005)
33. Lim, B., Lee, H., Kurnia, S.: Why did an electronic payment system fail? A case study from the system provider's perspective. Electron. Comm. Res. Appl. **7**, 202–213 (2006)
34. Barkhordari, M., Nourollah, Z., Mashayekhi, H., Mashayekhi, Y., Ahangar, M.S.: Factors influencing adoption of e-payment systems: an empirical study on Iranian customers. IseB **15**(1), 89–116 (2017)
35. The Federal Reserve: System, Board of Governors of the Federal Reserve -Consumers and Mobile Financial Services 2016 (2016). Accessed on 16 Aug 2020. Available at: https://www.federalreserve.gov/econresdata/mobile-devices/files/consumers-and-mobile-financial-services-report-201603.pdf

36. Ameme, B.K.: The impact of customer demographic variables on the adoption and use of internet banking in developing economies. J. Internet Bank. Comm. **20**(2) (2015)
37. Hsu, M.H., Chang, C.M., Yen, C.H.: Exploring the antecedents of trust in virtual communities. Behav. Inf. Technol. J. **30**(5), 587–601 (2011)
38. Bailey, A.A., Pentina, I., Mishra, A.S., Mimoun, M.S.B.: Mobile payment adoption by US consumers: an extended TAM. Int. J. Retail Distrib. Manag. **45**, 626–640 (2017)
39. Li, M., Dong, Z.Y., Chen, X.: Factors influencing the consumption experience of mobile commerce. Internet Res. J. **22**(2), 120–141 (2011)
40. Saaksjarvi, M.: Consumer adoptions of technological innovations. Eur. J. Innov. Manage. **6**(2), 90–100 (2003)
41. Yiga, C., Cha, K.J.: Toward understanding the importance of trust in influencing Internet banking adoption in Uganda. Inf. Dev. **32**, 622–636 (2016)
42. Fungácová, Z., Hasan, I., Weill, L.: Trust in banks. J. Econ. Behav. Organ. (2017)
43. Vejacka, M., Štofa, T.: Influence of security and trust on electronic banking adoption in Slovakia. E+M Ekon. A Manag. **20**, 135–150 (2017)
44. Ben Mansour, K.: An analysis of business' acceptance of internet banking: an integration of e-trust to the TAM. J. Bus. Ind. Mark. **31**, 982–994 (2016)
45. Hillman, S., Neustaedter, C.: Trust and mobile commerce in North America. Comput. Hum. Behav. **70**, 10–21 (2017)
46. Rogers EM (1962) Diffusion of innovations. Free Press of Glencoe, Macmillan Company
47. Anshari, M., Alas, Y., Sulaiman, E.: Smartphone addictions and nomophobia among youth. Vulnerable Children & Youth Studies. Int. Interdiscip. J. Res. Policy Care **I**(3), 242–247 (2019)
48. Yaokumah, W., Kumah, P., Okai, E.S.A.: Demographic influences on e-payment services. Int. J. E-Bus. Res. (IJEBR) **13**(1), 44–65 (2017)
49. Alalwan, A.A., Dwivedi, Y.K., Rana, N.P.: Factors influencing adoption of mobile banking by Jordanian bank customers: extending UTAUT2 with trust. Int. J. Inf. Manag. **37**, 99–110 (2017)
50. Hossain, M.A., Hossain, M.S., Jahan, N.: Predicting continuance usage intention of mobile payment: an experimental study of Bangladeshi customers. Asian Econ. Fin. Rev. **8**(4), 487 (2018)
51. El-Kasheir, D., Ashour, A.S., Yacout, O.M.: Factors affecting continued usage of internet banking among Egyptian customers. Commun. IBIMA **9**, 252–263 (2009)
52. Dehbi Dehbini, N., Birjandi, M., Birjandi, H.: Factors influencing the adoption of electronic payment cards in urban micro-payments. Res. J. Fin. Account. **6**(1), 39–47 (2015)
53. Hossain, M.A.: Security perception in the adoption of mobile payment and the moderating effect of gender. PSU Res. Rev. **3**(3), 179–190 (2019)
54. Lee, C.C., Hsieh, M.C.: The influence of mobile self-efficacy on attitude towards mobile advertising. In: 2009 International Conference on New Trends in Information and Service Science, pp. 1231–1236. IEEE (2009)
55. Gan, C., Clemes, M., Limsombunchai, V., Weng, A.: A Logit Analysis of electronic banking in New Zealand. Int. J. Bank Market. **24**(6), 360–383 (2006)
56. Wang, Y.S., Wang, Y.M., Lin, H.H., Tang, T.I.: Determinants of user acceptance of internet banking: an empirical study. Int. J. Serv. Ind. Manage. **14**(5), 501–519 (2003)
57. Alagheband, P.: Adoption of e-banking services by Iranian customers. Unpublished Master Thesis. The Lulea University of Technology (2006)
58. Izogo, E.E., Nnaemeka, O.C., Onuoha, A.O., Ezema, K.S.: Impact of demographic variables on consumers' adoption of e-banking in Nigeria: An empirical investigation. Eur. J. Bus. Manage. **4**(17), 27–39 (2012)
59. Connolly, S., Stavins, J.: Payment Instrument Adoption and Use in the United States, 2009–2013, by Consumers' Demographic Characteristics. Research Data Reports Paper (15–6) (2015)
60. Koulayev, S., Rysman, M., Schuh, S., Stavins, J.: Explaining adoption and use of payment instruments by US consumers. Rand J. Econ. **47**(2), 293–325 (2016)

61. Ozkan, S., Bindusara, G., Hackney, R.: Facilitating the adoption of e-payment systems: theoretical constructs and empirical analysis. J. Enterprise Inf. Manag. **23**(3), 305–325 (2010)
62. Luarn, P., Lin, H.H.: Toward an understanding of the behavioral intention to use mobile banking. Comput. Human Behav. J. **21**(6), 873–891 (2005)
63. Ghasemi, A., Zahediasl, S.: Normality tests for statistical analysis: a guide for non-statisticians. Int. J. Endocrinol. Metabolism **10**(2), 486 (2012)
64. Sekran, U., Bougie, R.: Research Methods for Business: A Skill Building Approach, 5th edn. Wiley, United State (2009)
65. Hair, J., Babin, B., Money, A., Samouel, P.: Essentials of business research methods. Wiley, United States (2003)
66. Venkatesh, V., Davis, F.D.: A theoretical extension of the technology acceptance model: four longitudinal field studies. Manage. Sci. J. **46**(2), 186–204 (2000)
67. Sumanjeet, S.: The emergence of a payment system in the age of electronic commerce: state of the art (2009). Available at: https://www.researchgate.net/publication/228309774_Emerge nce_of_Payment_Systems_in_the_Age_of_Electronic_CommerceThe_State_of_Art
68. Gbongli, K., Xu, Y., Amedjonekou, K.M.: Extended technology acceptance model to predict mobile-based money acceptance and sustainability: a multi-analytical structural equation modelling and neural network approach. Sustainability **11**(13), 36–39 (2019)
69. Poon, W.C.: Users' adoption of e-banking services: the Malaysian perspective. J. Bus. Ind. Market. **23**(1), 59–69 (2008)
70. Chen, K., Chen, J.V., Yen, D.C.: Dimensions of Self-efficacy in the study of smartphone acceptance. Comput. Standards Interfaces J. **33**(4), 422–431 (2011)
71. Shin, D.H.: Towards an understanding of the consumer acceptance of mobile wallet. Comput. Hum. Behav. J. **25**(6), 1343–1354 (2009)
72. Kalinic, Z., Marinkovic, V., Molinillo, S., Liébana-Cabanillas, F.: A multi-analytical approach to peer-to-peer mobile payment acceptance prediction. J. Retail. Consum. Serv. **49**, 143–153 (2019)
73. Hamza, A., Shah, A.: Gender and mobile payment system adoption among students of tertiary institutions in Nigeria. Int. J. Comput. Inf. Technol. **3**(1), 13–20 (2014)
74. Duane, A., O'Reilly, P., Andreev, P.: Realizing M-payments: modelling consumers' willingness to M-pay using smartphones. Behav. Inf. Technol. J. **33**(4), 318–334 (2014)

Customer's Decision to Shift to e-Purchase Through Social Media in COVID-19 Pandemic

Bahaa Mohammad Alhamad, Naseem Mohammad Twaissi, Zaid Ahmad Alabaddi, and Ra'ed Masa'deh ⓘ

Abstract The study aimed mainly to describe the role of COVID-19 pandemic in the shift of the customer's purchasing decision from the traditional purchase to e-purchase through social media. To achieve the objectives of the study, the analytical descriptive approach methodology and simple random sample technique were adopted, the researchers designed a questionnaire depend on previous studies, which consisting of (24) items to gather the information from the study sample which reached (740) respondents. The Statistical Package for Social Sciences (SPSS) program was used to analyze and examine the data and the hypotheses. The main results were: Customer's experience e-purchase experiment through social media was satisfying from the point of view during the COVID-19 pandemic and there is a statistically significant impact of that experiment on customer satisfaction. Also, that experiment has increased the level of customer trust toward e-purchasing through social media during the pandemic and there is a statistically significant impact of that experiment on customer trust. On the other hand, the research found a positive relation between customer satisfaction and customer trust of that experiment during the pandemic and the intention to continue e-purchase after the pandemic. Also, the research proves that experiment in COVID-19 pandemic affected the customers buying decision to shift to e-purchase through social media by strongest customers satisfaction and increased the level of customers trust and this shifting will continue to the future after the pandemic. The research recommends the stakeholders to exploit

B. M. Alhamad (✉)
Ma'an, Jordan

N. M. Twaissi · Z. A. Alabaddi
Business Administration Department, AHU University, Ma'an, Jordan
e-mail: n.twaissi@ahu.edu.jo

Z. A. Alabaddi
e-mail: Dr_zaid_abadi@ahu.edu.jo

R. Masa'deh
School of Business, The University of Jordan, Amman, Jordan
e-mail: r.masadeh@ju.edu.jo

the social media platforms in their marketing strategies due to their great advantages in such crises and pandemics.

Keywords e-purchasing through social media · Customer satisfaction · Customer trust · COVID-19

1 Introduction

Due to the increasing of global population and living more closer to the animal, the human population has become more exposed to the danger of new kinds of viruses. All societies around the world can take more preventive measures to prohibit any pandemic could occur, and should also make an effort to prepare our societies when such pandemics could occur [1]. In December 2019, a resident of Wuhan, China came down with a mysterious ailment, thus marking the beginning of a global pandemic that would come to define life in most countries across the globe in the early 2020s [2]. With the end of 2019 and the beginning of 2020, the world witnessed a new virus outbreak, which has called COVID-19, and as it is still in the middle of a pandemic outbreak, it is not easy to estimate its effects on business, economic, customer behavior, or even societal consequences as these aspects have not been studied to a great extent in the past with similar pandemics [3].

With strict governmental procedures and the implementation of martial law and lockdown procedures in most countries, the CONID-19 pandemic outbreak has forced many businesses to close, which leading to an unprecedented disruption of commerce in most industry sectors. Retailers and brands face many short-term challenges, such as those related to health and safety, the supply chain, the workforce, cash flow, consumer demand, sales, and marketing [1]. Many industry sectors and markets specially those engaged in tourism and hospitality, no longer exist due to the COVID-19 pandemic.

In fact, it is the time to all business organizations and firms to rearrangement of their prioritizes, so postpone tasks and spending more time have no value in this new environment. On the other hand, there are sectors have unprecedented growth such as online communication platform, e-commerce and online shopping platforms, which in turn provided new opportunities for business organizations to adapt with this new business environment. Another consequence of the current pandemic is the extreme increase in internet and social media usage. Previous research has indicated that humans who feel lonely tend to use social media more and, in some cases, even prefer social media over physical interaction [4, 5]. The Coronavirus pandemic has activated a new wave of commerce innovation in shifting SMEs to e-commerce. Where, new buying behaviors are forming that are likely to remain after the crisis has passed—and this presents opportunities to companies if such a pandemic could occur in the future. Especially, those who viewed digital commerce as a secondary channel, now need to reprioritize their business with a digital commerce focus [6]. The e-commerce experience through a social media platforms is becoming critical to customer decision-making, so many households are trying these new innovative services for the first time [6–9]. The focus of this paper to describe the role of the COVID-19 pandemic in the shift of the customer's purchasing decision from the traditional purchase to e-purchase through social media, identifying the extent to which the e-purchase experience through social media was satisfying in the pandemic, explaining how the experience of e-purchase through social media affects the level of customer trust towards e-purchase decision, and exploring if the customers will continue e-purchase through social media after this pandemic.

1.1 Study Questions

1. What is the role of the COVID-19 pandemic in the shift of the customer's purchasing decision from the traditional purchase to e-purchase through social media?
2. To what extent the e-purchase experience through social media was satisfying in the pandemic?
3. How does the experience of e-purchase through social media affect the level of customer trust towards e-purchase decision?
4. To what extent customers will continue e-purchase through social media after the pandemic?

1.2 Study Hypotheses

H1. Customer's experience e-purchase experiment through social media was satisfying from the point of view during the pandemic.

H2. The e-purchase experiment through social media increased the level of customer trust towards e-purchase decision during the pandemic.

H3. There is a relation between customer's satisfaction of the e-purchase experience through social media during the pandemic and the intention to continue e-purchase through social media after the pandemic.

H4. There is a relation between customer's trust in e-purchase through social media during the pandemic and the intention to continue e-purchase through social media after the pandemic.

H5. Customers will continue e-purchase through social media after the pandemic.

1.3 Study Model

Figure 1 represent the study model.

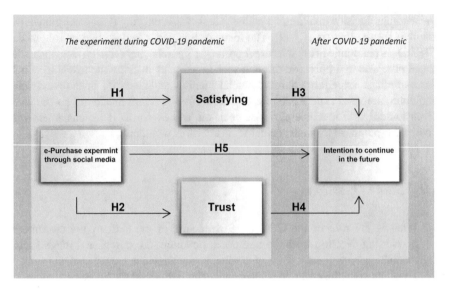

Fig. 1 The study model

2 Role of Social Media During COVID-19 Pandemic

Social media platforms played an important role during the COVID-19 pandemic in many fields, which produced a new opportunity to companies to take advantage of marketing and selling through social media during quarantine and lockdown law in the COVID-19 pandemic. Social media defines' as a communication platform between individuals that enables to creation of interactive content on the Internet to create a new environment of participation and cooperation between the individuals themselves and between individuals and companies [10], and consider as one of the most powerful tools of networking across the internet that has connected the social with economic aspects in the real world [11, 12].

The main role that social media platforms have played during the COVID-19 pandemic outbreak, they were a source of information to the majority. In recent years, traditional sources of information have been replaced by social media [13, 14]. So, finding a trusted source of information in such pandemic and circumstances is extremely important. Especially information that spread on the official and governmental accounts on social media. Instagram, for example, announced to only include COVID-19 related posts and stories in their recommendation section that are published by official health organizations [15]. So, such information can help keep us safe, providing us with a better understanding of what is occurring and how it might impact us and those we love [16]. COVID-19 has increased the use of social media on Facebook, Instagram, WhatsApp, Twitter, and Zoom. They are generating an enormous amount of data on word of mouth [17].

This increasing in the internet and social media usage caught the attention of some companies to take their chance of exploiting the social media platforms in marketing to sell their products, acquire more customer's, strongest their brands, and increasing their customers loyalty, trust, and satisfaction. De Valck [15] explained that, companies can use social media for commercial purposes or for communal purposes. In other words, companies use social media to brand, sell, market their business (which is close to traditional marketing efforts using mass-media) versus using social media to connect with and co-create with customers and—more importantly—to provide a platform to customers to bond together.

3 Customer's e-Purchase Experiment During COVID-19 Pandemic

Customer purchase-decision process come as an extent to customer behavior science, it can identify through these steps: desire, information search, alternatives evaluation, selection and post-purchasing evaluation. Basil et al. [18] defined the consumer decision-making process as "the phases that consumers go through to take a final purchase decision. The task of a marketer is to focus on the whole series of the purchasing process instead of emphasizing solely on a purchase decision, because consumers experience different phases before reaching a conclusion", all of that when talk about the traditional customer's purchase decision.

Whereas, since COVID-19 and in lockdown and quarantine law, a lot of consumers shifted their purchasing behavior and have begun to shop online more than before the pandemic, and the trend is likely to continue after the pandemic ends. As Bertsch [19] explained our survey found that 69% of respondents would be more willing to subscribe to a delivery service for essential items following the pandemic. Furthermore, Fig. 2 shows the results of a survey conducted in Germany, United Kingdom and United State about the shifting to online purchases because of the COVID-19 pandemic 2020, by category as spread in Statista website [20].

4 Research Methodology

4.1 Study Approach

To achieve the objectives of the study and answer its questions, the analytical descriptive approach methodology was adopted as the study has a specific clear problem statement and specific hypotheses [21], so it is the most suitable approaches for the nature of the study and its goals in terms of conducting data analysis procedures and subjecting them to the statistical analysis and tests its hypotheses.

	Germany	United Kingdom	United States
Restaurant delivery / takeaway	16%	19%	31%
Hygiene products (e.g. hand sanitizer, toilet paper)	13%	21%	27%
Clothing	25%	24%	26%
Household cleaning products	9%	17%	26%
Food and drink delivery (e.g. from supermarket)	10%	30%	24%
Health products (e.g. medicine)	15%	15%	21%
Books	14%	16%	15%
Hobby supplies	9%	14%	15%
Games	10%	13%	13%
Consumer electronics, household appliances, furniture	9%	11%	11%
Video	7%	7%	11%
Music	7%	8%	10%
Magazines & newspapers	6%	7%	6%
Financial products and services	4%	6%	6%
Other	4%	3%	3%
I have not shifted from offline to online purchases for any products or services	47%	35%	29%

Fig. 2 The result of survey

4.2 Study Population and Sample

The study adopted social media tools users among Jordanian population those above 18 years old, and targeted those who had actual online purchase experience through social media platforms specially in lockdown during COVID-19 pandemic period. The respondents were selected from those interested in online purchase through social media and posting comments on commercial ads on social media to ask about something (e.g. price). The simple random sample technique has adopted, so it is the most suitable sample technique as the nature of the study and its population. The questionnaire was loaded on the surveymonkey.com and the link was sent as massages through social media to 1340 respondents and 740 complete responses received. Table 1 below describes the demographic characteristics of the sample.

Table 1 Sample demographic characteristics

Sample characteristic					
Gender			Age		
Male	320	43.2%	18–29	260	35.1%
Female	420	56.8%	30–49	330	44.6%
			+50	150	20.3%

4.3 Questionnaire Design Processes

Questionnaire was designed to achieve the objectives of the study, and to answer its questions, as it is instrument to collect data from the sample to test study hypotheses. The first part of the questionnaire provides a general information to describe the demographic characteristics of the respondents as shows in Table 1. The second part of questionnaire represented by 22 paragraphs to elaborate the independent variable, moderator variables and dependent variable that would be tested in this research. All of those paragraphs adopted from [22–24] studies and restructured to correspond the goals and objectives of this study. Whereas, paragraph number 23 consists of question and 8 nominal options and adopted from survey spread in Statista website [20], and paragraph number 24 which consist of question and 5 nominal options to show the most popular social media tools in context of online marketing from the point of view of Jordanian users. A 5-point Likert scale anchored by "strongly disagree" (1) to "strongly disagree" (5) was used as the attitude measurement for the independent, moderator and dependent variables. The degree of approval was concluded through the following steps according to [25]:

- Extracting the scale range (5 − 1 = 4).
- Divide the average scale by the number of required levels, total of 3 (Low, Moderate, and High) 4 ÷ 3 = 1.33, which the category length, and thus resulting in three levels:

> 1.00-2.33 *low level* of approval.
> 2.34-3.67 *Moderate level* of approval.
> 3.68-5.0*High level* of approval.

5 Data Analysis and Result Discussion

5.1 Reliability Test

The reliability of a measure indicates the stability and consistency with which the instrument measures the concept and helps to assess the 'goodness' of a measure [26]. The reliability of the questionnaire was measured by extracting the Cronbach Alpha coefficient to measure the coefficient of the scale paragraphs' consistency. According

Table 2 Reliability of the study tool by Cronbach alpha

Cronbach alpha

Variable	Number of paragraphs	Coefficient of Cronbach alpha
Purchase experiment	5	0.780
Satisfying	7	0.884
Trust	5	0.899
Intention to continue	5	0.764

Table 3 Validity of the study by KMO and Bartlett's test

KMO and Bartlett's test

Kaiser–Meyer–Olkin measure of sampling adequacy		0.669
Bartlett's Test of sphericity	Approx. Chi-Square	16,648.022
	df	231
	Sig	0.000

to each variable, notes that all of them higher than the value of (0.7), which indicate the reliability of the study tool. Table 2 shows the study tool reliability of coefficients.

5.2 Validity Test

To verify the validity of the questionnaire questions used in the study, it had been presented to academic arbitrators specialized in this field and are working in a Jordanian Universities, in order to ensure the accuracy and validity of the paragraphs. To issue out their decision regarding the validity of the paragraphs, the soundness of their formulation and its relevance to the study subject. The questionnaire was assessed by four academic arbitrators specialized in business administration, marketing and strategic management. Furthermore, construct validity was adopted as validity measurement and factor analysis was used to measure the construct validity [26], the details of factor analysis as shows in Table 3 is appropriate because the KMO value is 0.669 is greater than 0.6 and approximate Chi-Square 16,648.022 with df 231 at the significant level 0.000 less than 0.05. Hence factor analysis indicate that it is appropriate for further data analysis.

5.3 Hypotheses Test

For this study, five hypotheses were tested using Simple Regression Linear and Pearson Correlation Factor as following:

Table 4 Regression factor for the first hypothesis

	Regression		(F) Test		(T) Test		R^2	R
	Std. Error	β	Sig	F-value	Sig	T-value		
Constant	0.105	1.098	0	370.005	0	10.44	0.334	0.578
CEE	0.031	0.591			0	19.237		

*Statistically significant at ($\alpha \leq 0.05$) level
CEE Customer e-purchase experiment

H1. Customer's experience e-purchase experiment through social media was satisfying from the point of view during the pandemic.

Table 4 shows that, the result from this table statistically accepted, where F-value (370.005) and it is statistically significant at the significance level ($\alpha \leq 0.05$), which confirms that, there is positively impact of e-purchase experiment through social media on the customer satisfaction, which indicate that the experiment was satisfying from the point of view during the pandemic, and the hypothesis supported, and with looking to Table 5 which represent the level of satisfying about that experiment notes that the average of total answers came with (3.05) which represent moderate to high level of satisfying, so this also support the hypotheses about that experiment was satisfying from point of view.

H2. The e-purchase experiment through social media increased the level of customer trust towards e-purchase decision during the pandemic.

Table 6 shows that, the result from this table statistically accepted, where F-value (182.301) and it is statistically significant at the significance level ($\alpha \leq 0.05$), which

Table 5 Level of satisfying about e-purchase experiment

Level of satisfying	Std. dev.	Mean	Paragraph	
Moderate	1.169	3.18	I felt comfort with that experiment	1
Moderate	1.248	2.77	I felt that experiment was easy to use	2
Moderate	1.105	3.57	My standards and expectations from the products/services I bought through social media were very high	3
Moderate	1.132	3.66	The performance of retailer fulfilled my expectation	4
Moderate	1.222	2.26	The retailer would not behave opportunistically (e.g., gaining money illegally)	5
Moderate	1.132	2.84	The retailer kept my best interests in mind	6
Moderate	1.004	3.11	I found better quality products/services on social media	7
Moderate	1.144	3.05	Mean of all paragraphs value	

Table 6 Regression factor for the second hypothesis

	Regression		(F) Test		(T) Test		R^2	R
	Std. Error	β	Sig	F-value	Sig	T-value		
Constant	0.134	1.408	0	182.301	0	10.498	0.198	0.445
CEE	0.039	0.529			0	13.502		

*Statistically significant at ($\alpha \leq 0.05$) level
CEE Customer e-purchase experiment

Table 7 Level of trust toward e-purchase experiment

Degree of satisfying	Std. dev.	Mean	Paragraph	
Moderate	1.196	2.80	The experiment was trustworthy and honest	1
Moderate	1.090	3.22	Marketers kept their promises and obligations	2
Moderate	1.407	3.13	The information provided by the marketers about the goods/services was plentiful and sufficient	3
Moderate	1.117	3.36	The retailer provided secure personal privacy	4
Moderate	1.226	3.28	It is important for me to buy products/services through social media retailers accounts with well-known brand names	5
Moderate	1.436	3.15	Mean of all paragraphs value	

confirms that, there is positively impact of e-purchase experiment through social media on the level of customer trust toward e-purchasing decision, which indicate that the experiment was increased the level of customer trust toward online purchase during the pandemic, and the hypothesis supported, and with looking to Table 7 which represent the level of trust about that experiment notes that the average of total answers came with (3.15) which represent moderate to high level of trust, so this also support the hypothesis about that experiment was increased the customer trust in online purchase from point of view during the pandemic.

H3. There is a relation between customer's satisfaction of the e-purchase experience through social media during the pandemic and the intention to continue e-purchase through social media after the pandemic.

Correlation coefficients shown in Table 8 indicates that, there is a statistically significant relationship between the customer satisfaction of the e-purchase experience through social media during the lockdown in the COVID-19 pandemic and intention to continue e-purchase through social media in the future after the pandemic, the value of the correlation coefficient has reached (0.711) at the significant level

Table 8 Pearson correlation factor to the third hypothesis

Correlations

		Satisfying	Intention to continue
Satisfying	Pearson Correlation	1	0.711**
	Sig. (2-tailed)		0.000
	N	740	740
Intention to continue e-purchase in the future	Pearson Correlation	0.711**	1
	Sig. (2-tailed)	0.000	
	N	740	740

**Correlation is significant at the 0.01 level (2-tailed)

(0.000), and represent a statistically significant positive relationships between the two variables, which support the third hypothesis.

> *H4. There is a relation between customer's trust in e-purchase through social media during the pandemic and the intention to continue e-purchase through social media after the pandemic.*

Correlation coefficients shown in Table 9 indicates that, there is a statistically significant relationship between the customer trust in the e-purchase experience through social media during the lockdown in the COVID-19 pandemic and intention to continue e-purchase through social media in the future after the pandemic, the value of the correlation coefficient has reached (0.668) at the significant level (0.000), and represent a statistically significant positive relationship between the two variables, which support the fourth hypothesis and this support the results of [22] study.

> *H5. Customers will continue e-purchase through social media after the pandemic.*

Table 10 shows that, the result from this table statistically accepted, where F-value (395.187) and it is statistically significant at the significance level ($\alpha \leq 0.05$),

Table 9 Pearson correlation factor to the fourth hypothesis

Correlations

		Trust	Intention to continue
Trust	Pearson Correlation	1	0.668**
	Sig. (2-tailed)		0.000
	N	740	740
Intention to continue e-purchase in the future	Pearson Correlation	0.668**	1
	Sig. (2-tailed)	0.000	
	N	740	740

**Correlation is significant at the 0.01 level (2-tailed)

Table 10 Regression factor for the fifth hypothesis

	Regression		(F) Test		(T) Test		R^2	R
	Std. Error	β	Sig	F-value	Sig	T-value		
Constant	0.09	1.979	0	395.187	0	21.929	0.349	0.591
CEE	0.026	0.524			0	19.879		

*Statistically significant at ($\alpha \leq 0.05$) level
CEE Customer e-purchase experiment

which confirms that, there is positively impact of e-purchase experiment through social media during the pandemic on the intention to continue e-purchasing through social media in the future after the pandemic, and the hypothesis supported, and corresponds with [6] and [19] results.

Figure 3 shows that, the existence of customer satisfaction and customer trust as a moderators variables found a partial mediation of the relationship between the e-purchase experiment through social media during the pandemic and the intention the continue e-purchase after the pandemic and contributes to significant result with regression factor (0.239) to the customer satisfaction variable at the significance level ($\alpha \leq 0.05$), and (0.324) to the customer trust variable at the significance level ($\alpha \leq 0.05$), which also support the fifth hypothesis.

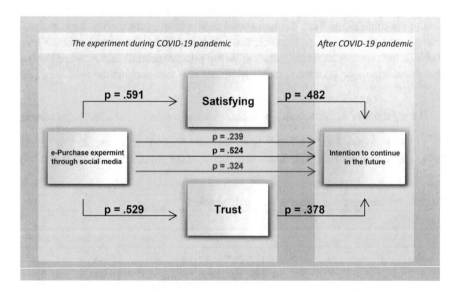

Fig. 3 Satisfaction and trust as moderators variables

Table 11 Ranking prioritizes of customers

	Product	Percentage (%)	Germany (%)	UK (%)	US (%)
1	Health products (e.g. *medicine*)	80.4	15	15	21
2	Food and drink delivery (e.g. *supermarket*)	68.9	10	30	24
3	Hygiene products (e.g. *hand sanitizer, toilet paper*)	61.5	13	21	27
4	Restaurant delivery/takeaway	60.1	16	19	31
5	Household cleaning products	48.6	9	17	26
6	Other	18.9	4	3	3
7	Clothing	8.8	25	24	26

5.4 Customer Prioritizes in e-Purchasing Through Social Media

Table 11 shows that, the ranking of answers of the question (23) in question-naire ("have you deliberately purchased any of these products or services online (through social media tools) instead of offline because of the COVID-19/coronavirus pandemic?"). In comparison to, [20] survey result in Fig. 2. Table shows that the highest rate came with (80.4%) percentage to the health products, whereas the minimum one came with (8.8%) to the clothing.

5.5 Customer Prioritizes in Social Media Tools in e-Purchase Decision

Figure 4 shows that, the ranking of answers of the question (24) in questionnaire ("please select the social media tools you prefer in online purchase decision."), notes that the highest rate came with (78.2%) percentage to Facebook, whereas the minimum one came with (8.3%) to Other.

6 Conclusion

6.1 Implication of the Research

The research findings provide significant implications to a wide range of stakeholders in various fields of industries. In terms of managerial decisions and implications, the research findings do some suggested solutions and insights to decision makers

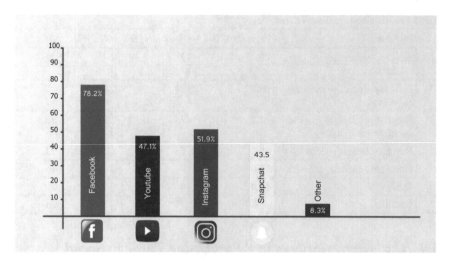

Fig. 4 Customer prioritizes of social media tools

in business organization and firms to rearrangement their prioritize and adopt new marketing strategies in crises such as lockdown period in COVID-19 pandemic, specially to those who viewed digital commerce as a secondary channel, so new buying behaviors are forming that are likely to remain after the crisis has passed [6]. Therefore, the research findings discovered that the customer's those have tried to e-purchase through social media platforms found that experiment was satisfying and increased theirs level of trust towards e-purchase through social media. In addition, the research findings prove that, the majority of those customer will continue to e-purchase through social media in the future after the pandemic which correspond with [6] and the results of [19] survey. The research discovered and proved that the customer's e-purchase through social media platforms experiment in COVID-19 pandemic affected the customers buying decision to shift to e-purchase through social media by strongest customer's satisfaction and increased the level of customer's trust and this shifting will continue to the future after the pandemic. So, the researchers recommended the stakeholders to exploit the social media platforms in marketing to sell their products, acquire more customer's, strongest their brands, and increasing their customer's loyalty, trust and satisfaction, so "postpone tasks and spending more time in the pandemic has no value, when there are another marketing strategies available".

6.2 Limitations of the Research and Future Recommendations

This research process provides some insights to the researchers, and these insights could be viewed in the shed light of limitations. This research focused on those customer's who have an experiment in e-purchase through social media and did not covered those who have an online purchasing in general, and did not take into consideration some moderating variables such as the customers loyalty and the impact of brand, and also the impact of gender and age differences. In addition, this study adopted the quantitative research approach and questionnaire study tool. Future study could combine the two approaches quantitative and qualitative, and interview as a study tool.

References

1. Donthu, N., Gustafsson, A.: Effects of COVID-19 on business and research. J. Bus. Res. **117**, 284–289 (2020)
2. Kirk, C., Rifkin, L.: I'll trade you diamonds for toilet paper: consumer reacting, coping and adapting behaviors in the COVID-19 pandemic. J. Bus. Res. **117**, 124–131 (2020)
3. Jorda, O., Singh, S.R., Taylor, A.M.: Longer-run economic consequences of pandemics (Report no. w26934). National Bureau of Economic Research (2020)
4. Nowland, R., Necka, E.A., Cacioppo, J.T.: Loneliness and social internet use: pathways to reconnection in a digital world? Perspect. Psychol. Sci. **13**(1), 70–87 (2018)
5. Al-Dmour, H., Masa'deh, R., Salman, A., Abuhashesh, M., Al-Dmour, R.: Influence of social media platforms on public health protection against the COVID-19 pandemic via the mediating effects of public health awareness and behavioral changes: Integrated model. J. Med. Internet Res. **22**(8) (2020). https://doi.org/10.2196/19996
6. Zealley, J.: Channel shift: prioritizing digital commerce (2020). Retrieved from https://www.accenture.com/_acnmedia/Thought-Leadership-Assets/PDF-2/Accenture-COVID-19-Channel-Shift-Prioritizing-Digital-Commerce.pdf. Accessed 9 Apr 2020
7. Hunaiti, Z., Mansour, M., Al-Nawafleh, A.: Electronic commerce adoption barriers in Small and Medium-Sized Enterprises (SMEs) in developing countries: The case of Libya. In: in the 11th IBIMA Conference on Innovation and Knowledge Management in Twin Track Economies, Cairo-Egypt, 4–6 January (2009)
8. Alshurideh, M., Alkurdi, B.: The effect of customer satisfaction upon customer retention in the Jordanian mobile market: an empirical investigation. Eur. J. Econ. Fin. Adm. Sci. **47**(April), 69–78 (2012)
9. Masa'deh, R., Shannak, R., Maqableh, M.: A structural equation modeling approach for determining antecedents and outcomes of students' attitude toward mobile commerce adoption. Life Sci. J. **10**(4), 2321–2333 (2013)
10. Alghizzawi, M., Salloum, S., Habes, M.: The role of social media in tourism marketing in Jordan. Int. J. Inf. Technol. Lang. Stud. **2**(3), 59–70 (2018)
11. Abd Al-samee, M.: Role of social media sites in marketing Egypt as an international touristic destination, Ministry of Tourism. Ministry of Tourism Contest for the Year 2012 (2012)
12. Altamony, H., Alshurideh, M., Obeidat, B.: Information systems for competitive advantage: implementation of an organisational strategic management process. In: Innovation and Sustainable Competitive Advantage: From Regional Development to World Economies-Proceedings of the 18th Inter-national Business Information Management Association Conference, vol. 1, pp. 583–592 (2012)

13. Shannak, R., Obeidat, B., Almajali, D.: Information technology investments: a literature review. In: 14th IBIMA Conference on Global Business Transformation through Innovation and Knowledge Management: An Academic Perspective, Istanbul-Turkey, 23–24 June, pp. 1356–1368 (2010)

14. Fotis, J., Buhalis, D., Rossides, N.: Social media impact on holiday travel planning. Int. J. Online Market. 1(4), 1–19 (2011)

15. De Valck, K.: What is the role of social media during the COVID-19 crisis? (2020). Retrieved from https://www.hec.edu/en/knowledge/instants/what-role-social-media-during-covid-19-crisis-0. Accessed 26 Apr 2020

16. Kushner, J.: The role of social media during a pandemic (2020). Retrieved from https://khoros.com/blog/social-medias-role-during-covid-19. Accessed 2 May 2020

17. Sheth, J.: Impact of Covid-19 on consumer behavior: will the old habits return or die? J. Bus. Res. 117, 280–283 (2020)

18. Basil, G., Etuk, E., Ebitu, E.T.: The marketing mix element as determinants of consumer's choice of made-in-Nigeria shoes in Cross River state. Eur. J. Bus. Manage. 5(6), 141–147 (2013)

19. Bertsch, E.: Consumers continue to shop during COVID-19. What are the biggest takeaways for retailers? (2020). Retrieved from https://www.mytotalretail.com/article/consumers-continue-to-shop-during-covid-19-what-are-the-biggest-takeaways-for-retailers/. Accessed 11 Aug 2020

20. Statista: Have you deliberately purchased any of these products or services online instead of offline because of the COVID-19/coronavirus pandemic? (2020). Retrieved from https://www.statista.com/statistics/1107859/shifting-to-online-purchases-because-of-the-covid-19-pandemic-by-category/#statisticContainer. Accessed 3 Aug 2020

21. Malhotra, N.K.: Marketing Research: An Applied Orientation, 4th edn. Prenticall-Hall, New Jersey (2004)

22. Ling, K., Chai, L., Piew, T.: The effects of shopping orientations, online trust and prior online purchase experience toward customers' online purchase intention. Int. Bus. Res. 3(3), 63–79 (2010)

23. Gehrt, K.C., Onzo, N., Fujita, K., Rajan, N.R.: The emergence of internet shopping in Japan: identification of shopping orientation-defined segment. J. Market. Theory Pract. 15(2), 167–177 (2007)

24. Yu-Hui, C., Barnes, S.: Initial trust and online buyer behavior. Ind. Manage. Data Syst. 107(3), 21–36 (2007)

25. Dancey, C., Reidy, J.: Statistics without Maths for Psychology, 3rd edn. Pearson Prentice Hall (2004)

26. Cavana, R.Y., Delahaye, B.L., Sekaran, U.: Applied Business Research: Qualitative and Quantitative Methods. Wiley, Queensland (2001)

The e-Learning of Students and University's Brand Image (Post COVID-19): How Successfully Al-Ain University Have Embraced the Paradigm Shift in Digital Learning

M. T. Nuseir⊙**, Ghaleb A. El-Refae**⊙**, and A. Aljumah**⊙

Abstract The COVID-19 has caused a state of emergency not only in the UAE but also around the globe. Following the preventive measures, the government of UAE has closed temporarily all types of educational institutes, replacing them by other institutes, thus shifting to e-learning. Therefore, this study aims at examining the role of information and communication technology, e-service quality, and e-information quality towards the brand image of universities by concentrating on students' e-learning. More specifically, this study investigates the impact of the information and communication ethnobiology post COVID-19 on the university's brand image of Alain University in UAE as one of the leading universities. The contribution of the study is twofold. First, the study is among the pioneering studies that have proposed a framework to develop an excellent online learning system, which can promote the university's brand image. Second, this study fills the gap by investigating the subject in the context of a Gulf country, which is among the most affected thirty countries due to COVID-19. It is, therefore, a critical contemporary issue for governments, higher education commissions, and universities to understand the antecedents of the online learning system and their relationships with the university's brand image.

Keywords Information and communication technology · Students' e-learning · University's brand image · COVID-19

M. T. Nuseir (✉)
Department of Business Administration, College of Business, Al Ain University, Abu Dhabi Campus, P.O. Box 112612, Abu Dhabi, UAE
e-mail: mohammed.nuseir@aau.ac.ae

G. A. El-Refae
Department of Business Administration, College of Business, Al Ain Campus, Al Ain University, P.O. Box: 64141, Al Ain, UAE
e-mail: ghalebelrefae@aau.ac.ae

A. Aljumah
Management Department, Emirates College of Technology, Abu Dhabi, P.O. Box 41009, Abu Dhabi, UAE
e-mail: aljumah37@gmail.com

© The Author(s), under exclusive license to Springer Nature Switzerland AG 2021   171
M. T. Alshurideh et al. (eds.), *The Effect of Coronavirus Disease (COVID-19) on Business Intelligence*, Studies in Systems, Decision and Control 334,
https://doi.org/10.1007/978-3-030-67151-8_10

1 Introduction

In the modern era of information technology, the e-learning system has revolutionized various sectors, including the education industry. This potential shift towards a contemporary teaching and training design will provide opportunities to the higher education institutions (HEIs) and represent at the same time a massive challenge for these institutions. For several years, an intense competition has been among face-to-face classes and online or virtual classes. Predominantly, face-to-face classes is a widely used method for teaching in all global universities. In spite of the several advantages associated with e-learning, such as perception, compatibility, and ease of use, the lack of communication is a significant obstacle in its adoption as the mainstream learning system [1], mainly due to its lack of responsiveness, reliability, empathy, assurance, and tangibility. Therefore, the current situation around the globe has compelled the entire education system to adopt the modern e-learning system.

The comparison of the online learning system with classical face-to-face education shows that e-learning provides better information and knowledge to the students, thus enhancing their learning ability. It also adequately delivers the quality of information and services to the students if proper information and communication technology (ICT) is established. In this regard, the advancement in ICT is a prominent factor that has landscaped several industries, including the education industry. Thus coupled with e-information quality and e-service quality, ICT brings several benefits for the university students, and several postgraduate and undergraduate university programs have started to explore the opportunities arising from this system. The existing literature concerning the relationship between students' e-learning and service quality indicates that e-learning is significantly determined by the service quality [2]. However, due to the scarcity of service providers and trained instructors, only limited research is conducted to observe the influence of e-information quality and e-service quality on students' learning. Besides other objectives, this study aims at analyzing the effects of e-information quality, e-service quality, and ICT on students' e-learning in the higher education institutions (HEIs) in Alain University in UAE.

Each teaching method particularly emphasizes on students' satisfaction, as it determines whether the quality of information and knowledge that is rendered to the students is in line with their expectations. Therefore, e-learning is assumed to enhance the students' learning efficacy and ultimately improve their efficiency. More specifically, E-learning with the highest quality of education may increase students' satisfaction [3]. However, the adoption of online teaching methods and maintaining accreditation standards and service quality are a massive challenge for UAE, since it has been undergoing a severe economic crisis due to COVID-19. Therefore, this study aims at analyzing the impact of e-learning on the students' satisfaction in the HEIs in the UAE.

The electronic Word Of Mouth emphasizes to have a direct social contact through the internet. eWOM is generally used in buzz marketing to make a particular message viral on the internet instantly. Although it is not a new concept, it is the speed of spreading a particular message, which makes it even more critical in marketing. Thus, eWOM is used as a marketing tool because people prefer to take advice online, and they like to share and trust the reviews of other users. EWOM is also assumed to have a significant relationship with online learning or any online service [4]. Students having some experience of online service or who have been affected by it usually share their negative or positive WOM, and this can instantly go viral on the internet or among the students community. Therefore, this study aims at analyzing how students' word of mouth affects the e-learning in HEIs in UAE.

Uppal et al. [19] defined the brand image as '*the unique bundle of association in a customer's mind.*' It is a specific image in the customers' eyes, which gradually develops over time. However, in the developing countries, universities have paid less attention to creating their brand image [23], and most HEIs emphasize on the average grades and reputation as a tool for attracting students. Due to COVID-19, the recent shift from classical to contemporary online learning has increased their vulnerability to eWOM. In other words, online tutorials and recorded videos with repetition and pause options may lead the students in determining the service quality provided by their universities, and it will be later translated into their views in the form of negative or positive eWOM [2, 4]. Therefore, creating and sustaining a specific brand image, particularly during the recent situation with a limited pool of resources is a significant challenge for HEIs in the UAE. As a result, this study aims to assess how student's satisfaction and eWOM affect the brand image of HEIs in the UAE.

This study contributes to the literature in two ways. First, it is a pioneering study because a comprehensive framework is proposed in this study for establishing a rigorous e-learning system. This framework is capable of promoting the university's brand image. Second, this study fills the existing research gap by addressing it in the context of the South Asian region, which is also among the list of top 30 countries that have been affected by COVID-19. Therefore, understanding the factors of the online learning system and determining their relationship with the universities' image is a critical issue in the higher education commission, universities, and the government.

2 Hypothesis Development

This section provides the literature concerning the relevant variables and relationships among them. Several prior studies have highlighted the significance of exchanging information among two parties and is considered one of the significant aspects of e-service [5]. Most scholars argue that the internet facility is commonly used by people to appease the need for information and satisfy their thirst for knowledge [6]. According to Verma and Illés [7], the comparison of online learning procedures and typical education systems reveals that better and more information and knowledge can be obtained through an online education system that may enhance the learning among students. That is, the information and knowledge acquisition is an essential online practice, and the continuous improvement in ICT will be likely to significantly contribute to the students' academic achievements [8]. In the education sector, information and communication technology (ICT) adoption has significantly revolutionized the learning procedures and teaching methods.

In addition, it has enabled access to new learning methods and teaching resources. Besides, it forms the basis of a new training method, i.e., e-learning in the education sector. The ICT quality depends on the way information and knowledge are rendered to the students. According to Zhou and Teo [9], information quality plays a significant role in the academic achievements of students. However, the recent global situation of Coronavirus breakdown has significantly augmented the need for ICT to facilitate learning. This breakdown has resulted in the discontinuation of physical contact among students and educational institutions, which gave rise to the demand for ICT to resume students' learning at a global level. Therefore, the following hypothesis is proposed:

H1: A positive relationship exists between students' e-learning and information and communication technology (ICT).

In the context of e-learning, the e-learning quality is of significant importance, followed by the quality of the course material, e-learning tutor, and the quality of the service support and e-learning administration [10]. In addition, the e-service quality is suggested to have a direct influence on the loyalty of e-learning students [11]. Besides the limited research on the website quality, a few studies have attempted to analyze the relationship between the academic achievements of students and the website quality. The qualitative literature concerning the online or web-based teaching and learning procedures have indicated that the website quality depends on the availability of a better education and learning requirements. The web-based education in its turn is related to the students' satisfaction with the teaching method [12]. More particularly, e-satisfaction is determined by online features, like consistency, convenience, conciseness, quality-based knowledge, cost reduction, security, and efficiency [13]. Based on the empirical studies, three website quality factors determine the positive behavior or positive WOM for education website, namely the organization, the entertainment, and the informativeness [4], whereby each factor differently affects the positive behavior towards the education website. In addition, these features of education website explain their significance compared to the classical face-to-face learning system. Hence, an increased adoption of web-based education system has appeared among the educational institutions around the globe, thus consequently leading to a shift from typical face-to-face learning to online education and is likely to influence the students' learning. This shift to a web-based education system has given rise to the demand to address this subject area. In order to address this area, this study aims at examining the institutional reputation and e-learning of students during the critical situation caused by the COVID-19. Therefore, it is necessary to examine the e-service quality for e-learning of students. As a consequence, the following hypothesis is proposed:

H2: A positive association exists among the e-learning of students and the e-services quality.

The empirical studies have proved that the provision of higher e-information quality improves the e-learning of students. In the view of [14] study, the comparison of web-based services and typical services forms the basis for the perspective that all service quality features do not have an equal influence on the whole service quality rendered to the students. In the current coronavirus situation, an enormous need arises for high-quality e-information, which may improve the students' e-learning. Further, the quality assurance is reported as the most significant feature of service quality for delivering the quality of lectures to the students [15]. In another study by Kong and Song [16], responsiveness and assurance were reported as the most key features in assessing the service quality. Thus, the information quality can be determined through assessing these measures. In the context of online education service, it has been mentioned by various studies that different service quality features are of different significance.

In the context of the web-based education system, it is driven by four elements, namely the use of the system, the entertainment, the information quality, and the system design [17]. Therefore, one of the vital elements of the e-learning session is the e-information quality, which improves the students' e-learning and is deemed as necessary for all institutions, particularly in the current situation arising from the coronavirus. E-information quality has also gained significant attention among the researchers. Thus, the following hypothesis is hypothesized:

H3: A positive association exists among the e-learning of students and e-information quality.

Because of the malfeasance, the environmental injustice, the carelessness, and maiming as the main obstacles in establishing face-to-face meeting, the Adoption Theory and the Technology Acceptance Theory state that it is intellectually and psychically impossible to spend too much time to share information with the world, thus giving rise to the need for technology. Such challenges demand the adoption of alternative means for handling such situations mainly because turning off the toxic producing system is not a wise option. Therefore, technology adoption is the best way to remove toxicity from the system.

Based on this theory, the present study aims at examining the ICT adoption in providing e-learning by the education institutions, which involve some levels of toxicity in their system due to coronavirus influence. At present, universities have adopted ICT as the only solution to provide e-learning and quality e-information to the students using high-quality e-services. The diffusion of innovation (DOI) model seeks to assess various types of innovation and provide four factors that influence the creation of a new idea, including channels, communication, innovation, and time. The DOI model is generally adopted at an organizational level and also forms the basis for the adoption of e-learning at the global level. Therefore, the DOI model particularly emphasizes on the environmental aspects, the organizational attributes, and the system characteristics that are likely to influence the individuals' interactions. In line with this theory, nowadays, the organizations are unable to perform their business operations due to the current COVID-19 pandemic, particularly the educational institutions that demand the creation of new ideas and procedures in the system. Therefore, most institutions have attempted to adopt innovative e-learning plans to continue their operations. Thus, the present study aims at assessing the effects of information and communication technology (ICT) adoption on students' satisfaction and e-learning, which would, in turn, affect the university's image [18].

Comparing the classical education (face-face education) with the online learning reveals that the later provides more and better knowledge and information, which improves the students' learning if a proper system of information and communication technology is followed and has the capacity to deliver the adequate quality of services and information to satisfy the students. One of the distinguishing factors is the advancement in information communication technology (ICT), which, like other industries, has landscaped the education industry. Hence, ICT, coupled with the e-service quality and e-information quality, offers many new benefits to students, whereby numerous undergraduate and postgraduate programs have begun to explore. The available literature on the relationship between the service quality and the student e-learning explains the service quality as one of the key determinants of the e-learning. However, limited service providers and a scarcity of trained instructors exist and the impact of the e-service quality and e-information quality on students' learning largely remains an unexplored area. Therefore, one of the objectives of the current study is to examine the impact of the ICT, e-service quality, and e-information quality on the e-learning of the students in the higher education institutes of UAE.

Student's satisfaction is at the heart of any method of teachings. That is, students' satisfaction is the judgment that the information and knowledge quality or the information and knowledge themselves cope with students' expectations. In this context, e-learning can improve the learning efficacy of students, thus increasing the efficiency of students. According to Oduma et al. [3], the highest quality of education associated with e-learning can help the universities in increasing the satisfaction of students. However, in countries like UAE, which is undergoing a deplorable economic situation, the sudden shift from face-to-face classes to an online method while maintaining the acceptable quality and accreditation standard is a real challenge. As a consequence, the study has envisaged examining the impact of this newly introduced mechanism of e-learning on the satisfaction of students studying at the higher education institutes of UAE.

The electronic word of mouth (eWOM) focuses on the explicit interpersonal contacts existing on the internet. eWOM is a kind of buzz marketing, where a message instantly can become viral. The concept itself is not new. However, the speed at which a message can be viral makes it essential. The eWOM, as people like to share, seek advice online and trust other users. Therefore, any online service, such as online learning, has a significant relationship with the eWOM. In other words, the influenced and affected students share their positive or negative word of mouth, which can be viral instantly. As a result, this study aims at examining the impact of e-learning on the word of mouth of students studying at the higher education institutes of UAE.

The brand image can be defined as a unique bundle of associations within the mind of the target customer [22]. It is an image in the eyes of customers, which develops over time. However, universities in developing countries have not always paid much attention to their brands [23], and many of the higher education institutions consider

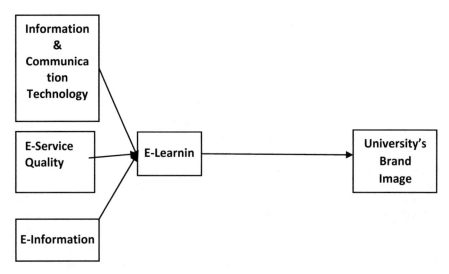

Fig. 1 Theoretical framework

the reputation and grade averages as a viable tool to attract the students. Accordingly, the following hypothesis was hypothesized (Fig. 1):

H4: **E-Learning** of students has a significant impact on the university's brand image.

3 Questionnaire and Measures

The questionnaire involved 59 items. All the instruments and measures are extracted from prior studies. In order to measure the questionnaire items, a five-point Likert scale was used ranging from 1 (strongly disagree) to 5 (strongly agree). In two studies conducted by [19, 24] who proposed dimensions that are integrated to measure the information technology and communication (ICT) orientation; these dimensions are ease of use, perception, compatibility, and advantage. This scale is mainly developed to measure the perspective of teachers regarding information and communication technology (ICT) usage in universities. In this study, all the items were modified in terms of students' perspectives about the ICT orientation for online learning. In addition, fifteen items were used to measure the ICT orientation construct, where 4 items were added for the advantage, 3 items for the compatibility, 5 items for the ease of use, and 3 items for the perception.

In this study, four dimensions including system availability, privacy, efficiency, and fulfillment were used in this study to operationalize the E-service quality. These dimensions were proposed by [25] by adopting the means-end approach. In their study, they developed and refined it as a scale with multiple items (ES-QUAL) to determine the e-service quality rendered to the online customers. Therefore, in order to explain the need to use two different scales as well as to develop it, the empirical data were collected at two stages to measure e-service quality. Thus, the scale for e-service quality was developed in this study, which consisted of four dimensions. These four dimensions further include 22 items that were adapted from prior studies. Four items were added for the system availability dimension, 3 items for the privacy, 8 items for the efficiency, and 7 items for fulfillment. In addition, nine factors were used for measuring the online information quality, and all these items were adapted from [26] study to develop a scale for information quality.

Alsabawy et al. [27] examined the increasing demand for distance learning programs. They mentioned that e-learning quality assessment is an important strategic issue for the survival of a program. Therefore, the current study adopted a 4-items e-learning scale from [27] study. Finally, an 8 item scale was adapted from [28] study to measure the university's brand image. This scale was mainly developed for measuring the university's brand image. For pre-testing, five experts, including 3 Assistant professors and 2 students were contacted who were well-informed about the online learning system. The pre-testing stage confirmed that no further amendments were required. Afterwards, the reliability of each measure was assessed by using the obtained responses. A total of 317 out of 470 questionnaires were received, and after data screening, 217 questionnaires were used for the data analysis.

3.1 Findings

For data analyses, Smart PLS 3 was employed in this study to perform PLS-SEM analysis, as it is advanced and one of the most popular estimation techniques, particularly in the tourism and hospitality industry [29]. To assess the structural and measurement models, the PLS algorithm and bootstrapping technique were also used.

3.2 Measurement Model Assessment

For the measurement model evaluation, the study assessed the convergent validity through the composite reliability, loadings, and average variance extracted. The measurement model is shown in Fig. 2.

Table 1 shows that for both countries, factor loadings are higher than 0.70 value, except for a few values. However, items such as ELQ3, ESQ2, ESQ3, ESQ11, ESQ15, ESQ22, IQ1, UBI1, UBI, and UBI9 with less than 0.70-factor loadings were deleted.

Similarly, in the case of composite reliability (CR), all the values were found above 0.70, which is the recommended range for acceptable CR. In this study, the AVE values for all the constructs were also found to be in line with the recommended range, i.e., above 0.50 [26, 29, 30] (Table 2).

For the discriminant validity, an advanced criterion (HTMT ratio) was proposed by Basheer et al. [30]. According to these scholars, the Fornell-Larcker criterion

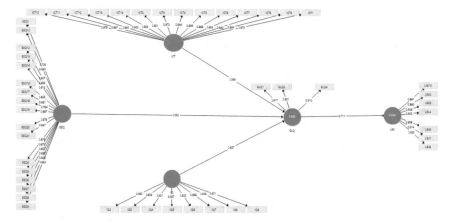

Fig. 2 Measurement model

can effectively evaluate the discriminant validity. However, in other studies, this approach has failed to detect inadequate discriminant validity. Thus in the present study, we used the HTMT ratio for assessing the constructs' discriminant validity. Table 3 shows all HTMT values in the context of the UAE. Table 3 also shows that all values are in line with the recommendations of several studies [30–32], i.e., below 0.90, thereby indicating that adequate discriminant validity is established for all the constructs.

To measure the amount of explained variance in an endogenous construct that is describable by the predicting variables, the coefficient of determination or R-square was observed following the recommendations of other studies [32, 33]. The R-square is defined as how much variance the independent variable is explainable by the predictor variables. It is regarded as a key measure to analyze if the model is a good fit, i.e., whether the measured regression line fits well with the observed items. The R^2 value usually lies between 0 and 1.

3.3   Structural Model Assessment (SEM)

After the measurement model evaluation, the structural model was assessed. The significance of the model was determined based on the t-values, standard errors, and path-coefficients. In Smart PLS 3, the bootstrapping procedure is performed to test the hypotheses and to determine the indirect and the main effects [30–32].

Table 4 and Fig. 3 present the results of the empirical testing of the direct and indirect hypotheses. The critical ratio (t > 1.645; P < 0.05) was used to support the hypotheses. Except few hypotheses, all the hypotheses were found to be supported based on the above criterion (Table 5).

Table 1 Cross loadings

	ELQ	ESQ	ICT	IQ	UBI
ELQ1	**0.917**	0.680	0.667	0.857	0.641
ELQ2	**0.901**	0.648	0.604	0.840	0.635
ELQ4	**0.910**	0.660	0.655	0.875	0.665
ESQ1	0.668	**0.735**	0.860	0.645	0.721
ESQ10	0.636	**0.835**	0.572	0.643	0.568
ESQ12	0.685	**0.907**	0.693	0.662	0.631
ESQ13	0.653	**0.895**	0.669	0.648	0.620
ESQ14	0.658	**0.912**	0.690	0.675	0.635
ESQ16	0.589	**0.855**	0.627	0.593	0.572
ESQ17	0.601	**0.831**	0.634	0.624	0.595
ESQ18	0.558	**0.794**	0.570	0.549	0.532
ESQ19	0.645	**0.887**	0.632	0.631	0.596
ESQ20	0.645	**0.879**	0.660	0.655	0.641
ESQ21	0.620	**0.837**	0.616	0.631	0.554
ESQ4	0.627	**0.879**	0.626	0.624	0.578
ESQ5	0.651	**0.871**	0.633	0.655	0.609
ESQ6	0.571	**0.857**	0.614	0.580	0.542
ESQ7	0.579	**0.857**	0.645	0.582	0.552
ESQ8	0.541	**0.800**	0.573	0.555	0.535
ESQ9	0.593	**0.853**	0.613	0.578	0.522
ICT10	0.619	0.648	**0.879**	0.585	0.707
ICT11	0.618	0.685	**0.887**	0.629	0.747
ICT12	0.623	0.668	**0.843**	0.582	0.662
ICT13	0.576	0.635	**0.822**	0.597	0.672
ICT14	0.649	0.698	**0.859**	0.640	0.722
ICT1	0.612	0.641	**0.875**	0.598	0.743
ICT2	0.576	0.646	**0.851**	0.561	0.671
ICT3	0.582	0.627	**0.873**	0.578	0.715
ICT4	0.615	0.637	**0.844**	0.594	0.710
ICT5	0.611	0.717	**0.899**	0.607	0.733
ICT6	0.615	0.638	**0.868**	0.612	0.732
ICT7	0.656	0.669	**0.893**	0.667	0.724
ICT8	0.567	0.583	**0.840**	0.578	0.712
ICT9	0.633	0.675	**0.894**	0.629	0.730
IQ2	0.771	0.579	0.529	**0.865**	0.559
IQ3	0.802	0.620	0.571	**0.856**	0.580

(continued)

Table 1 (continued)

	ELQ	ESQ	ICT	IQ	UBI
IQ4	0.864	0.674	0.639	**0.901**	0.644
IQ5	0.845	0.636	0.594	**0.907**	0.604
IQ6	0.905	0.696	0.677	**0.925**	0.711
IQ7	0.865	0.685	0.686	**0.896**	0.660
IQ8	0.855	0.648	0.624	**0.906**	0.634
IQ9	0.805	0.649	0.645	**0.871**	0.657
UBI10	0.601	0.566	0.684	0.611	**0.861**
UBI2	0.621	0.594	0.709	0.617	**0.880**
UBI3	0.646	0.617	0.717	0.629	**0.906**
UBI4	0.622	0.610	0.743	0.630	**0.905**
UBI6	0.686	0.692	0.789	0.682	**0.898**
UBI7	0.636	0.611	0.747	0.636	**0.914**
UBI8	0.657	0.660	0.781	0.655	**0.925**

The numbers in bold for each questions prove the correctness of their place for each variable, by obtain the highest value for each variable

Table 2 Reliability

	Cronbach's Alpha	rho_A	Composite reliability	Average variance extracted (AVE)
ELQ	0.895	0.896	0.935	0.827
ESQ	0.976	0.977	0.978	0.728
ICT	0.974	0.975	0.977	0.751
IQ	0.963	0.964	0.969	0.794
UBI	0.960	0.961	0.967	0.808

Table 3 Validity

	ELQ	ESQ	ICT	IQ	UBI
ELQ	0.899				
ESQ	0.729	0.853			
ICT	0.706	0.757	0.866		
IQ	0.743	0.729	0.698	0.891	
UBI	0.711	0.693	0.823	0.710	0.899

Table 4 R square

	R square
ELQ	0.895
UBI	0.506

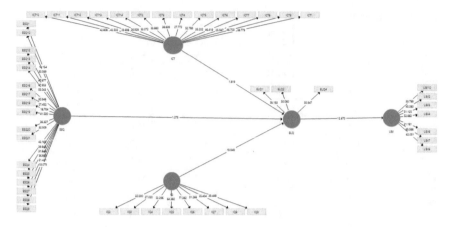

Fig. 3 Structural model

Table 5 Structural relationships

	Original sample (O)	Sample mean (M)	Standard deviation (STDEV)	T statistics (|O/STDEV|)	P values
ELQ -> UBI	0.711	0.713	0.056	12.670	0.000
ESQ -> ELQ	0.052	0.056	0.041	1.275	0.101
ICT -> ELQ	0.068	0.068	0.037	1.819	0.034
IQ -> ELQ	0.857	0.854	0.043	19.949	0.000

3.4 Blindfolding

Following [28], the predictive relevance (Q^2) test was conducted using a cross-validated redundancy approach (See Fig. 4).

As a result, the Q^2 value was calculated by performing a blindfolding method, which helps in assessing the predictive relevance ability of the constructs. In this approach, the path models and elements of the structural model are predicted. Further, by using the same procedures and criteria that were suggested by [33], the effect size was also analyzed for the predictive relevance [35] (Table 6).

4   Discussion and Conclusions

The current study was performed to assess the impact of e-information quality, e-service quality, and ICT on the e-learning of students in Alain university in the context of UAE. In this study, we proposed several hypotheses in the context of the COVID-19 effects on Alain university in the UAE to address the research objectives.

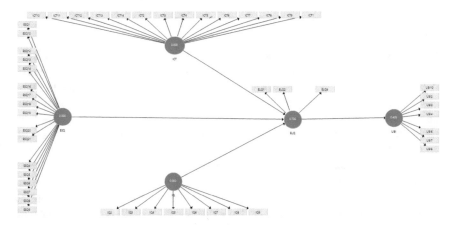

Fig. 4 Blindfolding

Table 6 Blindfolding

	SSO	SSE	Q^2 (=1 − SSE/SSO)
ELQ	651.000	172.039	0.736
UBI	1519.000	904.515	0.405

To achieve the research objectives in the current study, the data were collected through a survey and they were analyzed using different statistical tools for data analyses. The objectives were partially fulfilled, as indicated by the results. The findings also provided necessary implications for university management to adopt a digital online learning system due to the continuous operations during COVID-19.

Besides, students' e-learning is found as an essential element of online provision of education during crises or emergency situations, such as COVID-19. E-learning involves three main elements, namely e-information quality, e-service quality, and information and communication technology (ICT). Thus the findings suggest a positive role of information and communication technology in improving students' e-learning and this improved (ICT) can enhance the e-learning ability of students. Previous studies also support these results. For instance, Yadav and Tripathi [18] suggested that information and communication technology significantly promote e-learning among students. Other studies [9, 20, 21] also reported that information and communication technology significantly affect students' e-learning. However, students' e-learning in times of COVID-19 has significantly affected the universities. Thus the essential constituents of information and communication technology (ICT) are ease of use, perception, compatibility, and advantage.

In addition, the research findings also reveal the positive contribution of e-service quality in promoting e-learning of students. Thus a better e-service quality is capable of motivating students towards using the online system of learning. This finding is in line with [22] finding who discovered the direct effects of e-service quality on the loyalty of e-learning students'. Further, a better e-service quality enhances the effectiveness of the e-learning system. Similar findings were reported by several prior studies [34, 35] which reported the positive effects of e-service quality on students' e-learning. Thus, the present study confirmed the significant role of e-service quality in establishing a successful e-learning system during the current COVID-19 situation. The results also show that in order to promote the e-learning among students, the e-service quality must be reliable, assured, responsive, and tangible. Therefore, besides assurance, tangibility, responsiveness, and reliability, the universities are also required to focus on ICT while adopting e-learning.

A better provision of e-information quality must be ensured to achieve e-learning among students. The e-information quality is considered to play a vital role in developing an e-learning system and encouraging them to learn through the online learning system. As a consequence, the results obtained in this study are consistent with prior studies, which have provided empirical evidence about the positive role of e-information quality on students' e-learning [14, 36]. Thus, a significant e-information quality is needed for the effective adoption of e-learning among students.

5 Theoretical Implications

This study offers both theoretical and practical implications. Theoretically, this study is among the first ones that developed a framework for the adoption of the e-learning system by the students considering the post effects of COVID-19. Although numerous studies have examined the e-learning of students, the current study is one of the first ones to consider the effects of COVID-19 as well as providing a framework for establishing an online learning system that is capable of improving the brand image of the university. Thus the present study contributes to the existing body of literature by proposing a framework, keeping in view the effects of COVID-19. This study also contributes to the literature by addressing the existing research gap. This is achieved by analyzing the post-COVID-19 effects in South Asian countries, particularly those that are severely affected by COVID-19. Also, the present study attempted to prove that e-information quality, e-service quality, and information and communication technology (ICT) are the main determinants to achieve e-learning of students in the current COVID-19 condition [18]. Along with that, this study suggests that in order to launch e-learning effectively, universities are required to create a good brand image that can possibly be achieved through the students' satisfaction and positive word of mouth.

6 Practical Implications

Besides the theoretical implications, this study also provides some important practical implications for the universities concerning the adoption of the e-learning system during the recent COVID-19 situation. One of these implications suggests that universities can adopt the proposed framework to establish an online learning system. In the current global pandemic situation, universities cannot continue to perform their normal procedures. In this regard, a framework has been developed to adopt a rigorous e-learning system that could promote the brand image of the university. While developing an e-learning system, universities are currently faced by numerous issues. Therefore, this study suggests that these issues can be resolved through the proper implementation of ICT. Besides, various other unique implications have been proposed in this study for developing an e-learning platform by universities to continue their operations during the current situation. This study also offers implications for the university's image, whereby universities are required to promote the e-information quality, e-service quality, and information and communication technology for developing a rigorous e-learning system [37–51]. The results suggest that the students' satisfaction determine and improve the brand image [4]. Thus, this study provides unique insights for the universities, higher education commission, and government to develop understating about the determinants of the e-learning system and the relationship of these factors with the universities' brand image, as it may help universities in dealing with challenges arising from COVID-19.

References

1. Faqih, K.: Which is more important in e-learning adoption, perceived value, or perceived usefulness? Examining the moderating influence of perceived compatibility. In: 4th Global Summit on Education GSE (2016)
2. Carolina, V., Oktavianti, O., Handayani, R.: Tax avoidance & corporate risk: an empirical study in manufacturing company. JIA (Jurnal Ilmiah Akuntansi) 4(2), 291–300 (2019)
3. Oduma, C.A., Onyema, L.N., Akiti, N.: E-learning platforms in business education for skill acquisition. Nigerian J. Bus. Educ. (NIGJBED) 6(2), 104–112 (2019)
4. Kasemsap, K.: Promoting service quality and customer satisfaction in global business. In: Handbook of Research on Promotional Strategies and Consumer Influence in the Service Sector, pp. 247–276. IGI Global (2016)
5. Sezer, B.: Faculty of medicine students' attitudes towards electronic learning and their opinion for an example of distance learning application. Comput. Hum. Behav. 55, 932–939 (2016)
6. Hernández, J.B., Chalela, S., Arias, J.V.: Research trends in the study of ICT based learning communities: a bibliometric analysis. EURASIA J. Math. Sci. Technol. Educ. 13(5), 1539–1562 (2017)
7. Verma, C., Illés, Z.: Attitude prediction towards ICT and mobile technology for the real-time: an experimental study using machine learning. Paper presented at The International Scientific Conference eLearning and Software for Education (2019)
8. Singh:. Information and communication technology (ICT): its origin in India and pedagogical integration in the study of geography. i-Manager's J. Educ. Technol. 16(1), 1 (2019)
9. Zhou, M., Teo, T.: Exploring student voice in teachers' motivation to use ICT in higher education: qualitative evidence from a developing country. Int. J. Educ. Technol. 4(1), 26–33 (2017)
10. Kim, H.J., Hong, A.J.: The roles of academic engagement and digital readiness in students' achievements in university e-learning environments. Int. J. Educ. Technol. Higher Educ. 16(1), 21 (2019)

11. Nsamba, A.: Maturity levels of student support e-services within an open distance e-learning university. Int. Rev. Res. Open Distrib. Learn. **20**(4), 60–78 (2019)
12. Shahbazi, M., Farajpahlou, A., Osareh, F.: Development of a scale for data quality assessment in automated library systems. Lib. Inf. Sci. Res. **41**(1), 78–84 (2019)
13. Asogwa, B.E., Ugwu, C.I., Ugwuanyi, F.C.: Evaluation of electronic service infrastructures and quality of e-services in Nigerian academic libraries. Electron. Lib. **5**(2), 142–151 (2015)
14. Ilias, A., Sukib, N.: Determinants of the Intention to Re-use Internet Business Reporting (IBR): The Structural Equation Modelling Approach. Paper presented at the International Conference on Accounting Studies (ICAS) 2017 (2017)
15. Almisad, B.: A study of students' perceptions and attitudes toward the use of SMS to support learning and teaching at the Kuwait University (2015)
16. Kong, S.C., Song, Y.: An experience of personalized learning hub initiative embedding BYOD for reflective engagement in higher education. Comput. Educ. **88**, 227–240 (2015)
17. Luo, N., Zhang, M., Qi, D.: Effects of different interactions on students' sense of community in e-learning environment. Comput. Educ. **115**, 153–160 (2017)
18. Yadav, S.K., Tripathi, V.: A study on validating learning style questionnaire (LSQ) by measuring students learning style in an Asian context. J. Pract. Teach. Learn. **15**(2) (2017)
19. Uppal, M.A., Ali, S., Gulliver, S.R.: Factors determining e-learning service quality. Br. J. Educ. Technol. **49**(3), 412–426 (2018)
20. Supriadi, D., Sa'ud, U.: The effectiveness of implementing information and communication technology on student academic services (a case study in Bandung Institute of Technology for the 2015–2016 period). Int. J. Educ. **9**(2), 139–148 (2017)
21. Terry, C.: How can we evaluate the effectiveness of ICT resources in maximising student learning in Oman. Paper presented at the Proceedings of the Language Centre Forum (2016)
22. Lemoine, P.A., Sheeks, G.: Retention of online learners: The importance of support services. Int. J. Technol.-Enabled Student Support Serv. (IJTESSS) **9**(2), 28–38 (2019)
23. Jayakumar, G.D.S., Ali, M.S.D.: Modeling the brand image of cosmetics and its impact on customer satisfaction and loyalty intention with special reference to face creams. J. Impact Factor **7**(2), 226–236 (2016)
24. Bhat, S.B., Bashir, S.: Incipient iron deficiency in primary hypothyroidism. Thyroid Res. Pract. **15**(3), 138 (2018)
25. Kilburn, A., Kilburn, B., Hammond, K.: The role of quality in online higher education. J. Higher Educ. Theory Pract. **17**(7) (2017)
26. Hair, Jr., J.F., Hult, G.T.M., Ringle, C.: A Primer on Partial Least Squares Structural Equation Modeling (PLS-SEM). Sage (2016)
27. Alsabawy, A.Y., Steel, A., Soar, J.: Determinants of perceived usefulness of e-learning systems. Comput. Hum. Behav. **64**, 843–858 (2016)
28. Sultan, P., Wong, H.Y.: How service quality affects university brand performance, university brand image and behavioral intention: the mediating effects of satisfaction and trust and moderating roles of gender and study mode. J. Brand Manag.s **26**(3), 332–347 (2019)
29. Ringle, C.M., Sarstedt, M., Mitchell, R.: Partial least squares structural equation modeling in HRM research. Int. J. Hum. Resource Manag., 1–27 (2018)
30. Basheer, M.F., Hafeez, M.H., Hassan, S.G., Haroon, U.: Exploring the role of TQM and supply chain practices for firm supply performance in the presence of organizational learning capabilities: a case of textile firms in Pakistan. Paradigms **12**(2), 172–178 (2018)
31. Henseler, J., Hubona, G., Ray, P.A.: Using PLS path modeling in new technology research: updated guidelines. Industrial Management & Data Systems (2016)
32. Basheer, M.F., Siam, M., Awn, A., Hassan, S.: Exploring the role of TQM and supply chain practices for firm supply performance in the presence of information technology capabilities and supply chain technology adoption: a case of textile firms in Pakistan. Uncertain Supply Chain Manag. **7**(2), 275–288 (2019)
33. Hair, J.F., Sarstedt, M., Ringle, C.M.: Rethinking some of the rethinking of partial least squares. Eur. J. Market. (2019)

34. Khodadad Hoseiny, S.H., Noori, A., Zabihi, M.R.: E-learning acceptance in higher education: application of flow theory, technology acceptance model & e-service quality. Q. J. Res. Plan. Higher Educ. **19**(1), 111–136 (2013)
35. Ramayah, T., Cheah, J., Memon, M.: Partial least squares structural equation modeling (PLS-SEM) using SmartPLS 3.0. In An Updated Guide and Practical Guide to Statistical Analysis. Pearson (2018)
36. Mammo, Y., Ngulube, P.: Insights into e-information resources (e-journals) access models in higher learning institutions in Ethiopia. Int. Inf. Lib. Rev. **51**(1), 19–27 (2019)
37. Al Kurdi, B., Alshurideh, M., Salloum, S.A.: Investigating a theoretical framework for e-learning technology acceptance. Int. J. Electr. Comput. Eng. (IJECE) **10**(6), 6484–6496 (2020)
38. Alhashmi, S.F., Alshurideh, M., Al Kurdi, B., Salloum, S.A.: A systematic review of the factors affecting the artificial intelligence implementation in the health care sector. In: Joint European-US Workshop on Applications of Invariance in Computer Vision, pp. 37–49. Springer, Cham (2020)
39. Salloum, S.A., Alshurideh, M., Elnagar, A., Shaalan, K.: Mining in educational data: review and future directions. In: Joint European-US Workshop on Applications of Invariance in Computer Vision, pp. 92–102. Springer, Cham (2020)
40. Salloum, S.A., Alshurideh, M., Elnagar, A., Shaalan, K.: Machine learning and deep learning techniques for cybersecurity: a review. In: Joint European-US Workshop on Applications of Invariance in Computer Vision, pp. 50–57. Springer, Cham (2020)
41. AlMehrzi, A., et al.: Investigation of the key internal factors influencing knowledge management, employment, and organisational performance: a qualitative study of the UAE hospitality sector. Int. J. Innov. Creativity Change **14**(1), 1369–1394 (2020)
42. Al Kurdi, B., Alshurideh, M., Salloum, S., Obeidat, Z., Al-dweeri, R.: An Empirical Investigation into Examination of Factors Influencing University Students' Behavior towards Elearning Acceptance Using SEM Approach (2020)
43. Kurdi, B., Alshurideh, M.: Employee retention and organizational performance: evidence from banking industry. Manage. Sci. Lett. **10**(16), 3981–3990 (2020)
44. Kurdi, B., Alshurideh, M., Alnaser, A.: The impact of employee satisfaction on customer satisfaction: theoretical and empirical underpinning. Manage. Sci. Lett. **10**(15), 3561–3570 (2020)
45. Alshurideh, M.T., Salloum, S.A., Al Kurdi, B., Monem, A.A., Shaalan, K.: Understanding the quality determinants that influence the intention to use the mobile learning platforms: a practical study. Int. J. Interact. Mobile Technol. (IJIM) **13**(11), 157–183 (2019)
46. Alshurideh, M., Al Kurdi, B., Salloum, S.A.: Examining the main mobile learning system drivers' effects: a mix empirical examination of both the Expectation-Confirmation Model (ECM) and the Technology Acceptance Model (TAM). In: International Conference on Advanced Intelligent Systems and Informatics, pp. 406–417. Springer, Cham (2019)
47. Alshurideh, M., Salloum, S.A., Al Kurdi, B., Al-Emran, M.: Factors affecting the social networks acceptance: an empirical study using PLS-SEM approach. In: Proceedings of the 2019 8th International Conference on Software and Computer Applications, pp. 414–418 (2019)
48. Alshurideh, M., Kurdi, B.A., Shaltoni, A.M., Ghuff, S.S.: Determinants of pro-environmental behaviour in the context of emerging economies. Int. J. Sustain. Soc. **11**(4), 257–277 (2019)
49. Al-Dmour, H., Alshuraideh, M., Salehih, S.: A study of Jordanians' television viewers habits. Life Sci. J. **11**(6), 161–171 (2014)
50. Alshurideh, M.: The factors predicting students' satisfaction with universities' healthcare clinics' services: a case-study from the Jordanian higher education sector. Dirasat: Adm. Sci. **161**(1524), 1–36 (2014)
51. Hayajneh, N., Suifan, T., Obeidat, B., Abuhashesh, M., Alshurideh, M., Masa'deh, R.: The relationship between organizational changes and job satisfaction through the mediating role of job stress in the Jordanian telecommunication sector. Manag. Sci. Lett. **11**(1), 315–326 (2021)

Obstacles of E-Learning for Community College Students in the Light of Corona Pandemic

Saddam Rateb Darawsheh

Abstract This study aimed at identifying the obstacles of e-learning for the community college students in the light of Corona pandemic. The researcher followed the descriptive approach. The population of the sample was all the students at the University of Imam Abdul Rahman bin Faisal. The sample of the study consisted of (300) female students, who were chosen randomly from the community college students. The researcher designed a questionnaire entailing three domains. The first domain was the technical obstacles; the second domain was the financial obstacles, and the third domain was social obstacles. Data were processed by statistical methods. The findings revealed that the technical obstacles had got a mean score of (1.37) as it came in a low degree. The financial obstacles came in a moderate degree, with a mean score of (2.20). The social obstacles came in a high degree, with a mean score of (2.60). Thus, it is recommended to focus on the technical side in universities through supporting e-learning and introducing financial support to train specialists and technicians. Besides, it is important to hold educational courses for community members to enhance the status of e-learning.

Keywords Obstacles · E-Learning · Community college students · Corona pandemic

1 Introduction

University education today has faced several demands imposed on it by the simultaneous scientific and technological developments. There is a need to develop human cadres in order to face the increasing demand for university education and raise the level of its efficiency, effectiveness, quality in line with the requirements of the current era, as well as meet the labour sectors' needs and activate development plans. Therefore, the university education system should not be limited to the traditional teaching

S. R. Darawsheh (✉)
Department of Administrative Sciences, University of Imam Abdul Rahman Bin Fasil, Dammam, Saudi Arabia
e-mail: srdarawsehe@iau.edu.sa

© The Author(s), under exclusive license to Springer Nature Switzerland AG 2021 189
M. T. Alshurideh et al. (eds.), *The Effect of Coronavirus Disease (COVID-19) on Business Intelligence*, Studies in Systems, Decision and Control 334, https://doi.org/10.1007/978-3-030-67151-8_11

methods that are used by instructors in the classroom [1]. Rather, modern developments in technology must be employed and used to provide a pattern of learning that allows the instructors to introduce the curricula for their students at anytime, anywhere, and gradually vacating their seats for new students. This increases the capacity of universities and enables them to grant the necessary capabilities, skills and knowledge for the individuals' success in social and career life in the era of the cognitive revolution [2].

Educators are called to take advantage of everything new to keep pace with innovations while working to find modern concepts that are consistent with scientific progress and the flood of knowledge [3, 4]. Global, Arab and local interest of technology and technological education have increased in the light of changes that have been taken place at the level of technology, communication, invention, professionalism, and technical tools.

Thus, the universities in Saudi Arabia have started using the Learning Management System, which is currently known as the Eduwave portal, which is one of the technological tools in the learning communication process [5]. This helps in creating an effective learning process, enhancing research and learning capacity, as well as ensuring individuals' contribution in building a renewed knowledge-based economy that contributes to sustainable development [6].

The use of these technological tools contributes to raising the level of teaching and learning processes. It helps to break the imitation in learning, which is based on memorization, preserving information and retrieving to the vitality of learning resulting from exploration, research, analysis, and reasoning reaching to solve problems. This system has become one of the most prominent e-learning projects that has adopted by the Ministry of Education in Saudi Arabia. Also, it aims to improve and develop the teaching and learning processes to enhance the outcomes of the educational process [7].

Because of the significance of this system in the educational process, several Arab and foreign countries have adopted and started using the e-learning system in their universities [8]. These countries are benefiting from the Saudi experience in the field of e-learning. Using the e-learning portal is produced by the minds of distinguished Saudi Arabia, which places it at the forefront of countries that utilizes technology in education. Also, using this system contributes to improving the outputs of the educational process in Saudi universities and places it on the map of developed and modern countries, as well as exporting distinguished human competencies to compete regionally and worldwide [9].

The educational system has faced many academic, educational, social and economic problems. One of the most important problems is the electronic problems and how to communicate with students, especially in the conditions of Corona pandemic, which is disrupted the traditional learning to be replaced by e-learning. This learning is an integrated learning system between the teacher and the student from the first side, as well as curricula, the Internet, social and physical factors from the second side. All of these affect the virtual learning process [10].

The e-learning has faced many obstacles, especially in universities in the light of Corona pandemic. These obstacles include technical, physical, financial, and social obstacles that prevent the student from accessing information, academic courses and a faculty member. For these reasons, this study aims to investigate the obstacles of e-learning for the community college students in the light of Corona pandemic.

Many students have suffered from several academic problems, especially in e-learning in the light of Corona pandemic (COVID-19), which has afflicted the global system as a whole. Specifically, this study seeks to answer the following main question:

What are the obstacles of e-learning for community college students in light of Corona pandemic?

The study seeks to identify the most important obstacles of e-learning for community college students in the light of Corona pandemic.

This study stems its importance from the subject addressed by examining the most important obstacles of e-learning in the light of Corona pandemic to overcome them. Also, it introduces to the libraries, students and researchers a study that Corona pandemic is one of its variables. Besides, it tries to shed light on the importance of e-learning and how to deal with it in the light of Corona pandemic.

2 Theoretical Literature and Related Studies

2.1 Theoretical Literature

There are many definitions of e-learning. "E-learning is the use of electronic and computer media in the process of transferring and communicating information to the learners from one to another" [11, 37]. It is also a method of teaching or training that enables the learner to obtain instruction and apply it at anytime and anywhere in the world. This can be done through the interactive communication and information technologies synchronously in the classroom and as an asynchronous manner from a distance depending on self-learning and the interaction between the teacher and the learner [12]. Also, it is an interactive system that relies on an integrated electronic environment. It aims to build curricula in an ideal way that can be easily communicated by electronic networks relying on programs and applications that provide an ideal environment for integrating text with sound and image, and it also provides the ability to enrich information through links on sources of information in different locations [13]. Besides, it is as learning that does not depend on the presence of the professor and the student together inside the classroom with the interaction between them so that the place is cancelled [14]. Also, e-learning is defined as the use of the Internet and digital technology to create an experience and educate learners [15].

Based on all the above definitions, it can be concluded that all definitions focused on and agreed that e-learning is a method of learning by using or applying communication networks and the Internet through the tools by which e-learning is taught.

Also, it is a modern technology because the information is transmitted to all places and times. Accordingly, e-learning is a modern method that relies on electronic technologies (e.g., the Internet, computer, electronic library, e-book, etc.) in delivering information without the presence of the teacher and the student to the classroom.

2.2   Types of E-Learning

There are types for e-learning; (1) **Synchronous E-Learning**. It is a learning that requires the presence of learners at the same time to conduct a conversation between the students themselves and the teacher through the discussion room or receive lessons during the semester [16]. Also, it is over the air learning that requires the presence of learners at the same time in front of computers to conduct discussion and conversation between students themselves and the teacher through chat rooms [17], (2) **Asynchronous E-Learning**. It is a learning that does not require the presence of learners at the same time or place. It is carried out through the techniques and applications of e-learning; the learner selects the times and places in a way that suits him/her [18]. Also, it is indirect learning that does not require the presence of learners at the same time or the same place. It is done through some e-learning techniques, such as e-mail which information is exchanged among students themselves and the teacher in successive times. It is believed that the World Wide Web combined between synchronous and asynchronous learning by storing teaching lessons and courses to be available at any time according to the learner's conditions and times [19], and (3) **Blended Learning**. These types of learning include many learning tools, such as instant virtual collaborative learning software online courses, self-learning courses, and electronic performance support systems. It also mixes multiple activity-based events that include learning in the traditional classroom where the teacher meets with students face to face and self-learning as well [20].

2.3   Obstacles of Applying E-Learning

There are many obstacles to applying e-learning, including:

– Technical obstacles are among the main barriers in applying e-learning. They are represented by learners' needs for the equipment to be used in e-learning, such as computers, printers, and the Internet [21]. Reasons, such as the loss or non-establishment of technical infrastructure for e-learning from installing and using computers and phones lines are important reasons which lead to non-spread of e-learning. Also, the lack of a high-capacity network to ensure the speedy downloading of educational curricula and software leads to a reduction in the spread of e-learning [22].

– Financial Obstacles are also one of the barriers that contribute to limit the use of e-learning. These are represented in learners' need to have sufficient experience to use the computer and the Internet. If learners do not have the experience, educational institutions should hold qualifying training courses for learners to teach them how to use the computer [23]. This leads to that the organizations are responsible for some additional costs. Also, the non-spread of computers, the limited Internet coverage, its relative slowness and the high price have a major impact on the difficulty of applying this type of learning [24].

– Human Obstacles are among the obstacles that stand in front of the applying e-learning for the following reasons: (1) some educational institutions do not take into account the human cadre when applying e-learning through their obtaining intensive studies in the English language. This is represented as an obstacle in implementing e-learning [25]. (2) The lack of human and educational cadres who can advance this type of learning at the individual and community level. (3) The lack of training courses held by the educational institutions for their members and students, as there are a small number of faculty members who have e-learning skills in an integrated manner. Some students may feel frustrated because they believe that e-learning is not important, and it has no value. There are negative attitudes for some faculty members against e-learning. Also, the high workload for faculty members assigned to them by the administration preventing them to apply e-learning. Also, some students believe that e-learning is unprofessional because there is no direct supervision of learners' learning [26–28].

A Coronavirus is a group of viruses that can cause diseases, such as colds, severe acute respiratory syndrome (SARS) and the Middle East Respiratory Syndrome (MERS). A new type of coronavirus has been discovered after it had identified as the cause of the spread of one of the diseases that started in China in 2019. The virus is known as the severe acute respiratory syndrome virus Corona 2 (SARS Cove 2). The disease that is resulting from it called Coronavirus 2019 (COVID-19). In March 2020, the World Health Organization (WHO) is announced that it is listed the Coronavirus 2019 (COVID-19) as a pandemic. Public health groups, such as the United States Centers for Disease Control and Prevention (CDC) and the WHO, are monitoring the pandemic and spreading updates on their websites. These groups have also made recommendations about disease prevention and treatment (WHO, 2020).

2.4 Related Studies and Concluding Remarks

Al-Shehri [19] studied the obstacles that limit applying the e-learning programs. The findings showed that there were major obstacles in applying e-learning, such as technical and learning obstacles.

Souman and Hamza [18] aimed to identify the potential obstacles in the use of e-learning portal (EDUWAVE) from the Jordanian government schools teachers' point of view in Amman and their attitudes towards them. The results indicated that there was a satisfaction of male and female teachers about the e-learning portal site in terms of its design and components. The results also showed that there were several technical problems, and also there were negative attitudes towards using the e-learning portal in education administration.

Qahwan and Huthaili [27] studied the obstacles of university e-learning in Yemen. The findings revealed that there were obstacles in university education that were summarized as follows: Obstacles related to academics, learners, physical, technical, and community.

Haseeb and Al-Abaade [14] investigated the obstacles of applying e-learning at Hadba University College. The results of the study revealed that there were various obstacles of e-learning, such as technical, financial, and human obstacles.

Ghayad [3] aimed at identifying e-learning as a strategic choice for Algerian universities. The results showed that there was a necessary need to improve the efficiency and quality of university education. The results also indicated that there was a need to benefit from the experiences of global countries in the field of e-learning to overcome their obstacles.

Qamishi [2] aimed at identifying the obstacles of e-learning. The findings revealed that there was a necessity of reforming the educational system, especially e-learning because it faces changes in its inputs and outputs. The findings also showed that the most important obstacle facing e-learning was the financial deficit.

Al-Bedawi [9] aimed at identifying the obstacles of using e-learning from the faculty members' viewpoint at Qaseem University in the light of some variables. The questionnaire was prepared by the researcher after the validity and reliability were taken into account. The findings revealed that the limited number of available laboratories to implement e-learning, the weakness for the faculty member's in using e-learning, the lack of technicians and specialists to solve technical problems, students' lack of motivation and direct support by professors are the most important obstacles of using e-learning.

Alothman and Almassaad [28] aimed at revealing the incentives and obstacles related to the use of available e-learning systems and techniques at King Saud University from the faculty members' viewpoint in the College of Education. The results indicated that everything related to the motivating factors of the e-learning system came very high. Also, one of the most prominent obstacles were represented in the need for technical skills within the programs of preparing the faculty member with a moderate degree, and there were differences in the incentives domain due to the impact of gender in favour of males.

Boras and Al-Ashi [16] aimed at identifying the importance of e-learning in achieving human development. The findings indicated that different changes in the information technology have brought about a qualitative shift or what is known as the global transformations that have affected all the educational processes, especially those related to teaching methods. It has become easy to use and adapt communication and information technology to reduce social and cultural differences, as well as overcome time and space restrictions and scarce human resources. There were many methods of learning and e-learning. Also, the need has increased for the necessity of drawing future visions for the philosophy of e-learning, linked to the utilizing information and communication technology and their uses in all areas of the education system.

Al Mubarak [15] aimed at identifying the reality of applying e-learning in Sudanese universities in the light of some recent trends of comprehensive quality. The results indicated that applying the comprehensive quality standards in e-learning of the university education is achieved with a high degree. The results also indicated that the goals of the university electronic curriculum, which takes into account the overall quality standards, were high. Also, evaluating the university electronic curriculum adheres to the total quality standards with a high mean score.

Bachir and Ammar [17] aimed at identifying the reality of applying e-learning in Algerian universities at the University of Biskra. The results showed that using the computers for a long period increases students' burden in high costs and tests, as well as the high cost of maintenance of computers and the lack of technicians are the most prominent obstacles of e-learning.

Iskandar [1] aimed at investigating the reality of the quality of e-learning services provided to masters students with identifying the aspects of this learning system. The results showed that some financial constraints faced the students in e-learning.

Al Hayani and Ibtisam [26] studied the possibility of using e-learning to address students' learning problems from a faculty member' and students' point of views. The results showed that the faculty member and students had positive views on the use of e-learning.

Diab and Bruce [29] aimed at revealing the obstacles that faced digital learning in Algerian schools. The results showed that physical constraints were one of the most important obstacles that affect digital learning.

These studies were summarized as follows.

Obstacles related to the students' and teachers' competencies for the use of e-learning as in studies (e.g., [15, 16, 19, 26]).
The lack of adequate computer equipment and physical obstacles as in studies of (e.g., [3, 9, 27]).
Studies (e.g., [18, 28, 29]) studied students' perceptions about e-learning and social obstacles.

None of the previous studies investigated the obstacles in the use of e-learning portal at the Community College of Imam Abdul Rahman bin Faisal University. Thus, the present study seeks to conduct a study to cover this aspect.

2.5   Operational Definition of Terms

Obstacles: are defined operationally in this study as a set of technical, physical, administrative and supervisory problems or difficulties that prevent students from using e-learning in different educational situations.

E-learning: is defined as a learning system in an interactive environment that offers Saudi universities to their students through the media, electronic innovations, and virtual sessions.

Community College Students: they are attending the community college at Imam Abdul Rahman bin Faisal University in Saudi Arabia for the academic year 1440/1441. They are studying at the college in the three departments: Department of Administrative Sciences, Department of Financial Sciences, and Department of Computer Science.

Corona Pandemic: is an infectious disease caused by the newly discovered Coronavirus. There was no enough knowledge of this virus and this emerging disease before the outbreak of it in the Chinese city of Yuhan in December 2019.

3 Method and Procedures

The researcher followed the descriptive analytical approach through studying the phenomenon or the research problem and describing it scientifically to reach logical interpretations. This gives the researcher the ability to define frameworks for the research problem, which contributes to defining the research results.

3.1 Population and Sample of the Study

The population of the study consisted of all students at Imam Abdul Rahman bin Faisal University 2019/2020. A simple random sample of the study population was chosen from the Community College students. The sample of the study was (300 female students) who were chosen from the college students in Dammam branch.

3.2 Instrument of the Study

After the literature review was reviewed, a questionnaire was designed to identify the obstacles in the use of e-learning in the community college at Imam Abdul Rahman bin Faisal University. The questionnaire consisted of (22) items with three domains. The first domain was the technical obstacles with (9) items. The second domain was the financial obstacles with (7) items. The third domain was the social obstacles with (6) items.

Table 1 The consistency coefficient of the Cronbach alpha

Domain	Cronbach alpha	No. of Items
Technical obstacles	0.766	9
Finical obstacles	0.847	7
Social obstacles	0.863	6
Total	0.816	22

3.3 Validity and Reliability of the Instrument of the Study

To establish the validity of the instrument, it was given to a jury of experts in different Saudi universities. The jury was asked to review the items of the questionnaire as a tool to collect the data. The jury provided the researcher with some comments on it. After the needed modifications were made, the questionnaire became suitable for the study. To ensure the reliability of the instrument, the internal consistency was calculated according to the Cronbach Alpha equation. Table 1 shows the reliability index.

As shown in Table 1, the total value of Cronbach Alpha was (0.816). This result confirms that the instrument of the study was appropriate for this study.

3.4 Procedures of the Study

- The questionnaire used in this study was prepared after the literature review was reviewed.
- The researcher chose a random sample of the students, who were excluded from the study results to check the reliability of the questionnaire.
- The questionnaire was distributed on students of the selected community college in the second semester of the academic year 2019/2020.
- Analyzing the obtained results from the instrument of the current study.

3.5 Data Analysis

The data obtained from the questionnaire were processed statistically using the SPSS-Statistical Package for Social Sciences program. The descriptive statistic measures, including the frequency distributions and percentages, were used to answer the questions of the study.

The researcher used a 3-Point Likert Scale to correct the study instrument, by giving each of its items one degree out of the three degrees (high, moderate, low), which represents (3, 2, 1) respectively. The following scale was adopted to analyze and interpret the results. Table 2 presents the scale.

Table 2 The Statistical Model of Proportional Scaling

Degree	Means
Low	1–1.66
Moderate	1.67–2.33
High	2.34–3

4 Results of the Study

The study question is: *What are the obstacles of e-learning for community college students in the light of Corona pandemic?* Means and standard deviations were used to identify the degree of e-learning obstacles that community college students at Imam Abdul Rahman bin Faisal University face in the light of Corona pandemic. Table 3 presents the results.

Table 3 shows that the means score ranged between (1.37 and 2.60). The social obstacles had got a mean score of (2.60) as the first rank. The financial obstacles domain had got a mean score of (2.20) as the second rank, while the domain of technical obstacles had got a mean score of (1.37) as the last rank. The total mean score of the students' responses to the questionnaire was (2.05), with a standard deviation of (0.243). The means score and standard deviations for the study sample estimation were calculated for the items of each domain separately. They were as below.

4.1 The First Domain: Technical Obstacles

Table 4 shows that the means score ranged between (1.60–1.20). Item (5) *"I see there is a difficulty in dealing with e-learning systems, such as Blackboard and Zoom"* came in the first rank. Its mean score was (1.60) and its standard deviation was (0.295). Item (1) *"I see there is a slow communication on the Internet"* came in the last rank. The mean score of the item 1 was (1.20) and its standard deviation was (0.050). This is consistent with the findings of the studies of (e.g., [15, 16, 26]). In contrast, the findings are inconsistent with the findings of studies of (e.g., [19]).

Table 3 Means, standard deviations and ranks of the degree of e-learning obstacles for college community students

Rank	No	Domain	Mean	Std. dev	Degree
3	1	Technical obstacles	1.37	0.210	Low
2	2	Finical obstacles	2.20	0.320	Moderate
1	3	Social obstacles	2.60	0.358	High
Total			2.05	0.243	Moderate

Table 4 Means and standard deviations related to the domain of technical obstacles in a descending order according to the means score

Rank	No	Items	Mean	Std. dev	Degree
1	1	I see there is a slow communication on the Internet	1.20	0.050	Low
2	2	I see there is a lack of periodic maintenance of e-learning systems	1.22	0.078	Low
3	3	I feel there are many breakthroughs of the Internet in e-learning	1.30	0.070	Low
4	4	I see there are frequent interruptions on the Internet	1.33	0.187	Low
9	5	I see there is a difficulty in dealing with e-learning systems, such as Blackboard and Zoom	1.60	0.295	Moderate
5	6	I see the e-learning infrastructure is insufficient	1.35	0.192	Low
6	7	I feel weak in curriculum development in line with e-learning	1.40	0.194	Low
7	8	I feel there are no available e-learning applications in the Arabic language	1.45	0.195	Low
8	9	I see that students are not prepared for e-learning	1.50	0.290	Low
Total			1.37	0.190	Low

4.2 The Second Domain: Financial Obstacles

Table 5 shows that the means score ranged between (1.94 and 2.20). Item (13) *"The increased cost of using e-learning systems is outside the university"* came in the first rank. It had a mean score of (2.20) and a standard deviation of (0.350). Item (14) *"There are not enough incentives for students to use e-learning systems"* came lastly.

Table 5 Means, standard deviations related to the domain of financial obstacles in a descending order according to the means score

Rank	No	Items	Mean	Std. dev	Degree
1	13	The increased cost of using e-learning systems is outside the university	2.20	0.350	Moderate
2	12	The students have a high teaching load	2.18	0.408	Moderate
3	15	There is a need for human cadres and technicians in e-learning	2.17	0.453	Moderate
3	16	There is an increase in the prices of smart devices for applying e-learning	2.15	0.453	Moderate
5	10	There is a lack of laboratories and equipment for using in e-learning	2.12	0.458	Moderate
6	11	There is no laptop for every university student	2.05	0.426	Moderate
6	14	There are no enough incentives for students to use e-learning systems	1.94	0.463	Moderate
Total			**2.20**	**0.346**	**Moderate**

Table 6 Means and standard deviations related to the domain of social obstacles in a descending order according to the means score

Rank	No	Items	Mean	Std. dev	Degree
1	21	Learners from educated community members feel frustrated by their lack of e-learning proficiency	2.90	0.508	High
2	18	The community does not accept e-learning	2.85	0.515	High
3	17	There are no educational programs for the community about e-learning	2.70	0.550	High
4	19	I see there is a difference between e-learning and the culture of society	2.60	0.530	High
5	20	I feel there is a difficulty in understanding the idea of e-learning by society	2.50	0.533	High
6	22	I feel that body language and gestures are absent in e-learning, which hinders the communication process	2.40	0.535	High
Total			2.60	0.530	High

It had a mean score of (1.94) and a standard deviation of (0.463). The total mean score of the students' responses to this domain was (2.20), with a standard deviation of (0.346). These results are consistent with the findings of the studies of (e.g., [2, 3, 9, 14, 27]).

4.3 The Third Domain: Social Obstacles

Table 6 shows that the means score ranged between (2.40–2.90). Item (21) *"Learners from the educated community members feel frustrated by their lack of e-learning proficiency"* had got a mean score of (2.90) and a standard deviation of (0.508) as it came in the first rank. Item (22) *"I feel that body language and gestures are absent in e-learning, which hinders the communication process"* had got a mean score of (2.40) and a standard deviation of (0.535) as it came in the last rank. The total mean score of the students' responses to this domain was (2.60), with a standard deviation of (0.530). These results are consistent with the findings of the studies of (e.g., [28, 29]). However, the findings of this study are inconsistent with the findings of studies of (e.g., [18, 30]).

5 Conclusions, Limitations and Recommendations

1. The first domain was the technical obstacles, which came in a low degree, with a mean score of (1.37).
2. The second domain was the financial obstacles, which came in a moderate degree, with a mean score of (2.20).

3. The third domain (the social obstacles) had got a mean score of (2.60) as it came in a high degree.

The study is limited to the following limits:

- Temporal limits: In the academic year 1440/1441 Hijri.
- Spatial limits: Dammam Community College-Rakah. (Imam Abdul Rahman bin Faisal University).
- Objective limits: Obstacles of e-learning for community college students in the light of Corona pandemic
- Human limits: community college students (male and female) who are studying at the college now.

Different implications have emerged from the findings of the study. First, educators should find out solutions to the obstacles of e-learning in the learning institutions. Second, students should also train how to use e-learning applications in their learning to get better performance.

Many recommendations are presented as follows:

1. The need to focus on the technical side in universities through e-learning.
2. Financial support for e-learning through training the specialists and technicians.
3. Holding educational courses for community members to enhance the status of e-learning.

References

1. Iskandar, S.: Diagnosing the quality of the e-learning service provided to remote masters students: an exploratory study of a sample of Algerian University students. J. Econ. Human Dev. **10**(2), 238–252 (2019)
2. Qamishi, L.: Employing information and communication technology in e-learning. Al-Hikma J. Media Commun. Stud. **9**, 59–74 (2017)
3. Ghayad, K.: E-learning as a strategic choice for Algerian universities. Derasat **46**, 110–126 (2016)
4. Bani Yasin, B., Muhammad, M.: Barriers of using e-learning faced by teachers in Irbid first Directorate of Education. Palestinian J. College Educ. **3**(5), 115–136 (2011)
5. Zarrouq, B.: The obstacles facing the teacher of basic education in the employment of e-learning in teaching (Unpublished master's thesis). Sudan University of, Science and Technology, Sudan, Faculty of Education (2016)
6. Alzboon, M.: The effect of teaching using the electronic modules system (model) on the achievement of students of the University of Jordan with a course in computer skills and in developing their self-learning and social communication skills (Unpublished doctoral dissertation), University of Jordan, Amman, Jordan (2015)
7. Naseem, S.: The roles of kindergarten teachers in e-learning and obstacles in its applying from their viewpoint. Childhood Educ. J. **2**(2) (2010)
8. Al-Saeedi, O.: Attitudes of Majmaah University faculty members towards the employment of e-learning for the educational process. J. Hum. Adm. **9**, 30–57 (2016)
9. Al-Bedawi, S.: Obstacles in the use of e-learning from the faculty member's point of view at Qassim University in the light of some variables. J. Faculty Educ.-Assiut Univ. College Educ. **33**(7), 379–445 (2017)

10. Hamad, L.: The degree of using e-learning tools in teaching by faculty staff members in Jordanian universities and their attitudes toward it (Unpublished master's thesis). Middle East University, Amman, Jordan (2018)

11. Badawi, M.: Barriers in the use of faculty members at Menoufia University for management systems e-learning from their point of view. J. Psychol. Educ. Res. **30**(4), 96–146 (2015)

12. Eryilmaz, M.: The effectiveness of blended learning environments. Contemporary Issues Educ. Res. **8**(4), 251–256 (2015)

13. Khalifa, A.: Obstacles and problems in applying e-learning in university education from the viewpoint of faculty members at Jazan University in the light of some variables. J. College Educ. **20**(7), 129–154 (2011)

14. Haseeb, A., Al-Abaade, A.: E-learning application obstacles analytical study at AL-Hadbaa University College. Al-Rafidain Dev. J. **36**(116), 216–229 (2014)

15. Al Mubarak, H.: The effect of the application of electronic education at the Sudanese universities according to some of the current trends of comprehensive quality. J. Delta College Sci. Technol. Delta College **8**, 99–136 (2018)

16. Boras, F., Ashi, A.: The importance of e-learning in achieving the development of humanity. In: Research and Refereed International Conference: Electronic Management between Reality and Imperative, November, Jordan, Amman, pp. 40–46 (2017)

17. Bachir, B., Ammar, B.: The reality of the application of the e-education in Algerian universities: a case study in the university of Biskra. Int. J. Qual. Assurance **1**(1), 9–19 (2018)

18. Souman, A., Hamza, M.: The obstacles of utilizing EDUWAVE from the perspective of public schools teachers in Amman and their attitudes towards it. J. Educ. Sci. Stud. **38**(3), 917–930 (2011)

19. Al-Shehri, A.: Launching e-learning and training programs: a proposed model. Inst. Public Adm. **90**(3) (2010)

20. Radu, F., Radu, V., Croitoru, G.: The advantage of the new technologies in learning. In: Recent Researches in Artificial intelligence, Knowledge Engineering and Data Bases, pp. 150–155 (2011)

21. Qwaider, W.: Integrated of knowledge management and e-learning system. Int. J. Hybrid Inf. Technol. **4**(4), 59–70 (2011)

22. Kumpikaite, V., Duoba, K.: E-learning process: Students' perspective. In: 3rd International Conference on e-Education, e-Business, e-Management and e-Learning IPEDR Vol. 27 IACSIT Press, Singapore (2012)

23. Aggarwal, D.: Role of e-learning in a developing country like India. In: Proceedings of the 3rd National Conference; INDIACom-2009 Computing For Nation Development, February 26–27, 2009 Bharati Vidyapeeth's Institute of Computer Applications and Management, New Delhi (2009)

24. Alberdi, I., Iribas, A., Martin, A., Aginako, N.: Collaborative web platform for rich media educational material creation. World Acad. Sci. Eng. Technol. Int. J. Educ. Pedagog. Sci. **6**(5), 778–782 (2012)

25. Ahmadpour, A., Mirdamadi, M.: Determining challenges in the application of e-learning in agricultural extension services in Iran. Am.-Euras. J. Agric. Environ. Sci. **9**(3), 292–296 (2010)

26. Al-Hayani, S.B, Ibtisam, M.M.: Using e-learning to address students learning problems. Arab J. Educ. Psychol. Sci. **8**, 111–126 (2019)

27. Qahwan, M., Huthaili, S.: Obstacles of e-learning in university education in Yemen. In: The Twentieth Annual Scientific Conference, Education and Progress in Asian and Australian Countries, the Egyptian Association for Comparative Education and Educational Administration, Egypt, pp. 289–326 (2012)

28. Alothman, A., Almassaad, A.: Motivations and constraints of e-learning from the viewpoint of faculty members of the college of education, King Saud University. Int. J. Res. Educ. **41**(1), 96–132 (2017)

29. Diab, Z., Bruce, W.: Obstacles of digital education in the Algerian school. Arab J. Literat. Hum. Arab Found. Educ. Sci. Arts **7**, 153–178 (2019)

From Offline to Online Learning: A Qualitative Study of Challenges and Opportunities as a Response to the COVID-19 Pandemic in the UAE Higher Education Context

Shirley Leo, Nizar Mohammad Alsharari, Jainambu Abbas, and Muhammad Turki Alshurideh ⓘ

Abstract This paper aims to explore the challenges, opportunities and implications of COVID-19 pandemic in the UAE higher education context. It uses the qualitative study approach and websites methodologies to enrich the aims and results of the study. It focuses on the influence of COVID-19 pandemic on higher education institutions and their progressions respectively. As higher education institutions not only provide knowledge but also develop interpersonal skills and help mold an individual in all aspects to face real-world challenges. Institutions should, therefore, get back to traditional teaching methods as soon as possible by incorporating the benefits of virtual learning into traditional learning methods, as virtual learning cannot replace traditional methods. Consequently, over some time as the crisis continues globally, the higher educational institutions worldwide can take into development better modes of communication and continue to provide various forms of delivering their services, and as the COVID-19 with progress in medical fields can fade to a disease of history we can improve what was once traditional learning with the online learning experience from this time. As the COVID-19 crisis continues globally, the higher education institutions can develop better modes of communication

S. Leo · J. Abbas (✉)
Business Department, Higher Colleges of Technology, Abu Dhabi, UAE
e-mail: jabbas@hct.ac.ae

S. Leo
e-mail: shirley.leo@hct.ac.ae

N. M. Alsharari
Department of Accounting, Finance, and Entrepreneurship, College of Business, Jackson State University, Jackson, United States
e-mail: nizaralsharari@gmail.com

M. T. Alshurideh
Department of Management, College of Business Administration, University of Sharjah, Sharjah, UAE
e-mail: malshurideh@sharjah.ac.ae; m.alshurideh@ju.edu.jo

Department of Marketing, School of Business, The University of Jordan, Amman, Jordan

© The Author(s), under exclusive license to Springer Nature Switzerland AG 2021   203
M. T. Alshurideh et al. (eds.), *The Effect of Coronavirus Disease (COVID-19)
on Business Intelligence*, Studies in Systems, Decision and Control 334,
https://doi.org/10.1007/978-3-030-67151-8_12

and forms of providing their services and as the COVID 19 fades to a disease of history, so it can improve what was known as traditional learning into the online learning over the time. After substantiating the multiple benefits and consequences of having virtual learning in the higher education system, it is clear that there is a lot of room for improvement over the time. This study contributes to the COVID 19 concept and higher education tools as it introduces you to the various impacts of online educational tools used by higher education universities in the UAE during the spread of COVID-19. It can be considered as a pioneer study in UAE in exploration the relationship between COVID-19 and higher education field.

Keywords COVID-19 · Pandemic · Higher education · Distance learning · E-Learning · UAE

1 Introduction

The current world is facing an unprecedented global health, social and economic emergency with the COVID-19 pandemic. The pandemic affected almost all industries. The outbreak of the coronavirus not only impacted the physical health of those it affected but also its impact has been felt in every aspect of life—financially, socially, economically, psychologically and environmentally. While the long-term outlook of the COVID-19 pandemic is still highly uncertain, businesses and individuals are nevertheless having to adapt to new, troubling challenges such as severe supply and demand disruption, social distancing and serious health implications. Even though the COVID-19 crisis is concerning and disruptive, the way everyone tends to manage the crisis now will make a big difference in the future and for quick recovery. Education is also extremely affected by the pandemic and has uprooted higher education system globally. To cope with the crisis, educational institutions are forced to adopt virtual learning and hence digital transformation is no longer a luxury, but a necessity [1–3].

As we are looking into the current situation in the United Arab Emirates (UAE) and their response to COVID-19 it is clear to see that since the first reported case on January 29, 2020, there have been immediate drastic measures to ensure the safety on the citizens. As a night curfew and sterilization process was established in March, many businesses and institutes were advised to take measures to curb the growth of the disease [4]. Schools in the UAE decided to take education to an online platform and thus introduced distance learning initiative to make certain that education continued despite the closure of many educational institutions.

A pilot stage of closing schools with a spring break given to the students started on March 8th and led to all educational systems to start closing down face to face classes and giving virtual classes from March 22 and set to run till April 4th. While this helps contain the spread of COVID-19 it did propose many challenges to parents and the students alike, either with not having enough resources to maintain online classes or not having an efficient learning experience that they would have been provided had the classes been in a classroom setting.

Students took to online social media platforms to spread the hashtag "cancel the school year" in Arabic in order to take care of their health during these times and not be pressured to suddenly change in order to continue the education they deserve. However, most of the higher education institutions continued to issue online courses and encouraged their students to continue with their semester despite unrest and many mixed feedbacks. On April 2nd the Ministry of Education of the UAE announced that this form of eLearning would continue till the end of the semester for all public, private schools and universities across all the emirates [5].

This paper after introduction is divided into following sections. Section 2 introduces the objectives of the study. Section 3 shows the significant of the study to the literature. Section 4 discusses the literature review about COVID-19 influence and higher education field. Finally, the last section provides the study conclusions and main recommendations for future research.

2 The Study Objectives

This study aims to provide a brief explanation through a qualitative lens to the current COVID-19 pandemic and its various causes as well as challenges in day to day practices. Specifically, we will delve deeper into the higher education actions taken by various higher education institutions. Due to the pandemic many changes had to be adapted to the educational system, this included the shutting down of universities in many parts of the world which led to education being brought to an online platform in order to cater to the students and their growing demand for education during this time of a global shift. In this study, we will look into how changes are made to eLearning and how it has developed to a higher standard from the previous years of limited distance learning courses [6–8].

Through this research, we aim to provide the effects of both offline and online learning in the United Arab Emirates (UAE) in relation to higher education. The UAE has taken various actions to continue higher education and there are many stages of changes given through different higher education institutions in all of the emirates. The study will not only look into the challenges faced by higher education institutions and their students because of COVID-19 but will also explore the opportunities taken to continue providing education through new online tools and technology. Accommodating eLearning with a virtual classroom setting when it is a new avenue for most of the faculty and students to explore, leads to show how the prospect of improving education is plausible.

Online education, however, does not guarantee a positive reaction or result, this study also delivers the implications that are faced by different characters of this situation. Through this report, all the challenges and various opportunities that are witnessed during this time by the higher education institutions, the faculty, students, parents as well as the government and their actions can be witnessed. As the COVID-19 pandemic continues to grow, the data that can be obtained varies across time, this leads to show that technology always continues to improve and its increased use leads to different pathways that can be explored by higher education institutions in order to supply to the students in all diverse and distinctive scenarios.

3   The Significance of the Study

This study can be considered significant with current contributions where the impact of COVID-19 in higher education is discussed in terms of change in pedagogy, different platforms to deliver classes online and available technological tools also covers in detail the pros and cons due to transition from traditional classroom teaching to virtual learning. It can be considered as an innovative study in the UAE in terms of exploring the relationship between COVID-19 and the higher education field.

The factors that are to be evaluated while teaching online such as technology upgrade, students personal data security, psychological disturbances due to migration to different learning methods, cybercrime, the threat of job loss, financial issues like the additional cost to institutions towards technology upgrade versus expectation of parents for lesser fees due to non-utilization of campus facilities like labs, libraries etc., are discussed in detail.

Finally, this paper highlights the opportunities and challenges of virtual learning and suggests a solution, as possible improvements for the future by incorporating the advantages of eLearning experienced during this unprecedented pandemic situation to develop a hybrid pedagogy without compromising the advantages of traditional learning.

4   Literature Review

4.1   Covid-19 Pandemic and Education Field

As the world is slowly getting accustomed to the pandemic that started at the end of 2019, the novel coronavirus now known as Covid-19 has caused a global shift in the livelihood of all sectors of life. Many are faced with no choice but to turn towards technology and conform to virtual assistance in order to continue functioning. One of the biggest adjustments made as a result of COVID-19 is by the educational institutions on an international stage, where universities and colleges continue to teach during this time but online, thus following safety rules and regulations. Using distance learning technologies faculties have had to acclimatize and help students become familiarized so they can continue to learn. However, these days' students as well as faculty have not only adjusted, but are now treating virtual learning as a norm compared to the traditional learning which used to be standard [9–12]. The world focuses on factually and effectively communicating online to better help higher education so there are lesser risks at the end of the day [13–15]. This chapter takes up the concept and introduces you to the various impacts of online educational tools used by universities in the UAE during this time of COVID-19. With the use of many cyber apparatuses we will discern how we have overcome the difficulties of online teaching in higher education at this time of crisis.

Using the many platforms such as zoom, blackboard learn, teachers and professors can continue what they have been doing before the COVID-19 with not too much trouble. Although it is closest to their original form of teaching there are still many difficulties that they may face from their side or the students. One of those being disconnection or buffering due to the heavy load on the online platform and thus one method that was mentioned above is by using recorded lessons. With recorded lessons, professors are able to review and edit their content and take the time to capture all the material within a video that can be viewed by students in their own time. The students can take their time to understand the topic even taking time to rewind and understanding well unlike the traditional classroom setting [16].

Another form of teaching that started at the beginning of transitioning to virtual classrooms was blended learning. Some parts of the world still have the use of such an adaptation of education where there is more entry to students by enabling self-paced learning exercises [17–19]. Then by giving a distinctive alleyway of receiving education for each student through personal advising, thus improving computer-generated activities with proper feedback. Educational technology capabilities (ETC) hypothesize various perceptions to recognize the best method when providing education with technology-based implementations [20].

4.2 How COVID 19 Influence E-Learning in the UAE Context

As we have been able to look at the many virtual resources and the effectiveness of eLearning, it is crucial to see how effective the consequences of online learning are with the higher education system. Universities and colleges worldwide need to be able to do their best to provide better services during COVID-19 and with health issues, going online seemed to be the best and only solution. Nonetheless, can it be improved? Is it also possible to see the many advantages to distance learning instead of all the downfalls proven? There are many pros and cons to eLearning in the higher education field in the viewpoint of the students, workers as well as the organizations themselves [21–23].

With COVID-19 having more confirmed cases every day the most obvious pro to having eLearning in the educational field is the security of everyone's safety, as all students and faculty are asked to stay at home to avoid proximity with one another they are protected from the spread of the virus. Of course, with such an issue of safety being protected so as to avoid close contact there is a con that introduces isolation and the lack of interaction within the higher education community. This has a major psychological impact as people are called social beings and need to have general interactions, even requiring a community in colleges to promote a healthy competitive educational environment [24, 25].

Nowadays, moving further into the impact of virtual learning itself it is necessary to analyze the factor of fiscal footprint and time efficiency. First the advantages that eLearning provides with cost-effectiveness is that the schools will not have to spend as much money as they normally would in a regular semester on campus, these minor spending when gathered result in a large saving. On the other hand, universities have decided to reduce the fee for many of their majors due to the current crisis and this is a pro from the standpoint of students as they save money for their education. The budget impact that many universities will face in the upcoming future is inevitable and in some cases many majors or departments could be asked to shut down [26].

There is however a disadvantage that all employees and students will face when it comes to spending's as they will have to spend more money on their home necessities but most importantly on having stable internet connections and getting a plan that is suitable for their enter time taking lectures, undertaking assessments and doing tests. There is also the impact that moving online can have on many countries in the world and even some families that are not able to afford the requisites such as having a laptop and buying all the e-textbooks. Time efficiency is an umbrella that can be branched with many benefits and challenges that can be tackled by all people in this issue. The biggest benefit of virtual learning when it comes to time efficiency is that there is more time on the hands of faculty and students saved from the commute to the campus and the time spent within the campus grounds, all this time can add to spending it wisely on further research for classes or on other aspects of schooling whether it be teaching or studying.

With online classes, there is also a certain time frame given so professors can teach, and students can learn and then continue with the Q&A session, this is very similar to how traditional classrooms would follow a time schedule, but with recorded lessons also known as asynchronous learning, professors are allowed to take their time making sure they flexibly deliver all their material while taking time to modify and adjust how their content would like to be presented so as to not err. Students can also find this beneficial as they can take their time to understand what is being taught while doing it any time they prefer unlike a given class schedule, the recorded videos can also be paused and played back which is a huge plus for truly comprehending and learning. On the other hand, with synchronous learning or live classes there are various variables such as poor network connection or lack of quality with the professor or student's microphones or videos that can result in loss of time and only be a big challenge as it builds up over time [27].

Virtual learning has opened a world of new methods to the educational sphere and this has led to researchers, institutes, and businesses to come up with more platforms to continue the use of technology in an educational setting. This is a big advantage to the educational field as we have many types of eLearning such as webinars, distance learning, blended learning, and now with wanting a community online with students and teachers there have been introductions to personal learning environments or networks where peer learning and student queries are all addressed so as to create a joint space among students [28]. With a lot of help provided to students and such a community that helps students get all the benefits of the online learning environment there is a huge con that is present during the live online class. It is a matter of retention and inspecting to see if students truly do retain the information they have learned in class. In traditional settings, it is possible to check with the student through in-class participation which is also possible with online classes but as students are in their own rooms and are comfortable with their environment there is a higher chance of being distracted and not paying attention to the lecture which would not be the case if all students were present in an actual classroom [29].

With distractions have given prone to take place with online classrooms there is also a huge lack of control and there is a struggle in monitoring the actions of students, due to the deficiency of having smooth connections there is not an easy way to have complete control of the classroom by the professors. A pro to eLearning with the rise of technology is that now professors are given a chance to view assessments, announcements and if the students are indeed following the given instructions. There are opportunities to see each students' actions taken on the online classroom and traditional classroom settings would not guarantee that all students are able to perceive and attain information provided during lecture time, as they could be highly distracted. With online tracking, we are able to view all the students' activity, and this provides faculty with a system to have all students participate equally and to not miss out on any important notices.

An increase in more resources and opportunities in the IT world is helpful however this can be seen as a huge con as now professors of certain degrees are not experts in online pedagogy. Not many faculties are accustomed to using or educating virtually and lack training or practice to do so and a similar point can be said about students as well. There is a challenge of using this technology that is a challenge [30], in itself as many students will face difficulties in dealing with virtual learning and be exhausted with the actual course content that they would need to learn and on top of that this can cause a domino effect that could lead to a drop out of students in the future [31]. This is one of the major downfalls of having to move from offline to online on such short notice during this time of crisis.

4.3 E-Learning Tools Used in Higher Education

With COVID-19 affecting our daily lives, as well as the safety and health issues constantly being raised, to handle the continuation of day to day life almost every business or form of work has had to make drastic changes in order to continue. The international education sector has had to reconstruct and adapt to continue providing services. Higher education in general has had to go online and help students continue their education so as to not hinder their development during this time of crisis. Most measures taken by universities are common such as using online platforms to conduct virtual classes. With the increase in cases worldwide, we saw a global movement of more and more users online, students especially sought out various methods provided by their institutions. The UNESCO also provided various resources for distance learning to help many schools and organizations come to terms with continuing their work during COVID 19 [32].

According to Oranburg [33], some popular tools provided and used these days are Google classroom, Blackboard Learn, Ultra Collaborative, GoToMeeting, WebEx, Zoom, Respondus Monitor Lockdown Browser, and many more. There are also many instances where faculty and students communicate through social media platforms such as Facebook, WhatsApp, and many others that provide services that are more easily accessible and are accustomed. Going online however does not only require external apparatuses, as it is also a matter of finding different strategies to be able to teach to students on a virtual platform. Hence, the proposition of having the idea of synchronous learning and asynchronous learning being used in classrooms [34].

Through synchronous learning faculty are able to educate live using platforms to communicate and interact in real-time, while through asynchronous learning it is a mode of schooling done using recordings and channels used virtually without interaction and interruption. There are also many universities that provide a hybrid of the two synchronous learning and asynchronous learning in order to have the best experience for students, so they are interacting and learning at their own pace as well. There are also plenty of resources provided by institutions other than students' home universities where online lectures and classes are given with an unlimited number of participants and are for purposes only to lecture without too much of a classroom like nature. The massive open online course also known as MOOCs are a form of such online courses that have been around for a while now that specialize on a subject and provide adequate information for learners with feedback and have a growing online community. With a number of universities having to join together and work on providing for their students there have been cross-university collaboration settings afforded online, most of them using features such as seminars, virtual classes, collaboration on a specific project, peer-review assignments, exchange of personnel, to an exchange of pedagogical methods [35].

4.4 Covid-19 and E-Learning Implications

With the interference of the pandemic, many higher education institutions had to quickly change their method of teaching for the safety of all students and staff. With the adaptation of social distancing, it was a given that classroom settings were to be prohibited. Introducing universities to online tools and technologies was not a difficult procedure as most campuses in the twenty-first century was accustomed to having eLearning as an option for students either doing distance learning or blended learning. However, this time around virtual teaching was not provided as an option but the online method to continue providing education.

Many campuses quickly turned to provide existing students and upcoming students with previous material such as recorded lessons and online assignments so they can be distributed during the time of moving over from a form of traditional or blended learning in the beginning to a complete online or virtual teaching atmosphere. Providing prerecorded lessons (QS, 2020) as a form as asynchronous learning for students gave faculty a bit of a hard time to get control over the situation of giving assignments or conducting exams, but most importantly there was an issue of class participation and attendance, which is essential to provide proper education.

Therefore, the universities introduced Zoom, Blackboard Collaborate, WebEx, Google meet and other such online platforms to conduct classes and have students participate in online classes as well as lectures. These tools created a virtual classroom with interaction and provided the professors with a chance to conduct their lectures as they would in reality [36].

As these virtual classes were only for lectures, other applications such as Google drive, drop box and extra services were used to share files and have an easy exchange of materials online. This was a safe and effective way to have assessments provided to the students as well. Universities have come to learn and adapt to the different online pedagogies so as not to disturb the flow of regular education, but to swiftly abet the use of digital tools for both their staff and students in order to continue teaching and learning respectively [37].

Many challenges were presented with the introduction of online teaching but most of all was, acclimating to use the online tools in order to teach, which goes away from the norm of education. It can further be argued that this use of online tools and technologies is actually belated and if used earlier, it would have been easier to use now as it would have revolutionized and modernized education, as we now know it [38].

4.5   COVID-19 and E-Learning Challenges

The coronavirus, which surfaced at first in Wuhan, China in the year end of 2019, has now spread on a global scale and on 30th January 2020, the Director of the World Health Organization (WHO) declared this outbreak of Covid-19 to be a Public Health Emergency of International Concern [39]. On February WHO proposed an official name for the virus: Covid-19, an acronym that stands for the coronavirus disease of 2019. In March, UAE's educational ministry announced the closure of all schools and universities in the country and that students are expected to learn from home, avoiding all forms of meeting in public spaces to continue education. According to Clyde [40], the impact that Covid-19 had on higher education provided universities, faculties and students with a chance to explore the virtual classroom.

The most successful method found in keeping students retention power and maintaining access to learning during these times has been through the introduction of online courses. The closure of educational organizations to combat Covid-19, forced schools and universities to provide education through distance learning to ensure students continuous access to learning. Lectures were delivered using various online tools and technology such as Zoom, Black Board Learn ultra where students learn from their private home or office space. This shift from the traditional classroom setting to a virtual higher education experience proposed many challenges such as educational fees, employment and data privacy as well as cybercrime, which were all issues that had not been perceptibly present before the virus hit [40, 41].

One of the main issues that influenced many universities and the parents of the currently enrolled students and upcoming students was the tuition fees issue during this off-campus style of education. The parents feel that since all the services provided at campus are not consumed completely and also due to financial difficulties because of the economic situation being decelerated, the fees should be reduced [43]. On the other hand, universities argued that they are spending more than before because of the new use of technology and with the added regular outgo of salaries for their employees and rental for the lease premises.

The beginning of full time learning from home have shown results in the presence of students being online a lot of more which exposes students to more content on the internet other than the educational material provided by their professors. Also using online tools for having class interaction and participation requires sharing personal information online, which can be taken as a threat to every individual's data privacy. There is a huge possibility of an increase in cybercrimes during this pandemic scenario in all fields of education as well as work. Hackers are now prone to more active involvement in cybercrimes such as blackmail, issue of phishing emails, extortion and impersonation due to reduced online policing and control.

On the employment front, there has been a decrease in the utilization of labs, libraries and other non-academic services provided, because of the introduction of distance learning. Universities that have previously invested in technology and tools with well-equipped online systems and have trained their faculties have an advantage and can offer an increase in class size by allowing their students to register for courses with more access and even across campuses in different emirates as well as now the physical distance will not be an issue. This could have a positive impact for the students to get an education and experience no matter which campus or professor, but it also proposes a slightly negative impact on the employees of the teaching faculty. International universities of great repute who have branch campuses in the Emirates feel that there is a big possibility of surge in local candidates due to the growing fear over safety and travel bans [44].

4.6   COVID-19 and E-Learning Opportunities

With many applications and resources that are provided by the educational systems it was up to the educators to utilize all that was offered and continue to edify. However, with teaching it is not easy to jump from a traditional classroom to a virtual setting and convey lectures in a limited time as there are many aspects of online education that require intensive training and practice to avoid calamitous situations during lectures. Researchers have found that in order to articulately conduct eLearning there are different chambers of this industry that they must master, a leading researcher in eLearning has been able to provide an eight-dimensional framework with contents including; pedagogical, technological, institutional, interface design, evaluation, management, resource support and ethical. All these factors are needed to be developed by the university to competently conduct virtual learning [45].

Many effective measures are needed to be taken step by step in order to provide efficient services, not only do the providers of these many online tools give crucial information on how to operate their apparatus but students and personnel must learn to adapt to the new functions and find a way to do their best despite what little they can do. The transition from the original classroom to online is easiest with live lectures where professors are able to monitor their students and have lectures with interaction so the students can ask questions and get it solved right then and there [46].

To continue an uninterrupted learning procedure during this crisis the UNESCO even provided different regulations that can be followed by institutions all around the world. The first recommendation was to check the inclination of all faculty as well as students and choose appropriate and relevant tools to continue the course of the semester. With choices all in place it is up to the educational institute to have all sorts of privacy and security protocols put into place so that there is no issue with students or the professionals that could hinder their progress [47, 48].

Especially in the UAE it is very crucial to take into consideration all of the requirements of all personnel to be viewed by the board, not only are privacy related topics needed to be discussed but the security of data is a thread that concerns all users when it comes to anything done on an online platform. It is up to the organizations to consider how secure all the tests, assessments, exams are conducted, and to come up with ways to ensure there are no crude fabrications in the process of educating. There are also many psychological factors to consider before, during, and after teaching the class. Steps must be taken by the universities to ensure that all students and faculty are provided with the best help regarding issues outside of education as well as resources to help when it comes to pertaining knowledge provided by these online lessons.

Every student needs to be accounted for and it is of at most important to understand the psyche of all those involved and to prepare ahead of time to make sure all sources are given to help smoothen the procedure of lectures during this time of crisis. As advisors will need to be put into place to help everyone, it is essential to also have a study schedule that is uniformly accepted so that with the schedule there is a systematic way of handling the semester. Support provided must also include the suitable approaches for teachers to conduct their lessons while using the limited number of online tools or applications that they are asked to use.

Although we saw a huge list of platforms that are being enhanced every day for the use of eLearning, there must a solid decision to decrease the use of a huge number of applications so as not to complicate the student or faculty with the added stress of learning and teaching. Constant monitoring of students' progress is a given nonetheless faculty activity and learning methods, with recordings of the class sessions and targeted time-efficient tasks can help maintain the progression of a schedule. With the many protocols taken it is best to have a community on the side to help faculty and students remain in contact with one another so that they can still feel like the interaction is present online.

5 Conclusion and Recommendations

After substantiating the multiple benefits and consequences of having virtual learning in the higher education system, it is clear that there is a lot of room for improvement as time goes by. Through this study, we had a chance to witness the importance of technological advancement concerning the higher education area. COVID-19 not only had a huge impact on the day-to-day lives of all of humanity at the end of 2019 and for the better half of 2020, but with the pandemic, the world was forced to adapt to an online platform. These online media foundations were not only for human interaction and connection but a way of life and continuation of how humans survive with their work and livelihood.

Higher education institutions especially had to finally provide for the students and for the maintenance of their industry. Due to the continuation of education through virtual methods, there was a domino effect of schools shutting down and impeding the use of face to face classes and provide online classes for all of the courses and having students register and start to practice with more online tools. This brought a completely new audience to various applications and programs that could accommodate to the increasing requirements of the higher education institutes as well as the students and faculty.

Teachers, students and parents needed this opportunity to adapt to the latest technologies and tools in education. This resulted in a positive implication to a more comfortable procedure to learn and educate oneself. Another such benefit of this shift to virtual classrooms is that many institutions could enjoy a short-term financial advantage due to less expense on the estate and other such support systems and resources, but this should be considered as a temporary phenomenon to maintain continuity of learning during the pandemic situation as the primary role of the educational institution is to generate a literate society no matter what form of distribution provided. However, not all outcomes can be deemed positive and there are difficult challenges that are faced. The sudden shift to online methods of education did not allow schools and universities to conduct an experimentation period to judge whether this form of teaching can be successful or even amiable to all parties. Due to lack of research, even the boards in the institutions, as well as business that provide such online platforms, could not quickly develop changes to improve education. With platforms that are the most used and focused upon such as Google classroom, Blackboard Learn, Ultra Collaborative, Zoom and so on, there are difficulties in using them as a decoy for offline classrooms.

Faculty and students alike will find it more difficult to conduct regular interaction and to conduct assignments. Test-taking is guaranteed to be open book and will have to be monitored and reviewed by the faculty to make sure they are invigilated methodically. Cheating is much easier when the students have access to Wi-Fi and help from various websites and resources online as well as communicating with their classmates without the knowledge of the professors. Other more severe trials that both the institutions and students face is regarding privacy and protection from cyber-crime, this issue may not seem so precedent during these trying times but with more people online there is a huge market for criminal activities to grow. With a larger audience on the internet, there is also larger access to hackers who can access information that is private to the institutions or students alone and this can cause bigger problems that require immediate attention.

References

1. Alsharari, N.M.: Results Based Costing (RBC) System: Questioning the Unit of Analysis in ABC. Corporate Ownership and Control **13**(2), 587–603 (2016)
2. Alsharari, N.M.: The diffusion of accounting innovations in the new public sector as influenced by IMF reforms: actor-network theory. Int. J. Actor-Netw. Theory Technol. Innov. (IJANTTI) **8**(4), 26–51 (2016)
3. Alsharari, N.M.: Management accounting relevance: practice, variance and current research agenda. UOS J Hum. Soc. Sci., University of Sharjah
4. Hilton, T.: Coronavirus shuts schools in UAE, all you need to know about studying from home (2020). Retrieved from english.alarabiya.net/en/features/2020/03/05/Coronavirus-shuts-schools-in-UAE-all-you-need-to-know-about-studying-from-home
5. Ministry of Education (UAE): Distance learning system to continue to be applied till end of current academic year (2020). Retrieved from www.moe.gov.ae/En/MediaCenter/News/Pages/elearning3.aspx
6. Alsharari, N.M.: The development of accounting education and practice in an environment of socio-economic transformation in the Middle East: the case of Jordan. Int. J. Educ. Manage. **31**(6), 736–751 (2017)
7. Alsharari, N.M.: Internationalization of the higher education system: an interpretive analysis. Int. J. Educ. Manage. **32**(3), 359–381 (2018)
8. Alsharari, N.M.: Multilevel institutional analysis of accounting change in public management. Int. J. Organ. Anal. **26**(1), 91–106 (2018)
9. Alsharari, N.M.: Management accounting and organizational change: alternative perspectives. Int. J. Organ. Anal. **27**(4), 1124–1147 (2019)
10. Alsharari, N.M.: A comparative analysis of taxation and revenue trends in the Middle East and North Africa (MENA) Region. Pacific Account. Rev. **31**(4), 646–671 (2019)
11. Alsharari, N.M.: Accounting changes and beyond budgeting principles (BBP) in the public sector. Int. J. Public Sector Manage. **3**(2/3), 165–189 (2019)
12. Alsharari, N.M.: Internationalization market and higher education field: institutional perspectives. Int. J. Educ. Manage. **34**(2), 315–334 (2019)
13. Alhashmi, S.F., Alshurideh, M., Al Kurdi, B., Salloum, S.A.: A systematic review of the factors affecting the artificial intelligence implementation in the health care sector. In: Joint European-US workshop on applications of invariance in computer vision, pp. 37–49. Springer, Cham (2020)

14. Salloum, S.A., Alshurideh, M., Elnagar, A., Shaalan, K.: Mining in educational data: review and future directions. In: Joint European-US Workshop on Applications of Invariance in Computer Vision, pp. 92–102. Springer, Cham (2020)

15. Salloum, S.A., Alshurideh, M., Elnagar, A., Shaalan, K.: Machine Learning and deep learning techniques for cybersecurity: a review. In: Joint European-US workshop on applications of invariance in computer vision, pp. 50–57. Springer, Cham (2020)

16. Al Kurdi, B., Alshurideh, M., Salloum, S.A.: Investigating a theoretical framework for e-learning technology acceptance. Int. J. Electr. Comput. Eng. (IJECE) **10**(6), 6484–6496 (2020)

17. Al Kurdi, B., Alshurideh, M., Salloum, S., Obeidat, Z., Al-dweeri, R.: An Empirical investigation into examination of factors influencing university students' behavior towards Elearning acceptance using SEM approach. Int. J. Interact. Mobile Technol. **14**(2), 19–41 (2020)

18. Alshurideh, M.T., Salloum, S.A., Al Kurdi, B., Monem, A.A., Shaalan, K.: Understanding the quality determinants that influence the intention to use the mobile learning platforms: a practical study. Int. J. Interact. Mobile Technol. (IJIM) **13**(11), 157–183 (2019)

19. Alshurideh, M., Al Kurdi, B., Salloum, S.A.: Examining the main mobile learning system drivers' effects: A mix empirical examination of both the Expectation-Confirmation Model (ECM) and the Technology Acceptance Model (TAM). In: International Conference on Advanced Intelligent Systems and Informatics, pp. 406–417. Springer, Cham (2019)

20. Castro, R.: Blended learning in higher education: Trends and capabilities. Educ. Inf. Technol. **24**, 2523–2546 (2019)

21. Al-Duhaish, A., Alshurideh, M., Al-Zu'bi, Z.: The impact of the basic reference group usage on the purchasing decision of clothes (A field study of Saudi youth in Riyadh city). Dirasat: Administrative **41**(2), 205–221 (2014)

22. Al Shurideh, M., Alsharari, N.M., Al Kurdi, B.: Supply chain integration and customer relationship management in the airline logistics. Theoret. Econ. Lett. **9**(02), 392–414 (2019)

23. Al Dmour, H., Alshurideh, M., Shishan, F.: The influence of mobile application quality and attributes on the continuance intention of mobile shopping. Life Sci. J. **11**(10), 172–181 (2014)

24. Alshraideh, A.T.R., Al-Lozi, M., Alshurideh, M. T.: The impact of training strategy on organizational loyalty via the mediating variables of organizational satisfaction and organizational performance: an empirical study on Jordanian agricultural credit corporation staff. J. Soc. Sci. (COES&RJ-JSS) **6**(2), 383–394 (2017)

25. Alshurideh, M.: The factors predicting students' satisfaction with universities' healthcare clinics' services: a case-study from the Jordanian higher education sector. Dirasat: Adm. Sci. **161**(1524), 1–36 (2014)

26. Friedman, S., et al.: COVID-19 impact on higher education. Deloitte United States (2020). Retrieved from www2.deloitte.com/us/en/pages/public-sector/articles/covid-19-impact-on-higher-education.html

27. Murphy, M.P.: COVID-19 and emergency eLearning: consequences of the securitization of higher education for post-pandemic pedagogy. Contemporary Security Policy **41**(3), 1–14 (2020)

28. Tambouris, E., Panopoulou, E., Tarabanis, K., Ryberg, T., Buus, L., Peristeras, V., Lee, D., Porwol, L.: Enabling problem based learning through web 2.0 technologies: PBL 2.0. J. Educ. Technol. Soc. **15**(4), 238–251 (2012)

29. He, W., Xu, G., Kruck, S.E.: Online IS education for the 21st century. J. Inf. Syst. Educ. **25**(2), 101–106 (2019)

30. Rasheed, R.A., Kamsin, A., Abdullah, N.A.: Challenges in the online component of blended learning: a systematic review. Comput. Educ. **144**, 103701 (2020)

31. Harris, S.M., Larrier, Y.I., Castano-Bishop, M.: Development of the student expectations of online learning survey (SEOLS): a pilot study. Online J. Distance Learn. Adm. **14**(4), 1–12 (2011)

32. UNESCO: Distance Learning Solutions (2020). Retrieved from en.unesco.org/covid19/educationresponse/solutions

33. Oranburg, S.: Distance education in the time of coronavirus: quick and easy strategies for professors (March 13, 2020). Duquesne University School of Law Research Paper No. 2020-02 (2020). Available at SSRN: https://ssrn.com/abstract=3553911 or https://doi.org/10.2139/ssrn.3553911
34. Midkiff, S.F., DaSilva, L.A.: Leveraging the web for synchronous versus asynchronous distance learning. In: International Conference on Engineering Education, vol. 2000, pp. 14–18 (2000)
35. Pisoni, G., Marchese, M., Renouard, F.: Benefits and challenges of distributed student activities in online education settings: cross-university collaborations on a Pan-European level. In: 2019 IEEE Global Engineering Education Conference (EDUCON), pp. 1017–1021. IEEE (2019)
36. Abou Samra, R., Alsharari, N., & AlTunaiji, S.: Conceptual Model for challenges and succession opportunities for virtual project teams in the GCC. In: Future of Information and Communication Conference, pp. 328–340. Springer, Cham (2020)
37. Alshurideh, M., Nicholson, M., Xiao, S.: The effect of previous experience on mobile subscribers' repeat purchase behaviour. Eur. J. Soc. Sci. 30(3), 366–376 (2012)
38. Kandri, S.: COVID-19 is sparking a revolution in higher education (2020). Retrieved from www.weforum.org/agenda/2020/05/how-covid-19-is-sparking-a-revolution-in-higher-education/
39. Taylor, D.B.: How the Coronavirus Pandemic Unfolded: a Timeline (2020). Retrieved from www.nytimes.com/article/coronavirus-timeline.html
40. Clyde, C.: "Insight & Knowledge." COVID-19 Education: Ongoing Issues Affecting the Education Sector across the GCC (2020). Retrieved from www.clydeco.com/insight/article/covid-19-education-ongoing-issues-affecting-the-uae-ksa-and-gcc-wide-educat
41. Al-Shboul, M., Alsharari, N.: The dynamic behavior of evolving efficiency: evidence from the UAE stock markets. Q. Rev. Econ. Fin. 73, 119–135 (2018)
42. Adel Almasarwah, A.M., Omoush, N.A.: IFRS compliance and stock prices influence: evidence from Jordanian banks. Banks Bank Syst. 13(3), 24–35 (2018)
43. Alsharari, N.M., Eid, R., Assiri, A.: Institutional contradiction and BSC implementation: comparative organizational analysis. Int. J. Organ. Anal. 27(3), 414–440 (2019)
44. Masudi, F.: UAE Universities Expect More Local Applicants amid COVID-19 Uncertainty (2020). Retrieved from www.gulfnews.com/uae/education/uae-universities-expect-more-local-applicants-amid-covid-19-uncertainty-1.71786084
45. Khan, B.H.: Developing eLearning strategy in Universities of Bangladesh. In: ICT, pp. 78–106 (2014)
46. Lasyoud, A.A., Alsharari, N.M.: Towards an understanding of the dimensions and factors of management accounting change. Asia-Pacific Manage. Account. J. 12(1), 105–142 (2017)
47. UNESCO: 10 Recommendations to Plan Distance Learning Solution (2020)s. Retrieved from en.unesco.org/news/covid-19-10-recommendations-plan-distance-learning-solutions. Accessed: 17 Sep 2020. Available at: https://en.unesco.org/news/covid-19-10-recommendations-plan-distance-learning-solutions
48. Al-Qudah, A.A., Alsharari, N.: The US's debt ceiling crisis and budget constraints. Eur. J. Account. Audit. Fin. Res. 2(3), 31–39 (2014)

Business Innovation and Entrepreneurship and Corporate Governance

The Effect of Transformational Leadership on Achieving Effective Decisions in the Presence of Psychological Capital as an Intermediate Variable in Private Jordanian Universities in Light of the Corona Pandemic

A. A. Harahsheh, A. M. Abu Houssien, M. T. Alshurideh[ID], and AlMontaser Mohammad

Abstract This study aimed to discover the role of psychological capital in the impact of transformational leadership on taking effective decisions in Jordanian private universities in light of the Corona Pandemic. The sample included (108) respondents while the study population consisted of the lecturers of the (6) private Jordanian universities in the capital Amman. The researchers resorted to multiple and Stepwise linear regression in order to test the effect of transformational leadership on the effectiveness of management decisions. The researchers also used Hierarchical Regression to test the role of psychological capital in this effect as an intermediate variable. After analyzing the study data, it was revealed a high level of awareness of the importance of the leadership style. The results also showed interest in Jordanian universities in taking efficient decisions that contribute to addressing various types of crises, including the current crisis linked to the Corona pandemic. The results of the study also indicated that there is an effect of transformational leadership in terms

A. A. Harahsheh
Department of Educational Administration and Foundations, Amman Arab University, Amman, Jordan
e-mail: Alaa.harahsheh@aau.edu.jo

A. M. A. Houssien (✉)
Department of Management and Human Resources Management, Amman Arab University, Amman, Jordan
e-mail: Dr.hareth@aau.edu.jo

M. T. Alshurideh
Department of Marketing, School of Business, The University of Jordan, Amman, Jordan
e-mail: m.alshurideh@ju.edu.jo; malshurideh@sharjah.ac.ae

Department of Management, College of Business Administration, University of Sharjah, Sharjah, United Arab Emirates

A. Mohammad
Department Human Resources Management, University Malaysia Terengganu, Terengganu, Malaysia
e-mail: mo1993ntaser@gmail.com

M. T. Alshurideh et al. (eds.), *The Effect of Coronavirus Disease (COVID-19) on Business Intelligence*, Studies in Systems, Decision and Control 334,
https://doi.org/10.1007/978-3-030-67151-8_13

of its all dimensions (ideal effect, inspirational motivation, intellectual excitement, empowerment) on achieving the effectiveness of decisions in terms of its dimensions (quality of the decision, the timing of decision) in private Jordanian universities in light of the Corona pandemic. Given the direct and indirect impact of each dimension of transformational leadership, psychological capital has an intermediate role in the influence of transformational leadership style on the effectiveness of decisions in private Jordanian universities when studying the dimensions of the use of transformational leadership combined.

Keywords Transformational leadership · Effective decisions · Psychological capital · Jordanian universities · Corona pandemic

1 Introduction

Universities are keen to invest human skills and competencies as one of the most important inputs to their production system in order to achieve their goals effectively and to face the changes and challenges in the surrounding environment resulting from technological developments and globalization in all its forms, and this requires continuous modernization and development in administrative and operational ideas and methods. As a result of the proliferation of the new Coronavirus (COVID 19), change and the search for alternatives became an imperative for those responsible for the educational process in private universities and the trend towards adopting new leadership methods to keep pace with contemporary developments, including the Corona pandemic, which led to the spread of changing many methods and methods of leadership and management.

Transformational leadership focuses on leadership with motivation, where transformational leaders provide unusual incentives for their followers to raise morale and stimulate thinking towards new and creative ways and methods to make decisions and solve correct problems, which generates trust, loyalty, and respect for work, and motivates employees to do their best energies, efforts and demonstrate high positive behaviors, and building trust and respect between employees and management to achieve the desired goals. All organizational processes are affected by decisions, as the manner in which decisions and their accompanying effects are necessary is necessary in order to understand the organizational behavior of workers. In addition, decisions are the essence of the department and its compass to continue its administrative activities efficiently and effectively, especially as it depends mainly on anticipating the future and the complexities involved.

Psychological capital focuses on creating the causes of desired positive behaviors at work and bringing about positive change in organizational culture in order to improve organizational performance [1]. Whereas universities are among the organizations that give great attention to the distinguished performance of employees, empowering them and giving them the power and authority to perform their duties and make their decisions in a way that enables universities to achieve their operational and strategic goals in light of the intense competition environment. This study came to measure the role of psychological capital on the impact of transformational leadership on taking effective decisions in private Jordanian universities.

2 Study Problem

Private Jordanian universities are experiencing new competitive challenges with the spread of Corona Virus, and universities seek to achieve the highest levels of caution and caution in their activities and operations, especially in light of a very competitive environment that affects the performance of universities and puts them at stake in meeting the needs of students and their families and the community and still faces conditions Turbulent as a result of the change in the contents of the internal and external environments, which required it to take advantage of contemporary methods and approaches in analyzing problems and challenges and obtaining the decisions of the actors and emphasizing their quality, speed and timing.

The crisis resulting from the Corona pandemic has created sensitive environments in which leaders may be forced to make sudden and effective decisions using very little information, given that institutions have not gone through similar crises. In these types of environments, reason and logic may be easily overwhelmed by emotions and instincts. In order to reduce the impact of these difficult times, every competent leader must take a number of measures before, during, and after the spread of this pandemic. Leaders in educational institutions, including universities, are responsible for directing events in such complex situations. There are always risks in taking action; however, the scale of the crisis sometimes requires large-scale measures. So, university leaders need to have the authority and the qualifications to implement these measures. The leadership skills possessed by leaders of educational institutions in times of crisis may help resolve a crisis situation by eliminating panic and helping victims recover from a difficult situation more quickly. The human resources owned by private universities are considered one of the most important resources for achieving competitiveness and thus focused on enabling psychological capital, which includes the level of the individual's ability to rise and the level of his self-efficacy and flexibility to create the causes of positive behaviors desirable in the work and bring about positive change in organizational culture in order to improve organizational performance. Based on the foregoing, the researcher has determined that the main study problem lies in the question of what is the effect of transformational leadership on achieving the effectiveness of decisions in the presence of psychological capital as a mediating variable in private Jordanian universities in light of the Corona pandemic. According to the above, the problem of the study lies in answering the following questions:

1. Is there an effect of transformational leadership in terms of its dimensions (ideal effect, inspirational motivation, intellectual excitement, and empowerment) on

achieving the effectiveness of decisions in terms of its dimensions (quality of the decision, the timing of decision) in private Jordanian universities in light of the Corona pandemic?

2. Is there an effect of transformational leadership in terms of its dimensions (ideal effect, inspirational motivation, intellectual excitement, and empowerment) on psychological capital in Jordanian private universities?

3. Is there an effect of psychological capital on the effectiveness of decisions in terms of their dimensions (quality of the decision, the timing of decision) in private Jordanian universities?

4. Is there an effect of transformational leadership with their combined dimensions (ideal effect, inspirational motivation, intellectual excitement, empowerment) on achieving the effectiveness of decisions with their combined dimensions (quality of the decision, the timing of decision) When using the psychological capital as an intermediate variable in private Jordanian universities in light of the Corona pandemic?

3 Study Importance

The theoretical importance of this study shows that it is one of the important studies that examine the impact of transformational leadership on achieving the effectiveness of decisions in the presence of psychological capital as an intermediate variable in private Jordanian universities, especially in light of the challenges facing Jordanian private universities in the shadow of the Corona pandemic. In terms of application, this study is expected to provide a set of results for the study sample and the responses of its members, and to develop recommendations that correspond to and address each of the problems of the current study in light of the spread of Coronavirus, and the applied importance also is that it benefits private Jordanian universities in increasing flexibility to interact with The environment is to be supportive and proportional to critical decisions, problem-solving and transformation into important opportunities.

4 Study Hypotheses

According to the literature review, the researchers formulated the following hypotheses:

H01: There is no effect of transformational leadership in terms of its dimensions (the ideal effect, inspirational motivation, intellectual excitement, and empowerment) on achieving the effectiveness of decisions in terms of its dimensions (quality of the decision, the timing of decision) in private Jordanian universities in light of the Corona pandemic.

The researchers extracted the following sub-hypotheses from the main hypothesis:

H01-1: there is no effect of the ideal effect on achieving the effectiveness of decisions in terms of its dimensions (quality of the decision, the timing of decision) in private Jordanian universities in light of the Corona pandemic.

H01-2: there is no effect of inspirational motivation on achieving the effectiveness of decisions in terms of its dimensions (quality of the decision, the timing of decision) in private Jordanian universities in light of the Corona pandemic.

H01-3: there is no effect of intellectual excitement on achieving the effectiveness of decisions in terms of its dimensions (quality of the decision, the timing of decision) in private Jordanian universities in light of the Corona pandemic.

H01-4: there is no effect of empowerment on achieving the effectiveness of decisions in terms of its dimensions (quality of the decision, the timing of decision) in private Jordanian universities in light of the Corona pandemic.

H02: There is no effect of transformational leadership in terms of its dimensions (ideal effect, inspirational motivation, intellectual excitement, and empowerment) on the psychological capital in Jordanian private universities.

H03: There is no effect of the psychological capital on the effectiveness of decisions in terms of its dimensions (quality of the decision, the timing of decision) in private Jordanian universities in light of the Corona pandemic.

H04: There is no effect of transformational leadership in terms of its dimensions (ideal effect, inspirational motivation, intellectual excitement, and empowerment) on achieving the effectiveness of decisions in terms of its dimensions (quality of the decision, the timing of decision) in the presence of the psychological capital as an intermediate variable in private Jordanian universities in light of the Corona pandemic.

5 Literature Review

The concept of leadership is one of the behavioral concepts that sparked a lot of discussion and disagreement as the definitions differed in administrative thought, but they agree in their entirety about a set of steps or procedures that include employing the available resources to influence others in a way that supports their contribution to achieving specific goals [2]. Leadership is defined as the ability to influence a group in order to achieve a vision or set of goals [3]. Lussier et al. [4] also believes that leadership is the process of influencing leaders' behavior and behavior of subordinates to achieve organizational goals through change centered around some common interests, where the behavior of followers is directed by the leader, meaning that leadership occurs when Leads, persuades and influences others to work towards achieving the goals of the organization.

The concept of transformational leadership was first introduced by Burns [5] whose work focused on political leaders, and stated that the basis of transformational leadership is personal relationships, motivations, and values, as he referred to the leader who takes transformational leadership as his leadership as the leader who is looking for possible motivations for followers, seeks to meet their needs and gives them the opportunity to participate in the administrative process, which leads to mutual stimulation and upgrading that transforms followers into leaders. According to Osman and Siti Fatimah [6], transformational leadership generates incentives for subordinates and inspires them to reach beyond their expectations, which makes it an effective management method that school principals can pursue with the aim of effecting changes in the behavior of teachers and students. Transformational leadership is defined as "the ability the leader has to deliver any message to the organization, clearly articulating its future vision for the subordinates and working to motivate them by demonstrating high ethical behaviors and building trust and respect between the two parties to achieve the goals of the organization. It gives them the self-confidence to provide the best service or good with efficient use of resources while clearly supporting their social and emotional needs [7].

The importance of transformational leadership is evident in its influence with decisions, especially in the presence of crises such as the COVID-19 pandemic that transformational leadership works to empower individuals, develop their skills and enhance their self-confidence, and works to find self-reliant groups. Transformational leadership also causes a fundamental change in the behavior of followers and their values, in addition to that transformational leadership supports and sets values for the organization, changes in its culture and beliefs occur, and contributes to setting a standard of performance within an integrated and coherent system that makes it able to respond effectively to internal and external changes, in addition to that transformational leadership It helps workers in professional development and maintain a culture of professionalism and cooperation by talking to workers with each other, and through freedom of expression and planning in a collaborative manner, and encouraging joint responsibility, employee development, and effective communication.

Positive psychology emphasizes exploring the benefits and values of human psychology in a manner that enhances the flourishing quality of human life and contributes to the realization of optimal human potential for fully playing [8]. Luthans et al. [9] suggested a definition of psychological capital that claims to have four measurement combinations. The first structure is self-efficacy that requires an individual to make the necessary efforts to successfully accomplish difficult tasks. Followed by the second construction, it is optimism that attributes the positive attribution of the current and future victory. Third, it is the strength of an individual's will and hopes that he will be able to redefine methods to achieve goals. The latter relates to the flexibility that one can recover from a difficult situation quickly [10]. The content of psychological capital is related to the four types of positive psychological strength in unique positive psychology that can be operated in concrete terms—hope, self-efficacy, flexibility, and optimism. Thus, psychological capital is associated with positive psychological strength that leads to positive psychological energy. Otherwise, we can say that psychological capital is the positive psychological resources of the individual, which consist of structures (hope, self-efficacy, resilience, and optimism) or positive psychological state that can be raised and developed, rather than relatively stable features and difficult to be variable [11].

Authors in [12] concluded that psychological capital was directly related to performance, and indirectly to a performance by adapting with an emphasis on meaning and contentment. Additionally, psychological capital was directly related to contentment and highlighted the importance of this psychological construct in academia. Understanding the role played by adaptation and meaning-focused satisfaction in the relationship between psychological capital and performance may be useful for researchers and lecturers to design optimal evidence-based interventions to increase wellbeing and academic achievement [13].

Authors in [14] found that there is an effect of transformational leadership on employee motivation, and the study also found that motivation can be achieved by engaging workers in informal activities of social activities. The study also found a positive relationship between transformational leadership and other dependent variables. Authors in [15] concluded that there is an effect of the characteristics (sympathy, patience, intuition, equanimity) in developing confidence (self-efficacy) in private education schools, and the presence of a statistically significant effect of the characteristics of female leadership on the development of psychological capital and the study recommended the strengthening of interest in Ras Psychological money in its dimensions (hope, flexibility, optimism, self-efficacy).

Authors in [16] concluded that functional combustion is strongly linked to the performance of faculty members and that psychological capital increases his achievement and raises morale. The study recommended an increased interest in the personality traits of individuals, which leads to increased performance, stress relief, and emotional immobility. Authors in [17] concluded that project managers with high emotional intelligence enjoy managerial competencies and demonstrate transformational leadership behavior. The study recommended the necessity of developing and empowering project managers to increase their mental capabilities and administrative skills, which in turn are reflected in the project performance and success.

Authors in [18] concluded that there is a direct relationship of psychological capital to the level of the feeling of the faculty member and the quality of worklife, and there is no direct role for psychological capital in predicting organizational commitment. The study recommended the need to focus attention on the level of psychological capital among employees, and to develop a flexible and fair system of rewards and appreciation. Authors in [19] concluded that there is a strong positive relationship between teachers' perceptions of the manager's leadership practices and the ability of subordinates to make appropriate decisions. Authors in [20] concluded that there is a positive correlation between the level of school principals' practice of transformational leadership and the creative behavior of teachers and their ability to make decisions, and the hierarchical linear regression analysis indicated that increased teacher commitment affects the level of the relationship between transformational leadership and the creative behavior of teachers.

6   The Methodology of the Study

This type of studies is classified as an applied study, aimed at dealing with field problems, with regard to behavioral areas. In this study, researchers collected qualitative data, and then analyzed it in a quantitative manner. The researchers resorted to the descriptive approach as well as the analytical approach, with the aim of describing the role of psychological capital in the impact of transformational leadership on achieving effective decisions in private Jordanian universities in the light of the Corona epidemic. The study used many statistical methods in order to measure the variables of the study model.

6.1   The Population and the Sample of the Study

The population of the study included consist of the lecturers of the (6) private Jordanian universities in the capital Amman, the universities were chosen for Al-Ahliyya Amman University, the Middle East University, and Al-Ahliyya Amman University. Because of the researchers' inability to conduct a comprehensive survey as a result of the Corona epidemic, the researchers selected a simple random sample consisting of (130) lecturers at Jordanian private universities. The number of questionnaires that the researchers were able to retrieve was (125) questionnaires, which included (108) questionnaires that were complete and appropriate for the analysis.

6.2   The Sources of Data

The researchers relied on collecting the necessary data for this study on two sources, the first source representing secondary data through books and articles related to the study variables. The second source was the primary data that researchers obtained through the study administration, the questionnaire, which was distributed to university lecturers at Jordanian private universities.

6.3   Reliability Test

The researchers used the Cronbach alpha coefficient test in order to test the stability of the study tool, where the result is statistically acceptable if the Cronbach alpha value is greater than (0.60) [21], and the researchers prepared Table 1, which shows Cronbach's values Alpha, which was as follows:

The values of the Cronbach alpha coefficient for the instrument as a whole (0.936), and the value of the Cronbach alpha in relation to questionnaire questions related to

Table 1 Reliability test of study tool

Number	Variable	Reliability coefficient (α)
1	Ideal effect	0.717
2	Inspirational motivation	0.752
3	Intellectual excitement	0.743
4	Empowerment	0.789
Transformational leadership		0.868
5	Quality of the decision	0.779
6	Timing of decision	0.823
Effectiveness of decisions		0.894
7	Psychological capital	0.755
All paragraphs		0.936

transformational leadership (0.868), while the Cronbach alpha for the effectiveness of the decision (0.894), and for the psychological capital (0.755). None of the values was less than (0.70), and this leads researchers to judge that the study instrument was consistent and reliable.

7 The Study Analysis

7.1 Descriptive Statistics of the Demographic Characteristics of the Respondents

By reviewing Table 2, the following can be observed:

1. Males were the majority of the study sample, with a percentage of 87% of the total respondents. This is due to the tendency of males in Jordan more than females to graduate studies, which is reflected in the percentage of male faculty members in universities.
2. The members of the study sample within the age group who are over 50 years old formed the lowest group of the study sample, at a rate of (9.3%), which indicates that the members of the teaching staff in private universities are from the youth category.
3. The percentage of Ph.D. holders (90.75%) was among the respondents, while the respondents were holders of a master's degree (9.3%). The reason for this disparity is due to the requirements of the accreditation body for higher education institutions that do not allow the master's degree to exceed 20% of the total lecturers in Jordanian universities.
4. The percentage of respondents with experience ranging from 10 years to less than 15 years (60.2%). This result indicates that the study sample has sufficient

Table 2 Descriptive statistics of the demographic characteristics of the respondents

Variable	Group	Repetition	Percentage
Gender	Male	94	87
	Female	14	13
Age	30 ≤ 40	28	25.9
	40 ≤ 50	70	64.8
	50 ≥	10	9.3
Educational qualification	M.Sc.	10	9. 25
	Ph.D.	98	90.75
Years of experience	< 5 years	0.00	0.00
	5 ≤ 10	35	32.40
	10 ≤ 15	65	60.20
	15 ≥	8	7.40
	Total	108	100

experience to judge the effect of the prevailing leadership style on the quality of decisions.

7.2 Describe of Respondents Answers

The results in the Table 3 indicate:

Table 3 Mean, standard deviations, levels and relative importance of all dimensions of the study tool

Item		Mean	Standard deviation	Relative importance	Rank
Transformational Leadership	Ideal effect	3.96	0.486	High	3
	Inspirational motivation	3.86	0.469	High	4
	Intellectual excitement	4.20	0.426	High	1
	empowerment	4.14	0.477	High	2
	Transformational Leadership	4.04	0.368	High	
Effectiveness of decisions	Quality of the decision	4.13	0.469	High	1
	Timing of decision	4.01	0.521	High	2
	Effectiveness of decisions	4.07	0.420	High	
Psychological capital		4.06	0.504	High	

7.2.1   First: Transformational Leadership

The previous table showed high importance with regard to transformational leadership, the mean was 4.04 and the highest level of intellectual stimulation was then empowerment in the second place, the ideal effect in the third place, and the inspirational motivation in the fourth place and all dimensions were of the high relative importance.

7.2.2   Second: Effectiveness of Decisions

The previous table showed a high importance in relation to the effectiveness of administrative decisions in Jordanian private universities, the mean was 4.07 and the highest level of decision quality was then the timing of the decision in second place, and all dimensions were of the high relative importance.

7.2.3   Third: Psychological Capital

The previous table showed high importance with regard to psychological capital in Jordanian private universities, as the mean was 4.06.

7.3   Study Hypotheses Test

In order to test the study hypotheses, the researchers tested the first major hypothesis using "multiple and linear regression analysis". "Simple linear regression analysis" was also used to test the rest of the hypotheses and using "Path Analysis" to test the last major hypothesis.

H01: There is no effect of transformational leadership in terms of its dimensions (the ideal effect, inspirational motivation, intellectual excitement, and empowerment) on achieving the effectiveness of decisions in terms of its dimensions (quality of the decision, the timing of decision) in private Jordanian universities in light of the Corona pandemic.

The researchers used "simple linear regression analysis" to test the four sub-hypotheses, as shown in Table 4.

H01-1: there is no effect of the ideal effect on achieving the effectiveness of decisions in terms of its dimensions (quality of the decision, the timing of decision) in private Jordanian universities in light of the Corona pandemic.

Table 4 shows a positive correlation between the ideal effect and the effectiveness of decisions in Jordanian private universities. We also note that the ideal effect explains 52.4% of the variation in the effectiveness of decisions in Jordanian private universities, and based on the results of the analysis we can conclude that there is

Table 4 Results of testing the effect of the ideal effect on effectiveness of decisions private Jordanian universities

Dependent Variable	R	R^2	Calculated F	D.f	Sig F*	Variable	B	Standard error	Calculated T	Sig t*
Effectiveness of Decisions	0.724	0.524	177.588	1	0.000	ideal effect	0.534	0.040	13.326	0.000

*The effect is significant at the level ($p \leq 0.05$)

an important impact of the ideal effect on the effectiveness of decisions in Jordanian private universities in light of the Corona pandemic.

H01-2: there is no effect of inspirational motivation on achieving the effectiveness of decisions in terms of its dimensions (quality of the decision, the timing of decision) in private Jordanian universities in light of the Corona pandemic.

Table 5 shows a positive correlation between inspirational motivation and the effectiveness of decisions in Jordanian private universities. We also note that inspirational motivation explains 52.9% of the variation in the effectiveness of decisions in Jordanian private universities, and based on the results of the analysis we can conclude that there is an important impact of inspirational motivation on the effectiveness of decisions in Jordanian private universities in light of the Corona pandemic.

H01-3: there is no effect of intellectual excitement on achieving the effectiveness of decisions in terms of its dimensions (quality of the decision, the timing of decision) in private Jordanian universities in light of the Corona pandemic.

Table 6 shows a positive correlation between intellectual excitement and the effectiveness of decisions in Jordanian private universities. We also note that intellectual excitement explains 60% of the variation in the effectiveness of decisions in Jordanian private universities, and based on the results of the analysis we can conclude that there is an important impact of intellectual excitement on the effectiveness of decisions in Jordanian private universities in light of the Corona pandemic.

H01-4: there is no effect of empowerment on achieving the effectiveness of decisions in terms of its dimensions (quality of the decision, the timing of decision) in private Jordanian universities in light of the Corona pandemic.

Table 7 shows a positive correlation between empowerment and the effectiveness of decisions in Jordanian private universities. We also note that empowerment explains 32.2% of the variation in the effectiveness of decisions in Jordanian private universities, and based on the results of the analysis we can conclude that there is an important impact of empowerment on the effectiveness of decisions in Jordanian private universities in light of the Corona pandemic. The researcher used "multiple regression analysis" to test the effect of transformational leadership on achieving the effectiveness of decisions in private Jordanian universities in the light of the Corona pandemic, as shown in Table 8.

Table 8 shows a positive correlation between transformational leadership and the effectiveness of decisions in Jordanian private universities. We also note that transformational leadership explains 35.8% of the variation in the effectiveness of decisions in Jordanian private universities, and based on the results of the analysis we can conclude that there is an important impact of transformational leadership (the ideal effect, inspirational motivation, intellectual excitement, empowerment) on the effectiveness of decisions (quality of the decision, the timing of decision) in Jordanian private universities in light of the Corona pandemic.

H02: There is no effect of transformational leadership in terms of its dimensions (ideal effect, inspirational motivation, intellectual excitement, and empowerment) on the psychological capital in Jordanian private universities. The researcher used "multiple regression analysis" to test the effect of transformational leadership on

Table 5 Results of testing the effect of inspirational motivation on effectiveness of decisions private Jordanian universities

Dependent Variable	R	R^2	Calculated F	D.f	Sig F*	Variable	B	Standard error	Calculated T	Sig t*
Effectiveness of Decisions	0.727	0.529	180.890	1	0.000	inspirational motivation	0.717	0.053	13.450	0.000

*The effect is significant at the level ($p \leq 0.05$)

Table 6 Results of testing the effect of intellectual excitement on effectiveness of decisions private Jordanian universities

Dependent Variable	R	R²	Calculated F	D.f	Sig F*	Variable	B	Standard error	Calculated T	Sig t*
Effectiveness of decisions	0.775	0.600	241.987	1	0.000	Intellectual excitement	0.699	0.045	15.556	0.000

*The effect is significant at the level ($p \leq 0.05$)

Table 7 Results of testing the effect of empowerment on effectiveness of decisions private Jordanian universities

Dependent Variable	R	R^2	Calculated F	D.f	Sig F*	Variable	B	Standard error	Calculated T	Sig t*
Effectiveness of decisions	0.568	0.322	197.566	1	0.000	Empowerment	0.712	0.059	14.435	0.000

*The effect is significant at the level ($p \leq 0.05$)

Table 8 Results of the first main hypothesis test H01

Dependent Variable	R	R²	F	Sig F*	Transformational leadership	B	Standard error	Calculated	Sig t*
Effectiveness of Decisions	0.598	0.358	14.908	0.000	Ideal effect	0.195	0.073	2.680	0.009
					Inspirational motivation	0.100	0.113	0.883	0.379
					Intellectual excitement	0.356	0.098	3.634	0.000
					empowerment	0.216	0.099	2.174	0.032

*The effect is significant at level ($\alpha \leq 0.05$)

psychological capital in private Jordanian universities in the light of the Corona pandemic, as shown in Table 9.

Table 9 shows a positive correlation between transformational leadership and the psychological capital in Jordanian private universities. We also note that transformational leadership explains 59.9% of the variation in the psychological capital in Jordanian private universities, and based on the results of the analysis we can conclude that there is an important impact of transformational leadership on the psychological capital in Jordanian private universities in light of the Corona pandemic.

H03: There is no effect of the psychological capital on the effectiveness of decisions in terms of its dimensions (quality of the decision, the timing of decision) in private Jordanian universities in light of the Corona pandemic. The researcher used "multiple regression analysis" to test the effect of psychological capital on achieving on the effectiveness of decisions in private Jordanian universities in the light of the Corona pandemic, as shown in Table 10.

Table 10 shows a positive correlation between psychological capital and the effectiveness of decisions in Jordanian private universities. We also note that psychological capital explains 59.9% of the variation in the effectiveness of decisions in Jordanian private universities, and based on the results of the analysis we can conclude that there is an important impact of psychological capital on the effectiveness of decisions in Jordanian private universities in light of the Corona pandemic.

H04: There is no effect of transformational leadership in terms of its dimensions (ideal effect, inspirational motivation, intellectual excitement, and empowerment) on achieving the effectiveness of decisions in terms of its dimensions (quality of the decision, the timing of decision) in the presence of the psychological capital as an intermediate variable in private Jordanian universities in light of the Corona pandemic. The researcher used "path analysis" to test the direct and indirect impact of psychological capital in the relationship between the transformational leadership and the effectiveness of decisions in private Jordanian universities in the light of the Corona pandemic.

It is clear from Table 11 that (Chi2 = 13.993), and that Chi2 after dividing it by the degree of freedom is (4.664) which is less than the number 5 which means acceptance of the model, and that (GFI = 0.991) which is close to (1), and that (CFI = 0.982), which approximates (1), and (RAMSEA = 0.066) and approximates (0), which supports the quality of the model, and therefore approval of it.

It is clear from Table 12 that the direct impact of the ideal effect on psychological capital has reached (0.307) and that the direct effect of the inspiring motivation on psychological capital (0.209), and that the direct effect is intellectual excitement on psychological capital (0.401), and that the direct effect of empowerment on Psychological capital (0.294), and the direct effect of psychological capital on the effectiveness of decisions in Jordanian private universities (0.749). Which means that psychological capital positively affects the effectiveness of decisions in Jordanian private universities. Given that, the direct and indirect effect of each dimension of the transformational leadership is a significant effect and the psychological capital is a partial mediation. This confirms the existence of a positive role for psychological capital as an intermediate variable in the effect of the use of transformational

Table 9 Results of the test of the effect of transformational leadership on psychological capital in Jordanian private universities

Dependent variable	R	R^2 t	Calculated F	D.f	Sig F*	transformational leadership	B	Standard error	Calculated T	Sig t*
Psychological Capital	0.774	0.599	79.069	3	0.000	Ideal effect	0.329	0.068	4.865	0.000
						Inspirational motivation	0.197	0.091	2.159	0.032
						Intellectual excitement	0.355	0.076	4.665	0.000
						empowerment	0.299	0.082	3.756	0.000

*The effect is significant at the level ($p \leq 0.05$)

Table 10 Results of testing the effect of psychological capital on the effectiveness of decisions in private Jordanian universities

Dependent variable	R	R^2	Calculated F	D.F	Sig F*	Statement	B	Standard error	Calculated T	Sig t*
Effectiveness of Decisions	0.785	0.615	257.692	1	0.000	Psychological Capital	0.658	0.041	16.053	0.000

*The effect is significant at the level ($p \leq 0.05$)

Table 11 Results of the "path analysis test" for the direct and indirect effect of the transformational leadership on the effectiveness of decisions in the presence of psychological capital as an intermediate variable

Model Fit						
Statement	Chi^2	D.f	GFI	CFI	RAMSEA	Sig. level
The transformational leadership	13.993	3	0.991	0.982	0.061	0.000

GFI—"Goodness Fit index"
CFI—"Comparative Fit Index"
RAMSEA—"Root mean square error of approximation"

Table 12 Direct and indirect effects coefficients - H04

The estimates					
Path	The direct effect			The indirect effect	
	Direct effect factor	C.R	Sig level	indirect effect factor	Sig level
Ideal effect → Psychological capital	0.307	5.715	0.000	0.230	0.010
Inspirational motivation → Psychological capital	0.209	4.302	0.028	0.157	0.010
Intellectual excitement → Psychological capital	0.401	8.270	0.000	0.300	0.010
Empowerment → Psychological capital	0.294	5.113	0.000	0.218	0.010
Psychological capital → Effectiveness of decisions	0.749	25.068	0.000		

leadership on the effectiveness of decisions in private Jordanian universities when studying the dimensions of using transformational leadership combined.

8   Conclusions and Recommendation

The results of the tests revealed the interest of Jordanian private universities in the psychological capital of the lecturers. In addition, there is a high level of realization of the importance of the leadership style, in addition to the practice of transformational leadership on a large level among these universities. And the attention of Jordanian universities in taking efficient decisions that contribute to addressing various types of crises, including the current crisis linked to the Corona pandemic. Furthermore, results indicated that there is an effect of transformational leadership in terms of its all dimensions (ideal effect, inspirational motivation, intellectual excitement, and empowerment) on achieving the effectiveness of decisions in terms of its dimensions (quality of the decision, the timing of decision) in private Jordanian universities in light of the Corona pandemic. With regard to the role of psychological capital, the results of the study indicated that there is an indirect impact of the transformational leadership on the effectiveness of decisions in private Jordanian universities with the existing of psychological capital as an intermediate variable among in private Jordanian universities in light of the Corona pandemic. And given to the direct and indirect effect of each dimension of the transformational leadership is a significant effect, psychological capital is a partial mediation, and this confirms the existence of a positive role for psychological capital as an intermediate variable in the effect of the use of transformational leadership on the effectiveness of decisions in private Jordanian universities when studying the dimensions of using transformational leadership combined.

Transformational theories compare the responses of leaders to transactional and transformative situation. Transactional leaders do not have a wide-angle view; instead, they carry out their daily routine tasks in a specific way. A transformational leader works from the perspective of the big picture [22]. The current study has contributed to many practical effects that may be important in order to draw the attention of leaders in Jordanian private universities, as the study provides valuable inputs to leaders who have demonstrated their application of transformational leadership dimensions, and the extent of its contribution to containing crises, as it is expected that leaders will realize the role of style. Leadership and the role of taking care of psychological capital in dealing with crises in the future, as well as preparing leaders in a way that enables them to adapt to various health conditions and other conditions in the future.

This study recommends a set of recommendations, the most important of which is the need to benefit from the concepts of transformational leadership as a role model for managing private Jordanian universities in the light of the Corona pandemic. In addition to developing new programs in the management of private Jordanian universities that help to transform leaders' understanding of psychological capital to reach effective court decisions that seek to serve decision-making in universities, especially in times of crisis, as is the case in light of the Corona pandemic. Also, the necessity of improving the actual practice of leadership with its dimensions, the ideal effect, individual moral, inspirational motivation, and empowerment, in a way that contributes to enhancing the participation of workers in decision-making processes, which affects the psychological aspect of them and leads to access to efficient decisions.

References

1. Riolli, L., Savicki, V., Richards, J.: Psychological capital as a buffer to student stress. Psychology 3(12A), 1202–1207 (2012)
2. Ackerman, L.: Leadership vs. managership. Leadership Organ. Dev. J. 6(2), 17–19 (1985)
3. Robbins, S.P., Judge, T.A.: Organizational Behavior, 15th edn. Pearson, Upper Saddle River, New Jersey (2013)

4. Lussier, R.N., Achua, C.F.: Leadership: Theory, Application, Skill Development, 2nd edn. South-Western: Division of Thomson Learning (2004)
5. Burns, J.M.: Leadership. Harper & Row Publishers, New York (1978)
6. Osman, H.O., Siti Fatimah, B.: Transformational leadership and organizational commitment—A literature review. Sains Humanika 2(2), 127–133 (2014). https://doi.org/10.11113/sh.v2n 2.426
7. Cheung, M., Wong, C.: Transformational leadership, leader support and employee creativity. Leadership Organ. Dev. J. 32(7), 656–672 (2010). https://doi.org/10.1108/014377311111 69988
8. Åslund, C., Starrin, B., Nilsson, K.W.: Psychosomatic symptoms and low psychological well-being in relation to employment status: The influence of social capital in a large cross-sectional study in Sweden. Int. J. Equity Health 13(1), 1–19 (2014)
9. Luthans, F., Avolio, B.J., Avey, J.B., Norman, S.M.: Positive psychological capital: measurement and relationship with performance and satisfaction. Pers. Psychol. 60, 541–572 (2007)
10. Culbertson, S.S., Fullagar, C.J., Mills, M.J.: Feeling good and doing great: the relationship between psychological capital and well-being. J. Occup. Health Psychol. 15, 421–433 (2010)
11. Luthans, F., Youssef, C.M., Sweetman, D.S., Harms, P.D.: Meeting the leadership challenge of employee well-being through relationship PsyCap and health PsyCap. J. Leadership Organ. Stud. 20(1), 118–133 (2013). https://doi.org/10.1177/1548051812465893
12. Ortega-Maldonado, A., Salanova, M.: Psychological capital and performance among undergraduate students: the role of meaning-focused coping and satisfaction. Teach. Higher Educ. 23(3), 390–402 (2018). https://doi.org/10.1080/13562517.2017.1391199
13. Datu, J.A.D., Valdez, J.P.M.: Psychological Capital predicts academic engagement and well-being in Filipino high school students. Asia-Pac. Educ. Res. 25, 399–405 (2015). https://doi.org/10.1007/s40299-015-0254-1
14. Ahmad, F., Abbas, T., Latif, S., Rasheed, A.: Impact of Transformational leadership on employee motivation in telecommunication sector. J. Manage. Policies Pract. 2(2), 11–25 (2014)
15. Salih, A.A., Al-Dulaimi, M.S.: Women's leadership characteristics and its impact in developing the psychological capital: test of the moderating role of experience power (analytical study from the perspective of subordinates in private education schools in Amman). Int. J. Econ. Manage. Sci. 6, 1–10 (2017)
16. Rehman, S., Qingren, C., Latif, Y., Iqbal, P.: Impact of psychological capital on occupational burnout and performance of faculty members. Int. J. Educ. Manage. 31(4), 455–469 (2017)
17. Maqbool, R., Sudong, Y., Manzoor, N., Rashid, Y.: The impact of emotional intelligence, project managers' competencies, and transformational leadership on project success: an empirical perspective. Project Manage. J. 48(3), 58–75 (2017)
18. Abosaif, M.S.A.: The role of quality of work life in the relationship between psychological capital and organizational commitment among faculty members at the Faculty of Education. Taif University, Int. J. Res. Educ. Sci. 1(4), 109–181 (2018)
19. Saka, C.: A Study of the Relationship Between Teachers' Perceptions of Principals' Transformational Leadership Practices and Teacher Morale in Elementary Schools in Southern Ontario. Masters' theses of University of Windsor Scholarship at UWindsor, Canada (2018)
20. Ismail, A., Mydin, A.: The impact of transformational leadership and commitment on teachers' innovative behaviour. Adv. Soc. Sci. Educ. Hum. Res. (ASSEHR) 304, 426–430 (2018)
21. Sekaran, U.: Research Methods for Business, 6th edn. Wiley, Hoboken, NJ (2013)
22. Foster, D.T., Goertzen, B.J., Nollette, C., Nollette, F.P.: Emergency Services Leadership: A Contemporary Approach. Jones & Bartlett Learning, Burlington, MA (2011)

Using Innovation and Ecosystem to Transform Covide-19 from Crisis to an Opportunity: The Case of Jordanian Program "Faculty for Factory" (FFF)

Yousef Al Abdallat

Abstract The National Program for Linking Academic Institutes with Industrial Sectors (Faculty for Factory) is distinguished as the first of its kind in Jordan and the region, and one of the best programs in the world. Its idea is characterized as a non-traditional one, applicable to the mechanism of linking academic institutions to the productive community represented by the industrial sector. Development of an applied model of cooperation between industrial companies (SMEs) and academic institutions. During the life cycle of the program in 2020, Covid-19 virus appeared in the world, where the program worked to develop its goals and the nature of its work to confront the virus in all possible academic ways, starting from directing students and academics to work under one goal in proposing and adopting various ideas that contribute to supporting the Jordanian government in facing the virus. FFF also participated in various governmental tasks such as providing the crisis cell responsible for planning in the Crisis Management Center with the academics who are required to complete the planning process according to the best scientific practices, in addition to participating in the various committees formed by the Prime Minister and concerned with the promotion of manufacturing operations.

Keywords Linking academic institutions with industrial sectors and technology transfer · Smes support · Covid-19 · Corona-virus · Initiatives · Socio-economic

Y. Al Abdallat (✉)
The National Program for Linking Academic Institutes With Industrial Sectors "Faculty for Factory" Program, Department of Industrial Engineering, The University of Jordan, Amman 11942, Jordan
e-mail: abdallat@ju.edu.jo

© The Author(s), under exclusive license to Springer Nature Switzerland AG 2021   245
M. T. Alshurideh et al. (eds.), *The Effect of Coronavirus Disease (COVID-19) on Business Intelligence*, Studies in Systems, Decision and Control 334,
https://doi.org/10.1007/978-3-030-67151-8_14

1   Introduction

1.1   Faculty for Factory (FFF) Program

Faculty for factory program was created and launched in 2003 from the University of Jordan as a concept aimed to strengthen the role of applied scientific research carried out by academic institutions in cooperation with industry sector to serve the national economy, support the technological component in our pursuit of industrial development and enhance competitiveness [1]. The program has now turned from its small house to become a national program and effective tool to link industrial companies to academic institutions.

The FFF program has important and significant effect in strengthening the relationship between industry and academic institutions—approved by each party—as a mechanism able to clarify the desired true needs of the industry as well as overcoming the obstacles and difficulties that were always preventing development to the desired levels.

1.2   Program's Philosophy

By industrial companies to open their doors to benefit from the experiences of faculty members (participants) Throughout visits and studying the situations, opportunities are identified for cooperation projects and joint development between the industrial companies on the one hand and the participant on the other hand, during a period of three months; which is the summer holiday period of the faculty members supported by local and international organizations, in order to strengthen the links between academics and industry.

1.3   Program's Goals

Developing Practical model for cooperation between industry and universities to help participants (faculty members) from various Jordanian universities with different specialties and the Jordanian industrial companies from different sectors, creation of a genuine partnership between them.

Increasing the contribution of the participants in the preparation of the research projects and practical implementation in the Jordanian industrial companies according to their actual needs.

1.4 Mechanism of FFF

- Identifying a company and visiting it.
- Identifying problems.
- Preparing diagnostic study with the agreement of the parties.
- Assessing the diagnostic study.
- Implementing the project after approval.
- Reporting what has been accomplished.
- Assessing the final report.
- Paying the Allotment

1.5 Terms of Reference

Participant obtains his/her remuneration at the end of the project and if the final report with the sustainability of the project proposal is submitted in addition to a letter from the beneficiary company expresses its satisfaction with the performance of the participant.

The research and development project proposal to continue working with the company is considered to be indication of success for the project and, if it is not submitted, participant have to indicate the reasons and must be accepted by the Commission.

1.6 Through the Faculty-for-Factory Program

Faculty members will be recruited to apply their expertise to a specific SME's needs and will propose solutions, as well as introduce the SME to useful local sources of ongoing technical expertise or services.

Faculty may invite selected students to work on the project and will share their practical experience as case studies with students in their classrooms, stimulating interest in the real world of business.

Recent graduates may also be involved to capitalize on their current knowledge while exposing them to the business challenges and opportunities faced by Jordanian SMEs and industries. Both faculty and students will become more aware of and responsive to areas of applied research and employment that are in demand by the Jordanian private sector. SMEs benefit from the latest knowledge and applied research techniques.

1.7 Innovation, Entrepreneurship and Commercialization Culture of Research Products

Our era witnesses an exponential growth in data, information, knowledge and technology, avoiding participation in such venture; lacking in sources and lagging countless rounds behind the world will be the consequences of such denial. Lately, the globe is moving forward accelerating its progress by the aid of innovation and entrepreneurship in a diversity of fields and specializations [2]. This international trend has created—and still—new courses, specializations, and schools in many international, well-known and classified educational institutions [3]; to stimulate and quicken the creation and implementation of innovative and out-of-box solutions with higher effectiveness and efficiency than traditional and ordinary ones.

Jordan is well-known by its qualified and well-educated human resources, however; to keep up with the previously mentioned transaction; an urgent need to upgrade by enriching these resources and boosting their capacity to be continually updated with the modern methods and tactics.

Studies have proven that to boot the national production represented by Gross Demotic Production (GDP) and GDP per capita, a necessity for not only educating human resources, but also enabling them to initiate their own business – in the associated fields [2], advance new jobs, and generate revenues. Moreover, this modern technique has proven it effectiveness in the harmonization between local and global market needs and demand; with what is being taught in educational institutions that feed these markets to fulfill its needs.

Accordingly, Faculty for Factory program is now attempting to spread this culture in the Jordanian community by advancing a new course to be taught to sections in all faculties in University of Jordan; to take advantage of the newly developed ideas that would contribute in national progress; socially, environmentally and economically, and to utilize new generation's perspectives toward mitigating and maybe eliminating local and global unsolved issues.

2 Covid-19 Crisis

In just four months, the world has undergone dramatic transformations, many people died, and several thousands more have become sick, from a coronavirus that was already obscure before showing up in the city of Wuhan in December 2019. For many other people who have not come down with the illness, their whole lifestyle has changed by it, these individuals who have not been infected with the virus have changed their entire lifestyle, and have become afraid throughout the day of many things, including the possibility of being infected with the virus and exposure to symptoms they have seen through social media platforms or possibility to lost their jobs, including also the loss of members of their families.

During this short period, the world was subjected to fundamental transformations, including the complete disruption of air traffic and the disappearance of usually lifestyle, where most of the world-famous cities turned to ghost cities. Students missed their schools, as most countries, including weak countries, were transformed during a short period to completely e-learning school [4, 5] and all that's to aim reduce the spread of the pandemic.

Covid-19 virus worked as a business accelerator, but in a way that exceeds human capabilities. People's fear of infection with the virus, along with their fear for their future and their professional lives, and what the virus can lead to form high unemployment rates [6] and the displacement of many individuals, as happened in previous global crises, all of these led to the creation of pressure on mind of the entire workforce, and it was imperative for all countries to maintain their security and system, by develop creative solutions to confront this pandemic.

After declaring Covid-19 as a pandemic by the World Health Organization (WHO), a large number of countries-initiated restrictions on the movement of their citizens [7] in a way that disrupts all aspects of public life, including the wheel of production in many world countries which has a significant impact, especially on those industrialized countries, and this was apparent through the low levels of carbon dioxide in the atmosphere, where pollution levels in New York decreased by 50% compared to the same time of the year, and also 25% in China compared to the same time of the year, which is considered one of the most productive countries in the world [8].

3 Covid-19 Crisis in Jordan (Success Stories)

As (COVID-19) pandemic accelerates, global health care systems have become deals with many potentially infectious patients they are seeking testing and health care, to prevent infection for medical teams working to fight the pandemic, patients relies on effective use of personal protective equipment (PPE) (gloves, face masks, air-purifying respirators, goggles, face shields, respirators, and gowns). A critical shortage of all of these is projected to develop or has already developed in areas of high demand [9, 10]. In order to reduce the prevalence rates of infection, countries obligated their citizens to wear face masks, for the purpose of movement and to re-start the rotation of the economic wheel because it is became more important to preserve human life by avoiding economic systems collapsing that may have a greater impact on human lives than the pandemic itself [11]. These things combined have increased the demand for health protection equipment, specifically in face masks.

Re-arranging the process of opening different economic sectors and focusing on productive sectors with a high impact during the pandemic's spread, such as the pharmaceutical and health production sector, has become an important matter for many countries around the world. Different countries in the world have turned to innovative solutions through which they work to manufacture these different materials, and given the shortage of raw materials needed for industrialization, due to the monopoly of the various countries that produce these raw materials for themselves in order to preserve the health of their citizen, As concepts of innovation became necessary for different sectors of life in different countries of the world to face this pandemic [12].

Jordan's response to this pandemic was distinct compared to other countries around the world, as Jordan has sought from the first day to increase production of medical and sterile materials and achieve self-sufficiency among them for fear of a possible shortfall around the world, which resulted in an increase in the production of face masks to a million face masks per day in April according to the statement of Jordan's Minister of State for Media Affairs.

This increase in production came as a result of several measures taken by the Jordanian government in various fields with a view to maintaining the achievement of sustainable development goals (SDGs) Which is one of the concepts that is important to maintain throughout the pandemic period [13], Among these measures taken by the government is to facilitate the process of granting permits for masks production lines, in addition to facilitating the process of granting permits for clothing production lines to convert to the production of masks, before of these Jordan has guaranteed the agility of supply chains to obtain the largest possible amount of raw materials that are sufficient throughout the Covid-19 pandemic, as this is essential in managing any crisis around the world [14].

In line with the goals of SDGs, Jordan has not abandoned its role in ensuring the provision of medicines needed by different countries of the world to face Covid-19, since from the first moments of the announcement of (Hydroxychloroquine) as a drug that can be used for the purposes of mitigating the severity of Covid-19 [15], Al-Hikma Drug Company has worked to increase its production of this drug and provide it to a group of Arab countries, in addition to King Abdullah II Center for Design and Development (KADDB) also began producing high-quality medical protection materials such as disinfection tunnels and masks, The factories of the Fine and HiGeen also worked to increase their production of sterilization products in order to achieve internal sufficiency and also export to countries abroad, including China.

The various Jordanian factories began to develop their innovative capabilities to search for solutions on the possibility of manufacturing Ventilator and this came according to the government's directives to activate the process of scientific research related in this field through a set of scientific competitions held throughout the pandemic period.

This combination of measures that Jordan has taken throughout this crisis by focusing on achieving the values of marketing intelligence by confirming Jordan's role in producing medical and sterile materials, has contributed to putting Jordan on the map of countries that can become industrialized countries in this field after the end of the pandemic [16]. In addition to continuous follow-up by His Majesty King Abdullah II, through which his majesty focused on increasing production at the highest levels and preserving Jordan's role within the countries that share the world in combating the pandemic despite all the financial challenges Jordan faces today.

To reduce the economic effects of the Covid-19 crisis, the Jordanian government worked to form four specialized teams that include representatives from the public and private sectors, and representatives from the relevant military, security, and civil authorities, in follow-up to what His Majesty King Abdullah II ordered, to expand local capabilities to produce medicines, And the manufacture of medical equipment and supplies, and the promotion of food and agricultural industries, and ensure the continuity and sustainability of production.

According to this decision, four committees were formed as follows: Firstly, the Higher Steering Committee, headed by the Prime Minister, undertakes the tasks of directing the technical committees for national priorities, approving the national production plan, and providing the necessary requirements to implement this plan and the Pharmaceutical Manufacturing Committee, headed by the Director General of the General Food and Drug Corporation And it is responsible for determining the priorities of production of pharmaceutical products in the short and long term, and other related tasks and the Food Manufacturing Committee, headed by the Chairman of the Board of Directors of the Industrial Cities Company, and is responsible for setting the priorities for production of the product Food in the short and long term, and the associated tasks and other serious medical equipment and supplies manufacturing and sterilizers committee, headed by Chief Executive of the King Abdullah II Design and Development is responsible for determining the priorities of production of these materials, in the short and long term, and other tasks related to this process.

The continuation of Jordan in this way pushes us to invest the pandemic as a starting point after the end of this epidemic, especially as Jordan has maintained its health system from collapse despite the threats that have become apparent on the Jordanian economy after the end of the crisis, which can be avoided by anticipating the industrial steps as Jordan did during the production and export of various medical and sterile instruments.

As we previously talked about FFF programs, it becomes clear to us that this program plays an important role in dealing with the Corona crisis as part of the Jordanian committee. that stat from the first moment, which the program worked on adopting an initiative launched by graduate students, which aims to create a platform in which the owners of factories and companies facing challenges (due to the crisis caused by Corona) can meet with a group of experts and academics who are ready to provide support to decision makers to take the right decisions and actions based on science, studies and different opinions from a committee with local and international experiences in various field. This Jordanian youth initiative consists of 4 highly professional teams, where each team will include experts in a specific field.

These teams will include academics, engineers, technicians, and Jordanian with local and international experiences. In addition to specialized lawyers, experts in the Jordanian law, labor law, workers and property and other commercial laws, in addition to economic experts. The teams will be divided as the following order:

The First team: Is a group of engineers and technical experts. Members of the team will provide all the engineering and technical consultations and support to help overcome the challenge that are facing and will face the factories in Jordan during this difficult period and after the quarantine period.

The Second team: Is a team made up of economic experts? This team will provide financial and tax advice to the factories to avoid losses and pass this difficult period as it will be offering support in how to avoid the various financial challenges known to the industrial sector, such as dealing with banks and other companies, tax and more.

The third team: Is a group of legal experts who will provide all legal advice to the Jordanian factories, to avoid layoffs or exploiting employers? The experts will clarify all legal matters the bind these factories at the present time.

The fourth team: Is a youth team of experts in management in general, and risk management in particular? Members of the team will provide the best solution that they deem appropriate for each manufacturer in these difficult circumstances. They will also advise on how to get past this stage, restore desired productivity, and help solve various matters to reduce losses after this crisis.

FFF participated in the crisis cell responsible for planning the resources needed by the government during dealing with the Covid-19 crisis, through academics providing the necessary solutions using modern statistical techniques such as "System dynamics in forecasting the number of future cases", where the meetings were held periodically at the Jordanian Crisis Management Center, Participation in this working group resulted in the publication of a research paper by the working group under the title "Estimation of the probable outbreak size of novel coronavirus (COVID-19) in social gathering events and industrial activities" [17].

During the second semester of the year 2020/2020, the program also presented the "Entrepreneurship and Innovation" course at Department of Industrial Engineering level and at the university of Jordan level, where students were directed through the course to think about creative solutions that contribute to overcoming the Corona crisis and solutions were provided by students and these project currently under study and planning stage.

In addition to, the Higher Council for Science and Technology (HCST) did some of activities that related to COVID-19 and these activities summarized in the following points.

- **Higher Education Digital E-Learning Initiative (Teach Yourself)**

As a response to the urgent request of HRH Prince El-Hassan the chairman of the Higher Council for Science and Technology (HCST) an initiative was designed to enable university students in Jordan to continue learning through the internet while staying at home to salvage what remains of the current semester. Moreover, this program will prepare the stage to institutionalize digital learning within a "blended" system of distant-digital learning mixed with in-class learning for future semesters.

This initiative comes as a collaborative effort of the HCST, the Ministry of Higher Education and Scientific Research (MoHESR), the Accreditation and Quality Assurance Commission for Higher Education Institutions (AQACoHEI), and some of the Jordanian universities.

The main idea of this program is to utilize the already existing digital learning materials by many of the universities in addition to using open source materials available at the internet. Some of the Jordanian universities have already started this program for the current semester for their students. What is needed is to open up these university systems to all students for all universities.

- **HCST—Higher Council for Sciences and Technology**

In response to the global COVID 19 crises, the Industrial Research and Development Fund, which is affiliated to the HCST, has launched an initiative during the curfew to support the national and global response efforts. IRDF funded joint project proposals between the industry and academia.

The projects that have been supported within this initiative are as follows:

- A New COVID-19 Airborne Sanitizer.
- "Surfactant-Free" Safe Foaming Liquid Composition for Personal Care.
- A Novel Microbiological Safety Cabinet.
- COVID-19 Immunoassay Test Based on Lab on Disc Technology.

4 Conclusions, Limitation, and Future Directions

During the implementation of the FFF program, a true partnership between the Jordanian universities and industry has been achieved and a range of elements that may contribute to developing and sustaining partnerships between education and industry has been identified. An important issue related to this context has emerged, which is that factories always seek to develop their relationship with universities for the purpose of enhancing their practices and developing them to be optimized, and this has resulted in the need for universities to be available at all times to answer the inquiries of these factories within a specific communication range that does not conflict with the academic goals of universities.

The academia in Jordan has managed this by lunched Faculty-For-Factory (FFF) program that has a level of self-government, although it is still closely linked to universities' policies. Faculty-For-Factory offers regular communication, both formal and informal, between the universities and their industrial partners which play an integral role in developing an understanding the limitation, requirements and expectations of the two partners. In addition to, FFF also contributes to developing a long-term relationship between the two entities. The key finding from this study is that both parties agreed upon the program as a mechanism for promoting the relationship between academia and industry. Regardless of the explanation(s) for the faculty member/factory differences, the results provide support for the establishment of a formal needs/expectations process between prospective companies (factories) and prospective faculty members. FFF program is now attempting to spread Innovation, Entrepreneurship & Commercialization of research products culture in the Jordanian community, and technology transfer.

The program played an important role in combating the Covid-19 pandemic in Jordan through its approach to promoting the concepts of emphasis the participated process of student and academic in various workplaces, starting with motivating them to produce ideas through different academic subjects, and moving to participation in crisis cells and government committees that aim to bridge gaps and use different sciences to develop response processes.

Limitation

There are a set of limitation that faced the study, which are the rapid changes that accompanied the spread of Covid-19 virus worldwide and the accompanying governmental measures that determined the possibility of applying the program to some different ideas, and the accompanying difficulty of facing academics as a result of the sudden transformation to e-learning that made their efforts concentrated towards supporting students and direct them instead of the overall trend towards supporting the industry.

Future direction

1. Work on the idea of Industrial Professor to take advantage of existing expertise and artistic industry in Jordanian universities and open universities and research centers for industrial experience, according to the instructions and conditions must be available in the Industrial Professor.
2. Focus on industrial sectors Separately, as well as the training of students and prepare them for this specific sector thoughtfully selected for employment in accordance with the requirements of this sector is based on private sector and Supported by practical skills and knowledge required by this sector.
3. Commercialization of research products which is the graduation projects or Entrepreneurial ideas or support start-up projects and so on.
4. Building an integrated approach that enables different governments to confront any future pandemic.

References

1. Al-Abdallat, Y., Tutunji, T.A.: Faculty for Factory program: A University-industry link in Jordan. In: 2012 9th France-Japan & 7th Europe-Asia Congress on Mechatronics (MECATRONICS)/13th Int'l Workshop on Research and Education in Mechatronics (REM). IEEE (2012)
2. Schilling, M.A., Shankar, R.: Strategic Management of Technological Innovation. McGraw-Hill Education (2019)
3. د.ع.م. عثمان, واقع ومستقبل التعليم التقني في الاردن وعلاقته and العبداللات, ا.ي بالصناعة الوطنية. برامج التعليم التقني ومدى تحقيقها لمتطلبات الصناعة وحاجات سوق العمل, 2014
4. Hodges, C., et al.: The difference between emergency remote teaching and online learning. EDUCAUSE Review (2020). https://er.educause.edu/articles//3/the-difference-between-emergency-remote-teachingand-online-learning

5. Zhou, L., et al.: 'School's Out, But Class' On', The Largest Online Education in the World Today: Taking China's Practical Exploration During The COVID-19 Epidemic Prevention and Control As an Example. The Largest Online Education in the World Today: Taking China's Practical Exploration During The COVID-19 Epidemic Prevention Control As an Example (2020)
6. McKibbin, W.J., Fernando, R.: The global macroeconomic impacts of COVID-19: Seven scenarios (2020)
7. Prem, K., et al.: The effect of control strategies to reduce social mixing on outcomes of the COVID-19 epidemic in Wuhan, China: a modelling study (2020)
8. Henriques, M.J.B.N.: Will Covid-19 have a lasting impact on the environment (2020)
9. Livingston, E., Desai, A., Berkwits, M.J.J.: Sourcing personal protective equipment during the COVID-19 pandemic (2020)
10. Ranney, M.L., Griffeth, V., Jha, A.K.: Critical supply shortages—the need for ventilators and personal protective equipment during the Covid-19 pandemic (2020)
11. McKee, M., Stuckler, D.J.N.M.: If the world fails to protect the economy, COVID-19 will damage health not just now but also in the future, pp. 1–3 (2020)
12. World Health Organization: COVID 19 Public Health Emergency of International Concern (PHEIC). Global research and innovation forum: towards a research roadmap (2020)
13. Zhou, C., et al.: COVID-19: challenges to GIS with big data (2020)
14. Antoniadis, I., et al.: Business Intelligence during times of crisis: adoption and usage of ERP systems by SMEs. Procedia-Soc. Behav. Sci. **175**(Supplement C), 299–307 (2015)
15. Colson, P., et al.: Chloroquine and hydroxychloroquine as available weapons to fight COVID-19. Int. J. Antimicrob. Agents **105932**(10.1016) (2020)
16. Al-Tammemi, A.a.B.: The battle against COVID-19 in Jordan: an early overview of the Jordanian experience. Front Public Health **8**, 188 (2020)
17. Saidan, M.N., et al.: Estimation of the probable outbreak size of novel coronavirus (COVID-19) in social gathering events and industrial activities (2020)

Understanding the Influences of Cognitive Biases on Financial Decision Making During Normal and COVID-19 Pandemic Situation in the United Arab Emirates

S. F. Shah, M. T. Alshurideh, A. Al-Dmour, and R. Al-Dmour

Abstract The purpose of the study is to identify the effects of behavioral/psychological factors i.e. overconfidence, anchoring bias, loss aversion, herding effect on financial decision making, in both normal situation (NS) and COVID-19 pandemic uncertain situation (CVD-19) separately. This paper used a qualitative method by using semi-structured interview (virtual and physical) and all fifteen interviewers were based in the United Arab Emirates. Whereas, the results of the study show that in NS all the factors have a positive significant relationship with financial decision-making. But on the other hand, in the CVD-19 uncertain situation, majority of the factors has a negative effect on financial decision making, except for overconfidence, which shows positive effect. Though the limitation was a time constraint, limited factors, and CVD-19 itself is a stressful environment and people do not prefer to participate in interviews. Finally, the future research direction is to increase sample size and factors to understand the financial decision impact on performance.

Keywords Behavioral finance · Behavioral factor · Cognitive biases · Decision making · COVID-19 pandemic

S. F. Shah · M. T. Alshurideh (✉)
Department of Management, College of Business Administration, University of Sharjah, Sharjah, United Arab Emirates
e-mail: malshurideh@sharjah.ac.ae; m.alshurideh@ju.edu.jo

S. F. Shah
e-mail: U19106061@sharjah.ac.ae

M. T. Alshurideh
Department of Marketing, School of Business, The University of Jordan, Amman, Jordan

A. Al-Dmour
School of Business, Al Ahilyya Amman University, Amman, Jordan
e-mail: Ahmadmourh@gmail.com

A. Al-Dmour · R. Al-Dmour
School of Business, The University of Jordan, Amman, Jordan
e-mail: rand.aldmour@ju.edu.jo

1 Introduction

One of the essential parts in life is decision making, either small or big, every decision expose to a certain degree of risk. Also, the good decision does not give guarantee of good outcome. Such expediency yield good results as complexity is growing in risk management. A degree of understanding human behavior and technological improvement to support and imitate the cognitive process has improved decision making in numerous situations [1]. There are three types of uncertainty forms in decision making under uncertain situation, the first is insufficiency of understanding the situation, the second is absence of complete information and the third is confusing alternatives which are not properly distinguished [2, 3]. Whereas, Chhapra et al. [4] explains that firm's investment decision must rely on available data and analysis to determine and predict the financial position of the firm, yet sometimes the decision is made based on current news or mood. Similarly, Bakar and Yi [5] suggests that decision-makers must integrate all the essential available information correspondence to efficient market theory and should be unbiased in analyzing stock and markets' securities. However, human nature is subjective, where individual make systemic errors in judgment and decision making, i.e. overconfidence means investor heavily depends on an existing experience that leads to distortions [6]. Likely, [7–9] found the humans do not behave rationally in economics and stock market anomalies. So, the behavioral finance took place and became the most crucial field for research on psychological influence in human irrational decision making in the financial markets [5, 10–14]. The researchers try to understand how cognitive biases and emotions affect individual investor decision-making behaviors [15, 16].

In the first stage, this study took most consistently appeared cognitive bias (overconfidence, anchoring, and loss aversion) and herding impact on the decision-making process in a normal situation and the second stage will consider CVD-19 as a mediator to understand the impact of those biases on decision-making.

1.1 Research Questions

1. Are these psychological factors influencing investment decision making?
2. How are psychological factors affecting decision-making?

   (a) During NS
   (b) During CVD-19

3. What are the several aspects of investment decisions?

1.2 Objectives

To highlight the influence of different psychological biases on investment decision making.

1.3 Scope of the Study

Behavioral finance is a combination of psychology, economics, and finance knowledge. The outcome of this study could help investors and financial advisors to understand cognitive biases, able to identify and avoid them in order to improve investment decisions. Hence, it is essential to analyze these cognitive biases and distinguish the factors frequently influencing the financial decision-making. The rest of the paper is arranged as follows: The second part follows the literature review. The third part explains the model and propositions. The fourth part discusses and explains the results of the study. The fifth part is the limitations and future directions. The final part is the conclusion.

2 Literature Review

2.1 Overconfidence

It is an inclination to exaggerate the degree of how a person is accurate [17, 18]. It frequently appears in the decision-making process and many investors get a negative outcome in return due to flaws in this bias [19]. Similarly, overconfident investors are likely to overestimate the available information and potential gain, lead them to over trade which results in lower returns [20]. Besides, Oberlechner and Osler [21] disclosed that "investors are prone to be overconfident in two aspects: they underestimate ambiguity and overestimate their capabilities". Moreover, Barber and Odean [22] investigated that the common stock investment by using 35,000 households from a huge discount brokerage house and compare men's and women's overconfidence level, as a result, men traded overly and more overconfident than women. Also, many authors rely on past successes in business deals which makes investors overconfident and directs to highly increase trading volume in future periods [23].

2.2 Anchoring Bias

Anchoring is a cognitive bias in which decisions are made by undertaking an initial anchor [24]. Same as, the moment an individual makes an estimate from the beginning initial value and keep in mind till the final answer, whereas, an individual is not aware of the cognitive bias, so the correction is not enough. Any adjustment in starting value can cause a different estimate so the anchoring depends on initial values [25–27]. Likewise, expert predicts that releases are anchored bias using previous months releases [28]. Therefore, when people make judgment under uncertainty are more likely to use relative thinking rather than critical or absolute thinking [19]. The finding of [26] explains that the uncertainty level of information plays a significant role in anchoring behavior in real estate deals.

2.3 Loss Aversion

It is believed that loss brings disappointment and people try to prevent losses to avoid later regret [25]. Loss aversion is the mental penalty where investor put the same extent of losses or gains [29]. Although [30, 31] stated that loss effects are twice greater than gain, any attractive business deals involve potential loss might get rejected. The analysis of the intra-day trading reveals that loss in the morning leads to high risk-taking behavior in the market than an afternoon [32]. Finally, the market conditions influence on loss aversion, as investors become more loss averse in boom period (bull market) than in recessions (bear markets) [33].

2.4 Herding Effect

It is like a kind of imitation behavior, leading to an alignment of the individual's behavior [34–37]. According to scholars, a person conceals his or her own beliefs and imitate actions of others is known as herding behavior. Though, one of the reason is the investors imitate others as they believe that herding help to obtain valuable and reliable information [38]. The other reason for herding is the individual who has low intellectual level and self-belief, so prefer to imitate the high ability peers to improve professional performance evaluation and reputation [39]. Correspondingly, Anum [40] acknowledged that investors in stock markets favor their investment decision similar to the majority buying and selling transactions, on the contrary, rational investors behave oppositely. As a result of herding, the market becomes inefficient leads to create risky bubbles in the market.

2.5 COVID-19 Pandemic Uncertain Situation

It is SARS-CoV-2 disease declared as a pandemic, affected hundreds of people in more than 100 countries and it is almost unstoppable and created globally uncertain environment across the world [41]. World health organization [42] reported numbers

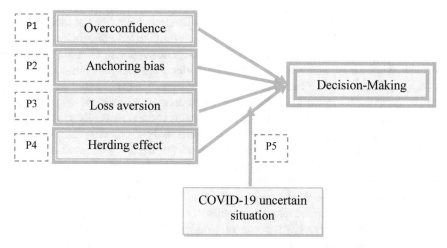

Fig. 1 The proposed new model

of fatality were aged between 60 and 89 years old people. The market in China dropped sharply due to additional Lunar new year holidays, the Shanghai composite index benchmark fell 7.7% dropping nearly $375 billion, which is sharpest fall in 1-day since August 2015 [43]. Craven et al. [44] imply global growth for 2020 is expected to slow down due to current COVID-19 pandemic situation, the reviewed two scenarios (1) Economic impact and recovery, and (2) Provide insight and best practice for business leaders to navigate the uncertain and rapidly changing situation. They also said that the coronavirus disaster is a story with an ambiguous ending. What is clear is that the human impact is by now tragic and that companies have an obligation to act right now to protect their employees, solve business issues and risks, and help to alleviate the outbreak in no matter what ways they can.

Based on the above explanation, the study model (Fig. 1) and suggested propositions can be seen below:

P1: The positive impact of overconfidence on investment decision-making.
P2: The positive effects of anchoring bias on investment decision-making.
P3: The negative effects of loss aversion on investment decision-making.
P4: The positive effects of herding effect on investment decision-making.
P5: The COVID-19 pandemic uncertain situation has negative impact on investment decision-making.

3 The Study methodology

The study implemented a qualitative method to collect primary data and analyzed the data by using the summative method. Once keywords, terms and research model is already identified through reviewing existing literature or individual interest in a particular area [45–47]. This research model forms through systemic review method by reviewing related prior literature papers. This mode supported to establish semi-structured questions, which concentrates on the selected factors only [48–51]. Moreover, [47, 52–54] explain that the quantification process helps to count exact word and by implying the meaning at the back of the interviews. Yet, [46, 55–57] said that coding is depending on the research question and it is done manually via a simple process. The qualitative research method used in this study was used by many scholars such as [58–60] to include, the first is "interviews (structured, semi-structured and unstructured) to study views, phenomenon, practice and personal perspective. The second is "focus group interview" to explore attitude, behaviors, and notions of normative behavior. The third is an analysis of documents, script or text, and reports i.e. official reports, financial reports, statistics, media or news journals, and websites to understand in extensively phenomenon [61–65]. Kvale [66] exclaimed an interview is the form of interaction between two or more people in which they exchange ideas, knowledge and focus on social events for research data on mutual interest and centrality. This study adopted a semi-structured interview to focus on the behavioral factors impact on decision-making and later stage understanding the influence CVD-19 (as a mediator) between factors and decision-making. The semi-structured interview has predetermined questions but the wordings are not fixed and the order of questions is changeable, it uses open response question [67].

3.1 The Frequency Table

Table 2 contains the study factors and the frequency counted positive and negative responses separately for each factor, which is the result of 15 interviews related field. The positive coded responses consider a positive opinion or statement similarly negative reflects a negative view of the study [46, 48].

Table 1 Summary of the factors and related codes of the study

No.	Factors	Code
1	Overconfidence	OVE
2	Anchoring bias	ANC
3	Loss aversion	LOA
4	Herding effect	HER
5	Decision making	DEM

Table 2 The incidence of positive and negative in frequency

Factors	Total	%	Positive (+)	Negative (−)	Balance of normal situation (+, −)	Total	%	Positive (+)	Negative (−)	Balance of COVID-19 uncertain situation (+, −)
Overconfidence	100	26	90	−10	80	100	28	55	−45	10
Anchoring bias	91	24	50	−41	9	98	27	30	−68	−38
Loss aversion	77	20	55	−22	33	85	24	25	−60	−35
Herding effect	115	30	75	−40	35	75	21	5	−70	−65
Total	383	100	270	−113	157	358	100	115	−243	−128

Table 3 Semi-structure interview questions and frequencies

	Subfactors	Normal situation			COVID-19 Pandemic Situation		
		Positive	Negative	Balance	Positive	Negative	Balance
OVE							
1. Do you agree that your skills, experience, and knowledge of business can help you to outperform the market? [68]	Skill, Experience, power, and knowledge	50	−5	45	15	−25	−10
2. What do you say about this statement, "I have the talent to evaluate the new information in the market"? [69]	Confidence on new information	40	−5	35	40	−20	20
ANC							
3. To what extent you agree on this statement. I am keen to look for new market information that is related to my investment plan and I use the new information to evaluate my investment decision	Value of Initial information and initial estimated value	20	−11	9	15	−45	−30

(continued)

Table 3 (continued)

	Subfactors	Normal situation			COVID-19 Pandemic Situation		
		Positive	Negative	Balance	Positive	Negative	Balance
4. Are you estimating potential business performance based on past business performance? [70]	Value of past information	30	-30	0	15	-23	-8
LOA							
5. When making stocks selling decisions, what are the consideration (s) drive your decisions of selling those stocks? [71]	Loss effects twice higher than gain	40	-6	34	10	-20	-10
6. Do Investors take a closer look at the growth potential than the loss potential of an investment?		15	-16	-1	15	-40	-25
HRE							
7. Do other investors' decisions of choosing stock types have impact on your investment decisions? [15]	Imitation and rely on market experts	40	-5	35	0	-30	-30
8. What do you think; investors follow the market trend is more soothing and reassuring?	Follow the majority	35	-35	0	5	-40	-35
Total Counts		270	-113	157	115	-243	-128

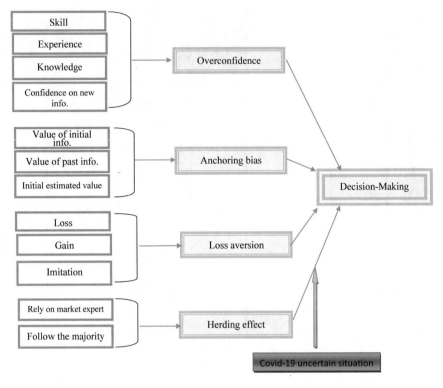

Fig. 2 Research Model

Based on the Tables 2 and 3, the study model can be drawn again in more details as seen below in Fig. 2.

4 Discussion

In NS the outcome of the above analysis in Tables 2 and 3 revealed that psychological factors (OVE, ANC, LOA and HER) has a positive effect on the financial DEM. Whereas, the total positive incidents were 270, while negative were −113. In the bar chart (Fig. 3) shows that OVE has a highly positive impact on DEM however, it got second place in the pie chart (26% in Fig. 4). Moreover, the HER represents the first place influencing on DEM because HER is common practice in the financial sector, where it is placed in the second position in effecting positively the DEM. Overall, the other factors ANC represents 24% and LOA 20% of total results but considering the positive effect LOA is in third and ANC comes last.

In CVD-19 situation, the outcome of the above analysis in Tables 2 and 3 revealed that psychological factors (OVE, ANC, LOA, and HER) has a negative impact on

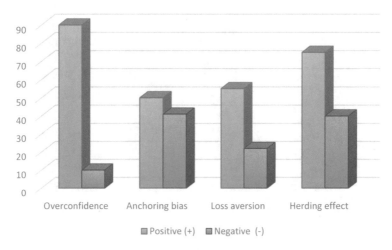

Fig. 3 The effects of factors in normal situation (Positive or Negative)

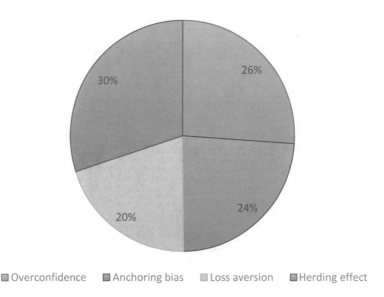

Fig. 4 Percentage influence of each factor in normal situation

the financial decision making. While, the total positive incidents were 115, while negative were −243. In the bar chart (Fig. 5) and pie chart (Fig. 6) ANC represents 27% of this part of the study and the first highest negative impact on the financial DEM. Whereas, HER effects show the second highest negative effect however it represents lowest percentage (21%) of this topic. Moreover, the OVE represent the first highest percentage (28%) of the topic but results in a positive impact on financial

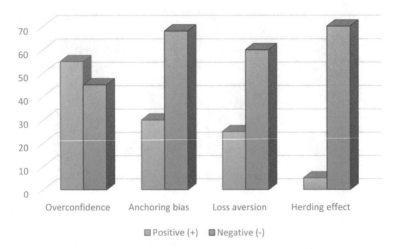

Fig. 5 The effects of factors in COVID-19 uncertain situation (Positive or Negative)

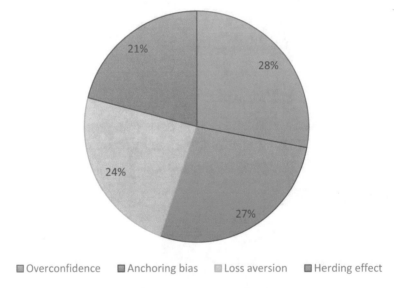

Fig. 6 Percentage influence of each factor in normal situation

DEM. Though the LOA comes second last (24%) and has negative impact on financial DEM, it has been noticed that people prefer gain over the loss.

4.1   Overconfidence (OVE)

It has been noticed that the interviews highly rely on personal skills, experience, and knowledge to perform better in the financial market, so the OVE represents a highly positive significant impact on financial DEM in a NS. Similarly, several authors found the same result [72–74]. The result supports the offered proposition "P1: The positive impact of overconfidence on investment decision making". On the other hand, during CVD-19, shows the same result as above mentioned but moderate impact. As it seems that people rely heavily on their skill, knowledge, and experience even in an uncertain situation. One of the participants said, *"COVID-19 situation appeared in news in January-2020, whereas intellectual and risk-taker investors immediately respond by promoting and investing in online businesses and took other initiatives related customer satisfaction to sustain in the business"*. In this case, investors highly relied on gut feeling rather than rationality in financial DEM.

4.2   Anchoring Bias (ANC)

Figure 3 illustrates that the ANC has a positive moderate impact on financial DEM, which reveals that participant relies moderately on initially presented information, value, and past information. In addition, ANC represents 24% (Fig. 4) of the total frequencies of the study in normal situation. A respondent said, *"I rely on initial information and value but also the investigation of the sources and reliability of information is important for me"*. The result supports this recommended proposition "P2: The positive effects of anchoring bias on investment decision making". On the contrary, in CVD-19 ANC reveals a highly negative impact on DEM among other factors and it also exemplifies the 27%, second highest of the total frequency of the responses. Whereas Costa et al. [75] claim that positive information enhance the impact on anchoring and negative information reduces the impact of anchoring.

4.3   Loss Aversion (LOA)

LOA represents the lowest part of the topic; similarly, like other factors it has a significant positive relationship with financial DEM. As a result, participants rely more on gain of the investment and avoid losses. One of the respondents stated, *"I believe in the law of supply and demand and prefer to study market-based knowledge. If I have stock which makes a loss and my investigation about that firm shows potential bankruptcy or financial constraint, I will definitely sell that stock"*. The result contradicts this proposed proposition "P3: The negative effects of loss aversion on investment decision making". On the opposing side, the impact of LOA is negatively related to financial DEM during CVD-19 and it contributes to third portion of CVD-19 situation of the study incidence. A participant indicated that "*I wouldn't sell any stock in this current situation; I will prefer to wait till things settle down and also purchase more stock with lower price for potential gain*". As Dominic and Gupta [25] tells that the effects of loss are twice larger than gain. Furthermore, Benartzi and Thaler [76] rely on loss aversion hypothesis and clarify many investors prefer to bond over the stocks because the bond is an agreement with low risk involved [77, 78].

4.4   Herding Effect (HER)

The HER is placed in the second positive significant influence in financial DEM and denotes the first highest (30%) of this area of the current study sample. The HER is commonly found in the financial institutions, where investors developed trust and avoid the risk of losing by finding alternative approach. Likewise, [79, 80] and [5, 81] found mixed results. A participant said, "*I prefer to follow intellectual, successful, trusted and close colleagues while making any financial decision*". The result supports the suggested proposition "P4: The positive effects of herding effect on investment decision making". While in CVD-19 situation, the result is completely opposite than in NS in DEM. Also, it illustrates the negative influence financial DEM and lowest of CVD-19 section of the study frequency.

4.5   Financial Decision-Making (DEM) in CVD-19 Situation

All the participants mentioned that the CVD-19 affected the process of decision-making. More prudently evaluate each decision to avoid any wastage of cash flow. Some believed that to sell the stocks could be a wise decision but on the other hand, a participant prefers to purchase more at a low price and wait until the market turns to stable. Though [82] divulge that financial distress started to be noticed and more are coming, the market has fallen into a stressful environment and keep on increasing the panic. In the current situation, investors do not try protecting themselves by selling and some are looking for a profit in this breakdown situation by taking short positions (potentially infinite). Therefore, people will turn towards central banks (infinite pockets), can potentially lend vast amounts. Moreover, the author said the CVD-19 continued spreading in the United States and caused huge shock to the real economy. It turned out to be harder for the businesses to earn export dollar, stability in financial markets, growth, and financing. Almost all the psychological factors have a negative relationship with financial DEM during CVD-19 situation except overconfidence. The result supports this proposition "P5: The COVID-19 pandemic uncertain situation has a negative impact on investment decision making".

5 Conclusion

This study has undertaken the most frequently appeared behavioral/psychological factors impact on financial DEM in both NS and CVD-19. The main factors are the Overconfidence (OVE), Anchoring Bias (ANC), Loss Aversion (LOA) and Herding effect (HER). The result reveals that in NS all the factors show significant positive relationship in financial DEM in different percentages. However, the second result during CVD-19 illustrates the negative influence on financial DEM, except "overconfidence" shows a positive effect. It can be observed that CVD-19 has changed the human behavior of purchasing and selling, especially in the financial sector.

5.1 Limitations and Future Research

This study offers a few significant contributions to the area. However, it has a few limitations. The first limitation was the time constraint, the second was few factors and the last was CVD-19 itself, which limits the sample size. For future research, it would be more valuable to increase the sample size and factors to get an in-depth understanding of this uncertain situation's impact on financial decision-making and outcomes.

References

1. Buchanan, L., Connell, A.O.: A brief history of decision making. Harv. Bus. Rev. **84**(1), 32 (2006)
2. Lipshitz, R., Klein, G., Orasanu, J., Salas, E.: Taking stock of naturalistic decision making. J. Behav. Decis. Mak. **14**(5), 331–352 (2001)
3. Lipshitz, R., Strauss, O.: Coping with uncertainty: a naturalistic decision-making analysis. Organ. Behav. Hum. Decis. Process. **69**(2), 149–163 (1997)
4. Chhapra, I.U., Kashif, M., Rehan, R., Bai, A.: An empirical investigation of investors behavioral biases on financial decision making. Asian J. Empir. Res. **8**(3), 99–109 (2018)
5. Bakar, S., Yi, A.N.C.: The impact of psychological factors on investors' decision making in Malaysian stock market: a case of Klang Valley and Pahang. Procedia Econ. Financ. **35**, 319–328 (2016)
6. Ritter, J.R.: Behavioral finance. Pacific-Basin Financ. J. **11**(4), 429–437 (2003)
7. Babajide, A.A., Adetiloye, K.A.: Investors' behavioural biases and the security market: an empirical study of the Nigerian security market. Account. Financ. Res. **1**(1), 219–229 (2012)
8. Bashir, T., Javed, A., Usman, A., Meer, U.I., Naseem, M.M.: Empirical testing of heuristics interrupting the investor's rational decision making. Eur. Sci. J. **9**(28) (2013)
9. Ashurideh, M., Al Kurdi, B., Abu Hussien, A., Alshaar, H.: Determining the main factors affecting consumers' acceptance of ethical advertising: a review of the Jordanian market. J. Mark. Commun. **23**(5), 513–532 (2017)
10. Bazerman, M.H., Moore, D.A.: Judgment in Managerial Decision Making. Wiley, New York (1994)
11. Duxbury, D.: Behavioral finance: insights from experiments II: biases, moods and emotions. Rev. Behav. Financ. (2015)

12. Duxbury, D.: Behavioral finance: insights from experiments I: theory and financial markets. Rev. Behav. Financ. (2015)
13. Hogarth, R.: Judgment and Choice. Wiley, Chichester, UK (1987)
14. Russo, J.E., Schoemaker, P. J. H.: Confident Decision Making: How to Make the Right Decision Every Time. Piatkus (1991)
15. Kengatharan, L., Kengatharan, N.: The influence of behavioral factors in making investment decisions and performance: Study on investors of Colombo Stock Exchange, Sri Lanka. Asian J. Financ. Account. **6**(1), 1 (2014)
16. Keswani, S., Dhingra, V., Wadhwa, B.: Impact of Behavioral Factors in Making Investment Decisions and Performance: Study on Investors of National Stock Exchange. Int. J. Econ. Financ. **11**(8), 1–80 (2019)
17. Fischhoff, B., Slovic, P., Lichtenstein, S.: Knowing with certainty: The appropriateness of extreme confidence. J. Exp. Psychol. Hum. Percept. Perform. **3**(4), 552 (1977)
18. Fabre, B., François-Heude, A.: Optimism and overconfidence investors' biases: a methodological note. Finance **30**(1), 79–119 (2009)
19. Fıçı, G.: Behavioural biases in investment decisions: a comparison between finance professionals and individual investors in Turkey. Bahçeşehir University Social Science Institute (2014)
20. Jain, R., Jain, P., Jain, C.: Behavioral biases in the decision making of individual investors. IUP J. Manag. Res. **14**(3), 7 (2015)
21. Oberlechner, T., Osler, C.L.: Overconfidence in currency markets. Available SSRN 1108787 (2008)
22. Barber, B.M., Odean, T.: Boys will be boys: gender, overconfidence, and common stock investment. Q. J. Econ. **116**(1), 261–292 (2001)
23. Gervais, S., Odean, T.: Learning to be overconfident. Rev. Financ. Stud. **14**(1), 1–27 (2001)
24. Dominic, C., Gupta, A.: Psychological Factors Affecting Investors Decision Making (2020)
25. Tversky, A., Kahneman, D.: Prospect theory: an analysis of decision under risk. Econometrica **47**(2), 263–291 (1979)
26. Chang, C.-C., Chao, C.-H., Yeh, J.-H.: The role of buy-side anchoring bias: Evidence from the real estate market. Pacific-Basin Financ. J. **38**, 34–58 (2016)
27. Al-Gasaymeh, A., Almahadin, A., Alshurideh, M., Al-Zoubid, N., Alzoubi, H.: The Role of economic freedom in economic growth: evidence from the MENA region. Int. J. Innov. Creat. Chang. **13**(10), 759–774 (2020)
28. Campbell, S.D., Sharpe, S.A.: Anchoring bias in consensus forecasts and its effect on market prices. J. Financ. Quant. Anal. **44**(2), 369–390 (2009)
29. Barberis, N., Huang, M.: Mental accounting, loss aversion, and individual stock returns. J. Finance **56**(4), 1247–1292 (2001)
30. Merkle, C.: Financial loss aversion illusion. Rev. Financ. **24**(2), 381–413 (2020)
31. Ngoc, L.T.B.: Behavior pattern of individual investors in stock market. Int. J. Bus. Manag. **9**(1), 1 (2014)
32. Coval, J.D., Shumway, T.: Do behavioral biases affect prices? J. Finance **60**(1), 1–34 (2005)
33. Hwang, S., Satchell, S.E.: How loss averse are investors in financial markets? J. Bank. Financ. **34**(10), 2425–2438 (2010)
34. Yang, J., Cashel-Cordo, P., Kang, J.G.: Empirical research on herding effects: case of real estate markets. J. Account. Financ. **20**(1) (2020)
35. Shefrin, H.: Beyond greed and fear: understanding behavioral finance and the psychology of investing. Oxford University Press on Demand (2002)
36. Welch, I.: Herding among security analysts. J. Financ. Econ. **58**(3), 369–396 (2000)
37. Musshoff, O., Hirschauer, N.: A behavioral economic analysis of bounded rationality in farm financing decisions. Agric. Financ. Rev. (2011)
38. Shabgou, M., Mousavi, A.: Behavioral finance: behavioral factors influencing investors' decisions making. Adv. Soc. Humanit. Manag. **3**(1), 1–6 (2016)
39. Kallinterakis, V., Munir, N., Radovic-Markovic, M.: Herd behaviour, illiquidity and extreme market states: evidence from Banja Luka. J. Emerg. Mark. Financ. **9**(3), 305–324 (2010)

40. Anum, B.A.: Behavioral Factors and their Impact on Individual Investors Decision Making and Investment Performance (2017)
41. Remuzzi, A., Remuzzi, G.: COVID-19 and Italy: what next? Lancet (2020)
42. World Health Organization: WHO director-general's opening remarks at the media briefing on covid-19—11 March 2020 (2020)
43. Ayittey, F.K., Ayittey, M.K., Chiwero, N.B., Kamasah, J.S., Dzuvor, C.: Economic impacts of Wuhan 2019-nCoV on China and the world. J. Med. Virol. (2020)
44. Craven, M., Liu, L., Mysore, M., Wilson, M.: COVID-19: Implications for Business. McKinsey & Company (2020). https://www.mckinsey.com/business-functions/risk/our
45. Al Kurdi, B., Alshurideh, M., Salloum, S.A., Obeidat, Z.M., Al-dweeri, R.M.: An empirical investigation into examination of factors influencing university students' behavior towards elearning acceptance using SEM approach. Int. J. Interact. Mob. Technol. 14(02), 19–41 (2020)
46. Alshurideh, D.M.T.: A qualitative analysis of customer repeat purchase behaviour in the UK mobile phone market. J. Manag. Res. 6(1), 109 (2013)
47. Hsieh, H.-F., Shannon, S.E.: Three approaches to qualitative content analysis. Qual. Health Res. 15(9), 1277–1288 (2005)
48. Ghannajeh, A.M., et al.: A qualitative analysis of product innovation in Jordan's pharmaceutical sector. Eur. Sci. J. 11(4), 474–503 (2015)
49. Alshurideh, M.: Do we care about what we buy or eat? A practical study of the healthy foods eaten by Jordanian youth. Int. J. Bus. Manag. 9(4), 65 (2014)
50. Alshurideh, M., Gasaymeh, A., Ahmed, G., Alzoubi, H., Kurd, B.: Loyalty program effectiveness: theoretical reviews and practical proofs. Uncertain Supply Chain Manag. 8(3), 599–612 (2020)
51. AlMehrzi, A., Alshurideh, A., Al Kurdi, B.: Investigation of the key internal factors influencing knowledge management, employment, and organisational performance: a qualitative study of the UAE hospitality sector. Int. J. Innov. Creat. Chang. 14(1), 1369–1394 (2020)
52. Alhashmi, S.F.S., Alshurideh, M., Al Kurdi, B., Salloum, S.A.: A systematic review of the factors affecting the artificial intelligence implementation in the health care sector. In: Joint European-US Workshop on Applications of Invariance in Computer Vision, pp. 37–49 (2020)
53. Al Kurdi, B., Alshurideh, M., Salloum, S.: Investigating a theoretical framework for e-learning technology acceptance. Int. J. Electr. Comput. Eng. 10(6), 6484–6496 (2020)
54. Kurdi, B., Alshurideh, M., Alnaser, A.: The impact of employee satisfaction on customer satisfaction: theoretical and empirical underpinning. Manag. Sci. Lett. 10(15), 3561–3570 (2020)
55. AlShurideh, M., Alsharari, N.M., Al Kurdi, B.: Supply chain integration and customer relationship management in the airline logistics, vol. 9, no. 02, pp. 392–414 (2019)
56. Ghannajeh, A., et al.: A qualitative analysis of product innovation in Jordan's pharmaceutical sector. Eur. Sci. J. 11(4), 1857–7881 (2015)
57. Yousuf, H., Zainal, A.Y., Alshurideh, M., Salloum, S.A.: Artificial intelligence models in power system analysis. In: Artificial Intelligence for Sustainable Development: Theory, Practice and Future Applications, Springer, pp. 231–242.
58. Ashurideh, M.: Customer Service Retention—A Behavioural Perspective of the UK Mobile Market. Durham University (2010)
59. Kurdi: Healthy-Food Choice and Purchasing Behaviour Analysis: An Exploratory Study of Families in the UK. Durham University (2016)
60. AlShamsi, M., Salloum, S.A., Alshurideh, M., Abdallah, S.: Artificial Intelligence and Blockchain for Transparency in Governance. In: Artificial Intelligence for Sustainable Development: Theory, Practice and Future Applications, pp. 219–230. Springer
61. Al-Jarrah, I.M., Al-Zu'bi, Z.M.F., Jaara, O.O., Alshurideh, M.: Evaluating the impact of financial development on economic growth in Jordan. Int. Res. J. Financ. Econ. 94, 123–139 (2012)
62. Assad, N.F., Alshurideh, M.T.: Investment in context of financial reporting quality: a systematic review. WAFFEN-UND Kostumkd. J. 11(3), 255–286 (2020)

63. Assad, N.F., Alshurideh, M.T.: Financial reporting quality, audit quality, and investment efficiency: evidence from GCC economies. WAFFEN-UND Kostumkd. J. **11**(3), 194–208 (2020)
64. Salloum, S.A., Alshurideh, M., Elnagar, A., Shaalan, K.: Machine learning and deep learning techniques for cybersecurity: a review. In: Joint European-US Workshop on Applications of Invariance in Computer Vision, pp. 50–57 (2020)
65. Salloum, S.A., Alshurideh, M., Elnagar, A., Shaalan, K.: Mining in educational data: review and future directions. In: Joint European-US Workshop on Applications of Invariance in Computer Vision, pp. 92–102 (2020)
66. Kvale, S.: InterViews: An Introduction to Qualitive Research Interviewing. Sage (1996)
67. Robson, C.: Real World Research: A Resource for Social Scientists and Practitioner-Researchers, vol. 2. Blackwell Oxford (2002)
68. Rahman, M., Gan, S.S.: Generation Y investment decision: an analysis using behavioural factors. Manag. Financ. (2020)
69. Metawa, N., Hassan, M.K., Metawa, S., Safa, M.F.: Impact of behavioral factors on investors' financial decisions: case of the Egyptian stock market. Int. J. Islam. Middle East. Financ. Manag. (2019)
70. Furnham, A., Boo, H.C.: A literature review of the anchoring effect. J. Socio. Econ. **40**(1), 35–42 (2011)
71. Le Luong, P., Thi Thu Ha, D.: Behavioral factors influencing individual investors' decision-making and performance: A survey at the Ho Chi Minh Stock Exchange (2011)
72. Kim, H.S., Jang, S.: CEO overconfidence and firm performance: the moderating effect of restaurant franchising. Cornell Hosp. Q., 1938965519899926 (2020)
73. Chen, Y.-R., Ho, K.-Y., Yeh, C.-W.: CEO overconfidence and corporate cash holdings. J. Corp. Financ., 101577 (2020)
74. Na, J., Kim, B., Sim, J.: COO's overconfidence and the firm's inventory performance. Prod. Plan. Control, 1–15 (2020)
75. Costa, D.F., de Melo Moreira, B.C., de Melo Carvalho, F., Silva, W.S.: Anchoring effect in managerial decision-making in accountants and managers: an experimental study. REBRAE **11**(3), 425–445 (2018)
76. Benartzi, S., Thaler, R.H.: Myopic loss aversion and the equity premium puzzle. Q. J. Econ. **110**(1), 73–92 (1995)
77. Mehra, R., Prescott, E.C.: The equity premium: a puzzle. J. Monet. Econ. **15**(2), 145–161 (1985)
78. Kadan, O., Tang, X.: A bound on expected stock returns. Rev. Financ. Stud. **33**(4), 1565–1617 (2020)
79. Chang, C.-H., Lin, S.-J.: The effects of national culture and behavioral pitfalls on investors' decision-making: Herding behavior in international stock markets. Int. Rev. Econ. Financ. **37**, 380–392 (2015)
80. Ghalandari, K., Ghahremanpour, J.: The effect of market variables and herding effect on investment decision as factor influencing investment performance in Iran. J. Basic Appl. Sci. Res. **3**(3), 313–318 (2013)
81. Dewan, P., Dharni, K.: Herding behaviour in investment decision making: a review. J. Econ. Manag. Trade, 1–12 (2019)
82. Gourinchas, P.-O.: Flattening the pandemic and recession curves. Mitigating COVID Econ. Cris. Act Fast Do Whatever, p. 31 (2020)

The Effect of Covid-19 Pandemic on Business Systems' Innovation and Entrepreneurship and How to Cope with It: A Theatrical View

K. A. Alameeri⬭, M. T. Alshurideh⬭, and Barween Al Kurdi⬭

Abstract This study investigates the issue of coronavirus effects on business intelligence systems regarding business innovation and entrepreneurship. Even though the Covid-19 pandemic has become the most significant crisis in the modern world and has negatively influenced numerous life spheres and business activities, it is possible to mention a few positive effects that might be used by different business organizations. Thus, the paper analyzes the relevant literature and shows that many positive ones follow short-term negative effects. The used the literature review approach to review many studies to answer the study questions and support its claims mentioned previously. The findings reveal that the coronavirus stimulates individuals and businesses to mobilize their efforts to overcome the existing Covid-19 pandemic challenge in the spheres of business innovation and entrepreneurship. The present paper also mentioned a set of advices related to how to cope with dealing with such a pandemic and mentioned a set of practical and theoretical implications for businesses in the near future.

Keywords Coronavirus · Innovation · Entrepreneurship · Business intelligence · Organizational change · Organizational learning · Professional people

K. A. Alameeri (✉) · M. T. Alshurideh
Department of Management, College of Business Administration, University of Sharjah, Sharjah, United Arab Emirates
e-mail: Kalameeri@hct.ac.ae

M. T. Alshurideh
e-mail: malshurideh@sharjah.ac.ae; m.alshurideh@ju.edu.jo

M. T. Alshurideh
Department of Marketing, School of Business, The University of Jordan, Amman, Jordan

B. Al Kurdi
Department of Management, Faculty of Economics and Administrative Sciences, The Hashemite University, Zarqa, Jordan
e-mail: barween@hu.edu.jo

© The Author(s), under exclusive license to Springer Nature Switzerland AG 2021
M. T. Alshurideh et al. (eds.), *The Effect of Coronavirus Disease (COVID-19) on Business Intelligence*, Studies in Systems, Decision and Control 334,
https://doi.org/10.1007/978-3-030-67151-8_16

1 Introduction

A coronavirus pandemic is one of the most significant events that has recently happened to the whole world. The outbreak started in Wuhan, China in 2019 and soon spread across the globe, affecting the people's health and paralyzing numerous life spheres. The pandemic has introduced new phenomena, including social distancing, self-isolation, and quarantine, and it is challenging to state whether these preventive measures will eventually disappear in the near future. That is why millions of people from various countries suffer from the pandemic and its consequences.

According to the information above, it is not a surprise that Covid-19 has affected every sphere of life, and business is not an exception. It is so because governments took appropriate measures to minimize social contacts to prevent the disease from spreading. Even though some industries suffered more than others did, it is impossible to find an economic sector that has not been influenced by the coronavirus. The new world order makes firms and companies of all sizes and specialties look for adequate ways to address the current challenges. The necessity to respond to the unprecedented economic conditions makes businesses rely on innovations, which motivates their growth and development. That is why the principal aim of this paper is to analyze the effect of the coronavirus disease on business intelligence systems, drawing specific attention to business innovation and entrepreneurship then the paper gives a set of approaches which might lead business organisations how to cope with it.

2 The Study Importance and Objectives

The principal aim of the study is to identify how Covid-19 pandemic affects the business sphere in the whole world. The paper analyzes the existing literature to determine that the current crisis is productive because it makes people do their best to address numerous modern problems. A significant objective is to show that a seemingly hopeless situation has the potential to result in positive outcomes. Accordingly, this study is planned to:

1. Gives an idea of what does BI systems means?
2. Gives an idea about business innovation and entrepreneurship.
3. Gives an idea if Covid-19 push organizations to innovate.
4. Gives an idea if Covid-19 push business entrepreneurship.
5. How business systems could cope with the Covid-19 pandemic.

3 Literature Review

3.1 Business Intelligence

Business intelligence (BI) usually refers to the use of new technology to conduct various business activities, playing a key role in informing firm functions, strategies, and efficiency. This broad definition stands for a process to explore and analyze structured information on an area "to discover tendencies or patterns from which to derive ideas and conclusion" [1] (p. 115). [2] supports this claim and adds that BI systems are the most valuable in those business areas where it is necessary to make decisions to create value. In addition to that, the researcher explains that the term denotes both a process and a product that includes technology and strategies used by companies [2]. As a result, BI has become a significant concept that allows many companies to deal with big data and benefit from it.

3.2 Business Innovation and Entrepreneurship

Since BI is an umbrella term, it has a direct relation to many significant phenomena, and business innovation and entrepreneurship are among them. According to [1], innovation is the introduction of a new product, process, or "organizational method in business practices, workplace organization, or external relations" (p. 114). The term also refers to the implementation of significantly improved goods and services. These phenomena are of significance when it comes to a dynamic business environment where companies need qualitative changes to achieve positive results. At the same time, entrepreneurship relates to the efforts of business persons "to generate value through the creation or expansion of economic activity" [1] (p. 114). Sometimes, these concepts are joined and create innovative entrepreneurship that stimulates economic growth because it denotes the use of new products to generate value.

3.3 Covid-19 and Innovation

Even though numerous examples can prove that innovations are often introduced against the background of the virus outbreak, some scholars still insist that coronavirus creates more problems. To begin with, [3] mention that being innovative means being resilient to continuous changes in the business environment. The scholars report that the current economic climate is unfavorable for innovation [3]. That is why these researchers stipulate that it is a rarity now that businesses willingly involve in the development and implementation of new products and strategies. Thus, the information below will describe that the current situation fosters the application of innovations in various business spheres. It is rational to mention that an element of truth is present in the article by [3]. It is so because the short-term effects of the Covid-19 outbreak are devastating, and both developed and developing economies have witnessed it. However, [4] considers this situation from a different point of view and states that innovations are needed. [5] also stipulates that approximately "75 percent of S&P 500 businesses were founded during a time of depression or catastrophe" (para. 4). [6] admit that innovations have become more significant in the modern world because they adapt business strategies to the new situation. The most typical innovation refers to the more active use of remote work. During the crisis, "companies developed various mechanisms which would enable them to better manage this process and these could be further deployed in order to cut costs" [6] (p. 606).

Simultaneously, [4] emphasizes the necessity for governments to implement innovative policies to provide businesses with necessary assistance and stimuli. The coronavirus outbreak has demonstrated that an economic sphere is sensitive to changes in the external environment. That is why the Covid-19 made governments actively participate in business by providing companies and firms with the necessary assistance. For example, the [7] understands the importance of this strategy and grants €314 to 36 innovative companies. In the European Union, they understand that those companies that enhance the implementation of innovations have the potential to lead to faster recovery from the crisis consequences [7]. Individual nations also provide their domestic companies with the required assistance. The United Kingdom is a suitable example here because the country understands its role. [8] offers a report that explains what kinds of support innovative companies can obtain to withstand the adverse impact of the coronavirus.

In addition to that, the Covid-19 outbreak has created suitable conditions for open innovations. [9] explain that this kind of innovation refers to finding new partners with complementary skills, unlocking potential in long-lasting relations, and others. These examples demonstrate that the crisis emphasizes the importance and benefits of cooperation. Even though collaboration is usually overlooked under normal circumstances, the coronavirus pandemic made companies collaborate in the sphere of innovation to achieve positive outcomes [9]. Companies understand that they should join their efforts and share knowledge to overcome the current challenges and make their businesses more profitable.

The information above explains the presence of significant business innovations, and the following examples describe them. Firstly, firms and companies cooperate in various spheres, and sharing employees is one of them. [10] explains that various firms exchange employees to avoid high unemployment rates. Secondly, the coronavirus outbreak has led to reshuffled priorities in the economy, making some jobs more significant than they were before. According to [10], keeping children entertained is a suitable example here because it is a challenge for parents to work remotely and manage children. That is why this business sphere is obtaining more attention now. Finally, the restaurant business is the one that has witnessed a significant portion of innovation. Delivery and pick up are the most typical options offered by restaurants that cannot serve visitors because of the virus [10].

The examples above are universal as businesses from numerous countries use and benefit from them. However, separate nations can also have specific trends in responding to coronavirus challenges. For example, India is one of them, and the business sector of this Asian country demonstrates specific responses to the pandemic. The new situation makes India invest funds and efforts in developing robots to prevent the disease from spreading, providing people with online consultation opportunities, crowdsourcing options, and others [11]. These innovations take place in India because the country is in a challenging humanitarian situation. Furthermore, India has many talented engineers who, in turn, have "a frugal innovation mindset to find hacks to problems with limited resources" [11].

The coronavirus outbreak has also influenced American business, which has made numerous companies look for possible innovations to respond to the new challenges. Back Forty Beer Co. is a suitable example of how the appropriate business innovation can lead to significant success. This alcohol producer anticipated the reduction of demand and decided to minimize its negative consequences by producing and selling hand sanitizer [5]. This production change has resulted in $1 million in benefits for the company [5]. Consequently, the given example demonstrates that the Covid-19 pandemic makes firms and companies reconsider their business strategies and find opportunities to introduce innovative decisions in these processes.

## 3.4	Covid-19 and Entrepreneurship

This section will focus on what specific effects the coronavirus disease has brought to entrepreneurship. To begin with, one should explain that numerous individuals and businesses have experienced essential losses. According to the [12], the crisis leads to the most significant downturn in the global economy. As for entrepreneurs, they "should be ready to lose 50% to 80% of their turnover" [12] (p. 7). Simultaneously, [13] stipulates that the negative effects on businesses differed depending on their owners. Thus, African American, Latin, immigrant, and female entrepreneurs are said to witnesses more negative outcomes in the United States [13]. In addition to that, [14] state that micro-enterprises experience the more significant impact of the pandemic because they are more subject to business environment changes.

Another adverse effect of the Covid-19 pandemic refers to the reduced volume of financing. For example, [15] stipulate that the availability of entrepreneurial financial sources has reduced by more than 50% since the beginning of the coronavirus outbreak. Two various phenomena can explain and justify this state of affairs. On the one hand, the current situation is uncertain, and numerous investors are afraid of providing entrepreneurs with finances. On the other hand, [16] analyze a British context and admit that funding volumes have reduced because investors and entrepreneurs cannot participate in face-to-face meetings to discuss financial matters.

This information denotes that the coronavirus preventive measures result in the fact that entrepreneurship lacks sufficient resources to run businesses.

With this in mind, governments understand the necessity to provide entrepreneurs with financial assistance. That is why appropriate loans and grants are the typical options to mitigate the adverse outcomes of the crisis [17]. Even though these measures seem positive and right, they also imply some negative outcomes. It refers to the fact that this assistance fails "to appreciate the potential of businesses and non-profits to help communities withstand and overcome crises" [17] (p. 1).

The thought above denotes that entrepreneurs require some freedom of action under the conditions of uncertainty, and this fact will manifest itself in changes that affect business intelligence systems. If it happens, entrepreneurs will do their best to implement a survival approach and develop recovery plans to improve the economy. Simultaneously, the current situation does not mean that the economy will not manage to overcome the difficulties. According to the [12], the given crisis indeed implies some positive features. It relates to the fact that the pandemic has an endpoint, and when it happens, entrepreneurship will become better and more sophisticated. It is so because entrepreneurs train their resilience and optimistic approach during the current situation. As a result, they are obtaining sufficient knowledge and skills that will manage to make a difference once the coronavirus pandemic ends.

4   Theoretical Framework

The study is investigating the effects of the coronavirus pandemic on business intelligence systems regarding business innovation and entrepreneurship. The model of the study was developed based on reviewing scientific literature on the topic. New facts of life, including social distancing, quarantine, and self-isolation, result in the fact that the business world should significantly change to meet the new requirements and operate within the new environment.

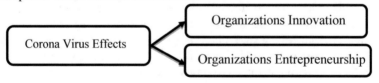

5   Propositions and Constructs

Based on the study model, the paper proposes two suggestions. Hypothesis 1 stipulates that the initial effects of the Covid-19 pandemic are adverse and paralyzing. However, Hypothesis 2 mentions that this crisis makes individuals and businesses more motivated to address current problems. Consequently, these two hypotheses suppose that the coronavirus will make the spheres of business innovation and entrepreneurship better by overcoming the existing challenges and issues.

The effect of Covid-19 can be stipulated by the following propositions:

P1: COVID 19 catalysts business organizations to be innovators.

P2: COVID 19 catalysts business organizations to be entrepreneurs.

6 Research Methodology

The study is using the literature method to discuss the topic under consideration. It relies on a set of articles and reports that analyze the coronavirus effects on business in general and organisations performance in specific. In addition to that, the paper draws attention to newspaper articles, official documents, and websites' reports of different organizations to suggest a set of approaches on how to cope with Covid 19 pandemic. This versatile nature of the sources allows the study article to address the issue from various sides, which contributes to arriving at reliable conclusions.

7 Research Findings: How to Cope with Covid-19 Pandemic

It has been proved that the Covid-19 global pandemic push organizations to change the way of doing things and leave the traditional business paradigms in addition to adapt new business themes and models. Such new business paradigms are relaying on believing in organizational changes, organizational learning, having more people that are professional. In addition, organizations should believe more in business innovation, collaboration, and even business integrations. Additional explanations about organisational change, organisational learning and the need for professional people are provided in the next three sections.

One can state that the coronavirus disease makes firms and companies invest more effort in the introduction of business innovations. Even though the pandemic's short-term effects are adverse overall, this situation demonstrates that it is necessary to rely on new products and strategies to overcome the current challenges. It is so because the crisis is unique, and it is impossible to address its issues with the help of traditional approaches. That is why businesses from the whole world implement appropriate innovations. Some of them are universally applicable, while others are specific for particular nations. In any case, the entire world is in a stage of increasing reliance on innovative solutions to the existing issues.

According to the information above, it is reasonable to comment on what specific effects and changes the Covid-19 pandemic brings to business intelligence systems within the sphere of entrepreneurship. Firstly, the existing challenges have indicated that entrepreneurship education is far from ideal. On the one hand, it is so because the available learning programs do not provide people with useful information on how to act during crises. In other words, entrepreneurial education has failed to prepare individuals to volatile economic conditions. On the other hand, traditional education means appear to be ineffective under the new conditions. [18] explain that the world currently needs online entrepreneurial education. [19] support this idea and mention that face-to-face lectures cannot be conducted under the current conditions. The necessity to rely on online education is the first step to implement to achieve the more widespread application of business intelligence systems in the entrepreneurship sector.

Secondly, new venture creation is another effect of the coronavirus on entrepreneurship. Even though the current situation is uncertain, new businesses tend to arise. [19] mention that this tendency is present because of the "creation, expansion, and substitution of demand" (p. 2). It happens because the world is changing, and new phenomena appear. For example, it is necessary to keep social distance, use hand sanitizers, work remotely, and others. This new reality means that individuals draw attention to activities that were not significant before. It relates to the reliance on innovations to meet the new demand. As a result, entrepreneurs see the opportunities and do their best to take advantage of them.

Thirdly, the Covid-19 pandemic is more likely to result in a significantly different entrepreneurship industry when the crisis ends. It relates to the fact that the current situation demonstrates that it is possible to run a business without many traditional phenomena. [19] admit that the new features will include the fragmentation of employment and an increase in home-based self-employment. In addition to that, it is reasonable to expect the gig economy to appear. The term denotes that a growing number of entrepreneurs will "insure against future disruptions by being more self-reliant and trusting less in the robustness of employment" [19] (p. 3). This fact demonstrates that the coronavirus teaches entrepreneurs to acquire more fragile and volatile skills and professions. At the same time, a different approach will be taken regarding cooperation with suppliers. Larger suppliers will be more requested because they will be less subject to changes in the business environment.

Finally, [20] admit that Covid-19 pandemic emphasizes social entrepreneurship's importance. The term "is an innovative, social value-creating research activity that can occur within or across the nonprofit, business, and public sectors" [20] (p. 177). This definition denotes that the business sector draws more attention to social value than an economic one. The virus has exposed ordinary people to many social issues, and it is necessary to take specific efforts to address them. Thus, the concept of social entrepreneurship means that entrepreneurs understand that they can make a difference within this sphere and start taking action to cope with numerous social problems [20]. Even though the study by [20] analyzes an Indonesian context, it demonstrates that the pandemic is a trigger for entrepreneurs to protect vulnerable parts of society.

In order to cope with Covid 19 pandemic, the study suggests three main business dimensions to approach, which are the need for organisational change, organisational learning and the need for professional people.

7.1 Organizational Change

Covid-19 has changed the way people live and work. Accordingly, organizations should change to cope with such changes. Peters (2020) mentioned that, nowadays, more than 50% of employees are conducting their business activities from home and 75% of them expected to increase demand for homeworking.

The extensive changes introduced by Covid-19 into the ways people used to live and work significantly influenced the fundamental business models applied by the companies and their organizational structures. Most businesses' primary reaction was to cut down the expenses by reducing the number of employees and optimizing the entire corporate structure. A survey conducted in the hard-hit areas of New York and New Jersey showed that 30 percent of firms reduced the staff numbers and over 40 percent of companies cut the work hours of existing staff [22]. In addition to cutting the workforce, many companies supported the increased amount of working from home instead of regular commuting to work. Within the United States, over 35% of those employed prior to Covid-19 switched to working from home [23]. Moreover, this state is supposed to remain after the pandemic since the companies invest in the assets required for remote work and develop the relevant organizational processes.

Along with the direct influence on the organizational structure, this crisis provided the companies with a unique opportunity of improving their business network by adjusting the market to their needs and desires. This is especially true for business intelligence system companies as these are already operating at evolving markets. In addition, forceful external occurrences are a substantial factor in giving "rise to deliberate shaping by individual actors or collectives" [24]. For the market-shaping to be successful, the company needs to develop a scalable vision of the market it desires and the image of a possible viable system that can support such a market. Then, using the marketing and advertisement tools, the company needs to drive changes in the characteristics of the existing market towards the desired one, along with adjusting its structure. Although this requires excellent cross-functional communication and significant management efforts, the implementation of market-shaping strategies is both possible and feasible during the Covid-19 crisis.

7.2 Organizational Learning

Besides adjusting to the current crisis and looking for opportunities to improve its overall performance, any company should learn lessons that may be helpful to overcome the possible future blows. In the case of Covid-19, this is especially important due to its unpredictable course and unknown time limits. A right solution is to deploy plan-ahead teams considering various aspects of company activities within time frames ranging from the immediate future to the next normal circumstances, which can be years ahead [24]. Such teams should develop different versions of future scenarios based on the current assessment of the company state. They should also define a portfolio of actions suitable for all likely scenarios, even if they are not the best ones for a specific case. These are particularly helpful given the unpredictable character of the crisis since they can be beneficial regardless of its course. Finally, identifying the improvements that can remain advantageous after the pandemic helps maximize the opportunities for further development. Consequently, by learning the lessons of the crisis, an organization can both overcome it and find ways to thrive from it.

7.3 Needs for Professional People

Professional people are seen as the core of business change and response [21, 25–27]. Although the value of organizational aspects in crisis management is hard to over-estimate, in many cases, the right staff is crucial for overall success. The dramatic changes in the lifestyle and work conditions require staff to adjust to them. A survey conducted among sales professionals in Italy just before the crisis showed that they were not accustomed to digital communication [28]. Moreover, the structural changes needed to ensure resilience require an increased level of teamwork and communication abilities. As a result, companies need more professionals that can be "innovative and learn new areas" and support the diversity among them as a way to "improve their flexibility and enhance the collaborative processes" [28] (p. 243). Therefore, increased professionalism of the staff and adjustments in the skills deserving highest appreciation are the methods largely used by the companies during the current crisis.

8 Study Conclusion

As for business intelligence systems and innovations, the Covid-19 pandemic has become a trigger for further development. It is so because the situation under consideration is unique, and the world does not have sufficient measures to solve the new challenges. As a result, there is an urgent necessity to introduce business innovations to make the economy work. Numerous governments understand it and provide innovative companies with financial assistance. This strategy results in various creative decisions, and some of them are universally applicable.

When it comes to entrepreneurship, the pandemic results in both negative and positive effects. An adverse feature refers to the reduced volume of financing because of preventive measures. However, uncertainty and the lack of sufficient resources make entrepreneurs accumulate their knowledge and skills, which leads to many profitable results. The paper shows that Covid-19 pandemic is likely to change entrepreneurship education and make it online. New venture creation with the help of business intelligence systems is another tendency within the entrepreneurship sector. Even though the whole world is afraid of the Covid-19 outbreak, it can indeed make business better.

9 Study Recommendations

Many scholars such as [29–51] have provided a set of suggestions which might help to cope properly with Covid-19 and post Covid-19. These suggestions are:

1. Business organizations need to use data and analytics techniques more than before.
2. Create implement new technology programs.
3. Adapt and facilitate online learning and working.
4. Start in creating and managing virtual business teams.
5. Working on maintaining work/life balance.
6. Provide remote support.
7. Search for new sales and distribution channels.
8. Provide more virtual communication and information supports.

10 Study Limitations

This study implies a single significant limitation that can influence the findings. It refers to the fact that the events under discussion are happening right now, meaning that there are no specific examples that can entirely prove or refute any of the paper's ideas. As a result, the article relies on the existing literature, but it is still possible to achieve unexpected results in the future.

11 Study Theoretical and Practical Implications

On the one hand, the theoretical implication arises since the study demonstrates that the coronavirus pandemic results in positive effects in the spheres of business innovations and entrepreneurship. On the other hand, the practical implication is represented by the motivation that individuals and businesses can extract from the study. This information will be useful to introduce more intelligence systems in business, which, in turn, will make this sphere better.

References

1. Giménez-Figueroa, R., Martín-Rojas, R., García-Morales, V.J.: Business intelligence: an innovative technological way to influence corporate entrepreneurship. In: Entrep. Tendencies Empir. Approach, pp. 113–132 (2018)
2. Trieu, V.-H.: Getting value from business intelligence systems: a review and research agenda. Decis. Support Syst. **93**, 111–124 (2017)
3. Kuckertz, A., et al.: Startups in times of crisis—a rapid response to the COVID-19 pandemic. J. Bus. Ventur. Insights **13**, 1–13 (2020)
4. Barua, S.:Understanding coronanomics: the economic implications of the coronavirus (COVID-19) pandemic. SSRN Electron. J. (2020). https//doi.org/10/ggq92n
5. Anderson, J.: COVID-19 fosters wave of business innovation, Back Forty Beer founder says. Hoover Sun (2020). (Online). Available: https://hooversun.com/businesses/covid-19-fosters-wave-of-business-innovation-back-forty-beer/
6. Dan, M.C., Goia, S.I.: Challenges for entrepreneurs amidst the COVID-19 crisis. In: Pamfilie, R., Dinu, V., Tăchiciu, L., Pleşea, D., Vasiliu, C. (eds.) 6th BASIQ International Conference on New Trends in Sustainable Business and Consumption. In: New Trends in Sustainable Business and Consumption, pp. 602–609 (2020)
7. European Commission: Coronavirus: EU grants €314 million to innovative companies to combat the virus and support recovery (2020)
8. G.-UK: Innovate UK Product Range in Support Package: What Products will Innovate UK Offer Through its New Funding? (2020)
9. Dahlander, L., Wallin, M.: Why now is the time for 'open innovation'. Harvard Business Rev (2020). (Online). Available: https://hbr.org/2020/06/why-now-is-the-time-for-open-innovation. Accessed 03-Aug-2020
10. Wunker, S.: Six Forms of Business Innovation During the Coronavirus Crises (2020). (Online). Available: https://www.forbes.com/sites/stephenwunker/2020/04/07/six-ways-to-innovate-during-the-coronavirus/#33d1bd461266
11. Sahasranamam, S.: India: How Coronavirus Sparked a Wave of Innovation (2020). (Online). Available: https://theconversation.com/india-how-coronavirus-sparked-a-wave-of-innovation-135715
12. Glion Institute of Higher Education. What are the Impacts of COVID-19 on Entrepreneurship? (2020)
13. Fairlie, R.W.:The Impact of Covid-19 on Small Business Owners: Evidence of Early-Stage Losses from the April 2020 Current Population Survey. National Bureau of Economic Research (2020)
14. Fabeil, N.F., Pazim, K.H., Langgat, J.: The impact of Covid-19 pandemic crisis on micro-enterprises: entrepreneurs' perspective on business continuity and recovery strategy. J. Econ. Bus. **3**(2), 837–844 (2020)
15. Brown, R., Rocha, A.: Entrepreneurial uncertainty during the Covid-19 crisis: Mapping the temporal dynamics of entrepreneurial finance. J. Bus. Ventur. Insights **14**, e00174 (2020)
16. Brown, R., Rocha, A., Cowling, M.: Financing entrepreneurship in times of crisis: exploring the impact of COVID-19 on the market for entrepreneurial finance in the United Kingdom. Int. Small Bus. J. Res. Entrep. (2020)
17. Haeffele, S., Hobson, A., Storr, V.H.: Coming back from COVID-19: lessons in entrepreneurship from disaster recovery research. Mercat. Spec. Ed. Policy Br. 1–7 (2020)
18. Liguori, E., Winkler, C.: From Offline to Online: Challenges and Opportunities for Entrepreneurship Education Following the COVID-19 Pandemic. SAGE Publications Sage CA, Los Angeles, CA (2020)
19. Maritz, A., Perenyi, A., de Waal, G., Buck, C.: Entrepreneurship as the unsung hero during the current COVID-19 economic crisis: Australian perspectives. Sustainability **12**(11), 1–9 (2020)
20. Nuringsih, K., Nuryasman, M.N., Amelinda, R.: The propensity for social entrepreneurship during the coronavirus outbreak. J. Manaj. **24**(2), 174–193 (2020)

21. Peters, R.: What lessons can COVID-19 teach us about organisational change? CIPD (2020). (Online). Available: https://peopleprofession.cipd.org/insights/articles/lessons-covid-19-organisational-change. Accessed: 08-Aug-2020

22. Abel, J.R., Bram, J., Deitz, R.: Businesses in the Tri-State Region Struggling to Weather the Coronavirus Outbreak. Federal Reserve Bank of New York (2020)

23. Brynjolfsson, E., Horton, J.J., Ozimek, A., Rock, D., Sharma, G., TuYe, H.-Y.: COVID-19 and Remote Work: An Early Look at US Data. National Bureau of Economic Research (2020)

24. Hirt, M., et al.: Getting ahead of the next stage of the coronavirus crisis. McKinsey Co. website. https//www.mckinsey.com/business-functions/strategy-and-corporate-finance/our-insights/getting-ahead-of-the-next-stage-of-the-coronavirus-crisis. Publ. April, vol. 2, pp. 1–11 (2020)

25. Alkalha, Z.S., Al-Zu'bi, Z.M.F., Al-Dmour, H., Alshurideh, M., Masa'deh, R.: Investigating the effects of human resource policies on organizational performance: an empirical study on commercial banks operating in Jordan. Eur. J. Econ. Financ. Adm. Sci. **51**, 44–64 (2012)

26. Alshurideh, M., Masa'deh, R.M.D.T., Alkurdi, B.: The effect of customer satisfaction upon customer retention in the Jordanian mobile market: an empirical investigation. Eur. J. Econ. Financ. Adm. Sci. **47**(47), 69–78 (2012)

27. ELSamen, A., Alshurideh, M.: The impact of internal marketing on internal service quality: a case study in a Jordanian pharmaceutical company. Int. J. Bus. Manag. **7**(19), 84–95 (2012)

28. Sharma, A., Rangarajan, D., Paesbrugghe, B.: Increasing resilience by creating an adaptive salesforce. Ind. Mark. Manag. **88**, 238–246 (2020)

29. Abu Zayyad, H.M., Obeidat, Z.M., Alshurideh, M.T., Abuhashesh, M., Maqableh, M., Masa'deh, R.: Corporate social responsibility and patronage intentions: the mediating effect of brand credibility. J. Mark. Commun. 1–24 (2020)

30. Al Kurdi, B., Alshurideh, M., Al afaishata, T.: Employee retention and organizational performance: evidence from banking industry. Manag. Sci. Lett. **10**(16), 3981–3990 (2020)

31. Alshurideh, M., Gasaymeh, A., Ahmed, G., Alzoubi, H., Kurd, B.A.: Loyalty program effectiveness: theoretical reviews and practical proofs. Uncertain Supply Chain Manag. **8**(3), 599–612 (2020)

32. Alzoubi, H., Alshurideh, M., Al Kurdi, B., Inairata, M.: Do perceived service value, quality, price fairness and service recovery shape customer satisfaction and delight? A practical study in the service telecommunication context. Uncertain Supply Chain Manag. **8**(3), 1–10 (2020)

33. Fawzi Assad, N., Alshurideh, M.: Investment in context of financial reporting quality: a systematic review. WAFFEN-UND Kostumkd. J. **11**(3), 255–286 (2020)

34. Assad, N.F., Alshurideh, M.T.: Financial reporting quality, audit quality, and investment efficiency: evidence from GCC economies. WAFFEN-UND Kostumkd. J. **11**(3), 194–208 (2020)

35. Hayajneh, N., Suifan, T., Obeidat, B., Abuhashesh, M., Alshurideh, M., Masa'deh, R.: The relationship between organizational changes and job satisfaction through the mediating role of job stress in the Jordanian telecommunication sector. Manag. Sci. Lett. **11**(1), 315–326 (2021)

36. Al Kurdi, B., Alshurideh, M., Salloum, S.A., Obeidat, Z.M., Al-dweeri, R.M.: An empirical investigation into examination of factors influencing university students' behavior towards elearning acceptance using SEM approach. Int. J. Interact. Mob. Technol. **14**(2), 19–41 (2020)

37. Kurdi, B., Alshurideh, M., Alnaser, A.: The impact of employee satisfaction on customer satisfaction: theoretical and empirical underpinning. Manag. Sci. Lett. **10**(15), 3561–3570 (2020)

38. Mehmood, T., Alzoubi, H., Alshurideh, M., Ahmed, G., Al-Gasaymeh, A.: Schumpeterian entrepreneurship theory: evolution and relevance. Acad. Entrep. J. **25**(4), 1–10 (2019)

39. Herrero, M., Thornton, P.: What can COVID-19 teach us about responding to climate change? Lancet Planet. Heal. **4**(5), e174 (2020)

40. Salloum, S.A., Alshurideh, M., Elnagar, A., Shaalan, K.: Machine learning and deep learning techniques for cybersecurity: a review. In: Joint European-US Workshop on Applications of Invariance in Computer Vision, pp. 50–57 (2020)

41. Al Kurdi, B., Alshurideh, M., Salloum, S.: Investigating a theoretical framework for e-learning technology acceptance. Int. J. Electr. Comput. Eng. **10**(6), 6484–6496 (2020)
42. Salloum, S.A., Alshurideh, M., Elnagar, A., Shaalan, K.: Mining in educational data: review and future directions. In: Joint European-US Workshop on Applications of Invariance in Computer Vision, pp. 92–102 (2020)
43. TSBSP. COVID-19 and entrepreneurs in the developing world: supporting business survival and recovery. TechnoServe Business Solution to Poverty (2020). (Online). Available: Supporting%0ABusiness Survival%0Aand Recovery. Accessed: 08-Aug-2020
44. AlMehrzi, A., Alshurideh, M., Al Kurdi, B.: Investigation of the key internal factors influencing knowledge management, employment, and organisational performance: a qualitative study of the UAE hospitality sector. Int. J. Innov. Creat. Chang. **14**(1), 1369–1394 (2020)
45. Alhashmi, S.F.S., Alshurideh, M., Al Kurdi, B., Salloum, S.A.: A systematic review of the factors affecting the artificial intelligence implementation in the health care sector. In: Joint European-US Workshop on Applications of Invariance in Computer Vision, pp. 37–49 (2020)
46. Alshraideh, A., Al-Lozi, M., Alshurideh, M.: The impact of training strategy on organizational loyalty via the mediating variables of organizational satisfaction and organizational performance: an empirical study on Jordanian agricultural credit corporation staff. J. Soc. Sci. **6**, 383–394 (2017)
47. Alshurideh, M., Al Kurdi, B., Abu Hussien, A., Alshaar, H.: Determining the main factors affecting consumers' acceptance of ethical advertising: a review of the Jordanian market. J. Mark. Commun. **23**(5), 513–532 (2017)
48. Alshurideh, M.: Do electronic loyalty programs still drive customer choice and repeat purchase behaviour? Int. J. Electron. Cust. Relatsh. Manag. **12**(1) (2019)
49. Alshurideh, M., Salloum, S.A., Al Kurdi, B., Monem, A.A., Shaalan, K.: Understanding the quality determinants that influence the intention to use the mobile learning platforms: a practical study. Int. J. Interact. Mob. Technol. **13**(11) (2019)
50. AlShurideh, M., Alsharari, N.M., Al Kurdi, B.: Supply Chain Integration and Customer Relationship Management in the Airline Logistics, vol. 9, no. 02, pp. 392–414 (2019)
51. Alshurideh, M., Al Kurdi, B., Shaltoni, A.M., Ghuff, S.S.: Determinants of pro-environmental behaviour in the context of emerging economies. Int. J. Sustain. Soc. **11**(4), 257–277 (2019)

Arab Potash Company (APC) Response Management Plan (RMP) to COVID-19

M. F. Nsour and S. A. AL-Rjoub

Abstract This chapter includes a Response Management Plan (RMP) for APC to support management functions, mitigate risk, and ensure continuous business operations during COVID-19. The company follows certain procedures and take necessary actions to quickly communicate decision making to different management levels. All responses, risks, risk management practices, and mitigation tools are reflected in the RMP. The actions and procedures taken by APC during the pandemic enrich the company experience in dealing with similar emergencies. Also, this chapter stresses the importance of planning ahead for uncertainties and the imminent need for incubating risk management thinking in the organization body to ensure business continuity.

Keywords Response management plan · COVID-19 · Risk management

1 Introduction

Since day one, the day of the first registered case of COVID-19 in Jordan, Arab Potash Company (APC) took decisions and actions to keep all production operations and other business activities operating smoothly without any kind of disruptions. These actions and decisions are accomplished in a way to ensure employees' safety and to comply with government laws and regulations.

M. F. Nsour
CEO and President of Arab Potash Company, Amman, Jordan
e-mail: Maen.n@ArabPotash.com

S. A. AL-Rjoub (✉)
Department of Banking and Finance, Hashemite University, Zarqa, Jordan
e-mail: salrjoub@hu.edu.jo

M. T. Alshurideh et al. (eds.), *The Effect of Coronavirus Disease (COVID-19) on Business Intelligence*, Studies in Systems, Decision and Control 334,
https://doi.org/10.1007/978-3-030-67151-8_17

APC fast and prompt response to COVID-19 result in minimizing the pandemic's effect on its operations; the objective is to secure business continuity and not only recovery. All departments, functions, and line managers, take necessary measures and actions to manage the emergency and provide response plans and initial crisis response activities to support business continuity and ensure incident recovery at the early stages of COVID-19 spread.

The company performs remarkably during the pandemic; production reached record numbers by achieving the highest daily production quantity of 8700 tons on April 12, 2020. Quarterly production is the highest in the company history; APC produces 645,430 tons in the second quarter of 2020 and utilizes only 26% of its employees. While achieving these records, the company ensures that all employees return safely to their families with no workplace incidents.

This chapter includes a Response Management Plan (RMP) for APC to support management functions, mitigate risk, and ensure smooth business operations. The company follows specific procedures and take necessary actions to transfer decision making to different management hierarchy. The response management requirements contribute effectively to the management of this contingency and participate in the early resolution of COVID-19 [1]. All responses, risks, risk management practices, and mitigation tools that APC took during the incident are reflected in the RMP.

1.1 APC

APC is one of the leading potash producers worldwide, rank as the eighth largest, and is the sole producer of potash in the Arab World. APC holds the best track records among other companies in work safety, corporate governance, corporate social responsibility, and environment preservation. The government of Jordan grants APC an exclusive concession to invest in Dead Sea minerals through extraction, manufacturing, and marketing minerals. The company's main product is potash but also invests in several downstream in related industries that extract other Dead Sea salts and minerals, including potassium nitrate, bromine, and other derivatives. APC employs more than 2000 workers across its three locations inside Jordan and is considered one of the major national corporations contributing to the national economy. The production site of APC consists basically of a Solar Evaporation Ponds System with a total area of 112 km^2 and three processing plants.

2 Scope, Objectives, and Methodology

2.1 Scope

This chapter identifies all the significant challenges that encounter APC and introduced by COVID-19, formulates business impact analysis, documents all response activities exercised, and discusses existing and developed controls in the RMP as part of APC's Business Continuity Plan (BCP).

2.2 Objectives

To identify and assess the set of countermeasures and response activities exercised by APC during coronavirus COVID-19 outbreak to ensure employees' safety, business recovery, and business continuity. We formulate an RMP for APC by analyzing the impact, defining challenges, spotting gaps, and suggesting better responses or rooms for improvement that need to be invested in or counted for in the future. Further, this chapter aims to raise the risk management maturity level either in APC or in other organizations with similar work scope. Risk management culture must be embedded within the strategic thinking of an organization.

2.3 Data and Methodology

We obtain information about the context and response actions from the company's top management, line managers, lowest level possible of employees, internally available data, and existing emergency plans. Most of the crisis data is usually unstructured and poorly documented, so data about pandemic order laws and details are from published news in the local and international media. We categorize all responses according to different production lines and functional departments after conducting a business impact analysis. In the final step, we conduct a gap analysis and write recommendations. We present the methodology in Fig. 1.

3 National Epidemic Briefing

Jordan reports the very first COVID-19 infected case early on March 2 of 2020. Since then, cases have propagated gradually, reaching 808 on June 7. The growth of confirmed cases propagated steadily but at a much lower rate than the global average with around 8 cases per day. Death rates of COVID-19 have also been below the global average, where 9 cases have died since the identification of case zero, which

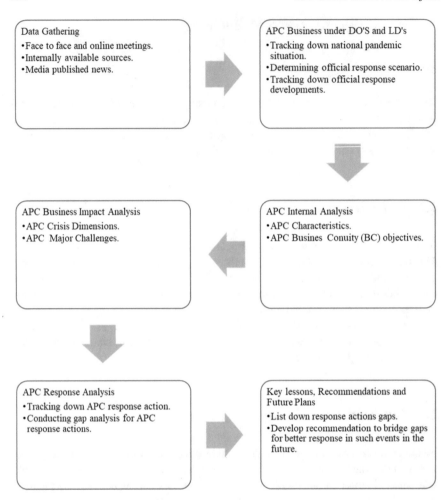

Fig. 1 Methodology framework

makes Jordan's most recent COVID-19 death rate around 1.13%. Figure 2 illustrates the behavior of COVID-19 cases in terms of propagation, recovery, and fatality as published by the Jordan Ministry of Health (MOH).

Figure 2 shows that Jordan is doing an exceptional job controlling and containing COVID-19 compared to other countries worldwide.

4   Official Response Timeline

To examine the effect of government actions on APC's operation and hence responses, we first list the timeline of different Defense Orders (DO) and Defense Laws (DL)

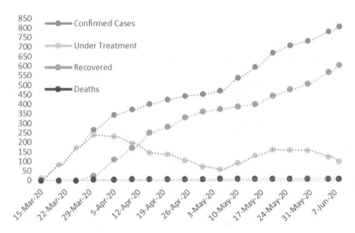

Fig. 2 Case overview

issued by the Government of Jordan (GOJ) to deal with coronavirus. Jordan's official response to COVID-19 is mainly applied through a series of DO's and official body resolutions in the light of situation developments and available information. Table 1 lists the major orders, DO's, and DL's issued by the government toward the coronavirus outbreak after Identifying case **Zero** in Jordan on March 2, 2020.

5 APC Business Under DO's and DL's

Precise information about the virus in the early days of the outbreak is scarce. Hence, APC top management took several gradual steps to ensure the safety of its employees. These steps include but are not limited to securing personal protective items like masks, gloves, and sanitizers, and enforcing social distancing, in addition to proactively preparing the company for what may come ahead.

The preparation phase starts by mimicking the experience of first infested countries like China, South Korea, and Italy in dealing with the disease, and by following the government of Jordan response actions to contain the propagation of COVID-19. APC top management realizes early the drastic nature of governmental response actions that are yet to come, including quarantine, remote working, movement restrictions, and the challenges they would pose. The second part of the preparation phase includes identifying gaps in the critical response actions to APC challenges (brought up by governmental efforts to contain the outbreak), different control measures, and desired responses. One of the identified gaps, for example, is the inability of transport employees to the company premises. APC resolves this by preparing the needed residence for employees near the site, determining employees who can work remotely, and providing them with needed laptops and software to work from a distance. Another gap is banning meetings, which are essential to organize and coordinate

Table 1 DO's and DL's timeline

March 10	• Travel restriction from and to several countries that either had a COVID-19 outbreak or have insufficient controls over the virus • Enforce 14 days of self-quarantine for Jordanians returning from these countries
March 14	• Halt all educational institutions operations starting the morning of March 15 for two weeks • Shut all land/sea border crossing for passenger movement • Halt all public events, gatherings, prayers in mosques and churches
March 17	• DL came into force • Issue DO No. 1, suspending and replacing some of the Social Security Law • Suspension of work and closure of public/private sectors
March 19	• The Central Bank of Jordan decided to take a set of measures to support the national economy: • allow banks to restructure the loans of individuals and companies • Inject over 550 million Jordan Dinars (JD's) to the national economy by reducing compulsory reserve • Support the Jordan Loan Guarantee Corporation procedures by reducing loan guarantee commissions and increasing the local sales guarantee program coverage
March 20	• Issue DO No. 2, imposing a nationwide lockdown and curfew
March 26	• Issue DO No. 3, tightening curfew regulations by announcing curfew violation punishments, punishments included, Fines, Imprisonment, and Vehicle Seizing for citizens
March 31	• Issue DO No. 4, establishing a national fund named "Himmat Watan," to support the Kingdom's efforts to eradicate coronavirus • Issue DO No. 5, suspending legal periods and submission dates before courts and other legal entities
April 8	• Issue DO No. 6, outlining measures related to employment conditions and wages
April 15	• Issue DO No. 7, regulating distance learning in schools and universities • Issue DO No. 8, outlining specific actions and obligations for citizens to protect public health and limit the spread of COVID-19
April 17	• Issue DO No. 9, launching, in cooperation with the Social Security Corporation, a program protecting businesses, freelance workers and daily wage workers
May 3	• Issue DO No. 10, related to the submission of tax returns • Issue DO No. 11, related to social distancing practices

efforts. APC resolve this gap by employing video conferencing software. The goal of RMP is to create response actions that prevent, prepare, respond, and recover the business from any operational disruptions caused by the pandemic [2].

Before DL came into force, several government instructions started to affect APC business and operations; first, there was a partial travel restriction on March 10, followed by closing the borders four days later. The company was no longer able to provide its sales team with the trips they need to close their deals or provide its operations team with necessary means to conduct their usual inspection field visits to suppliers and vendors worldwide.

However, the scene gets more severe on March 17, when DL came into force, followed by issuing DO No.1 requiring companies to suspend work and force closure of both public and private sectors. APC was one of the first companies to communicate with the government's official bodies, presenting their business case and asking for an exemption from this DO. The next significant development occurred on March 20 when the government issued DO No. 2, imposing a nationwide lockdown and curfew, and restricting the movement of employees from and to their workplace. At this point, APC top management identifies several procedural guidelines, conducts human resources planning, and re-engineered its processes to match the available workforce. APC categorizes employees into two main categories: employees whose physical presence is essential at the workplace to ensure production continuity, and those who can work remotely and need the necessary hardware and software to conduct business. The utilization of information technology (IT) improve business continuity and support the company's goals and give APC resilience to possible risks [3].

6   APC Characteristics and Business Continuity (BC) Objectives

The formulation of BC and RMP requires considering APC's current characteristics and their implications on APC's operations.

Table 2 list the main characteristics and their implications on APC operations. For example, employees' demographics have the main implication on essential factors such as the fatality rate, which is highly likely among employees, especially the elders, given the fact that employees' average age is above 42 years. On the other hand, the fact that operations are generally standardized enables management to respond promptly and effectively to a wide range of scenarios. The length and sequence dependency of the operation process requires a high degree of coordination and care in managing BCP.

Moreover, the geographic location which is distant from main cities will impose some difficulties in managing and monitoring response plans. The site activities like production, documentation, and employees' transportations are highly uncertain since any blockage at any point will block the entire operation. Finally, the low level of outsourcing in core business operations helps APC to manage its responses better internally.

7   APC Incident Impact Analysis (IIA)

The COVID-19 pandemic outbreak results in a comprehensive national action (mainly presented by a set of LD's and DO's) that impacted APC business and operations. In these impacts, we identify the impact of losing business functions or resources. The IIA is formally a Business Impact Analysis (BIA), but we tailored it to

Table 2 APC Main characteristics

Area	Characteristic	Implication
Process and Technology	Operations are generally standardized and adequately defined	Enable management to respond effectively to a wide range of scenarios
	Operations include Carnalite[a] pumping from the Dead Sea to the Carnalite Pans, Harvesting Carnalite, pumping Carnalite to the plants, processing Carnalite into final potash product, storage of the final product into the warehouses, selling the product, transporting the product to Aqaba site (The only shipping port in Jordan) and shipping it to customers worldwide	The length and sequence dependency of the process requires a high degree of coordination and care in managing BCP
	APC has a relatively low outsourcing level, especially when it comes to its core operations	It enhances its position and capability to manage its responses internally
Geographic Location	Core operations cover wild areas in Ghore Site (south of the Dead Sea), Aqaba Site	It imposes some difficulties in managing and monitoring response plans
	Company has three locations: – Amman Head Office	The site activities like production, documentation, and employees' transportations are highly uncertain since any blockage at any point will block the entire operation
	[a] A saline from which potash can be extracted. – Ghore Site – Aqaba Site	
Employees Demographics	APC has around 2000 employees	Disease transmission is highly likely among employees
	Employee's come from their residence from all over the Kingdom	It results in exposing APC to the risk of disease spreading
	Employee's average age is 42.3 years	Elderly employees are more exposed to infection risks
Corporate Values	– Safety – Customer Centricity – Sustainability and Corporate Social Responsibility (CSR)	– Employee's safety and wellbeing are maintained – Customers' needs must be satisfied – Support the National call

Table 3 COVID-19 outbreak impact analysis

Situation Characteristics	Key Business Impacts
Curfew and Movement Restrictions	• An employee's physical presence might be difficult • The operations sequence might get disrupted • Employees, goods, and documents movement might get disrupted
Economic Sectors Closure	• Production might shut down • Risk of interrupting off-site activities such as equipment manufacturing and steel fabrication • Risk of leaving construction sites unsecured • Risk of delays in project engineering and construction activities • Risk of extra costs related to contract deferrals • Risk of losing warranties for delivered equipment • Risk of delays in inspections, commissioning, and testing due to travel restrictions • Fail to deliver goods/ services from local and global vendors
Risk of Disease Transmission	• The disease spread among employees • Fail to safeguard the local community by failing to contain the disease • Shut down or production cut
National Call for Support	• Support governmental efforts that require financial resources

fit the COVID-19 incident. For executive management, the impact and priorities are apparent, and no need for significant exercise (like the use of detailed forms or questionnaires). Instead, our IIA involves interviews and meetings with senior managers and other employees at different levels. Most of the discussion was on quantifying the impact losses and prioritizing them in order of importance and risk [4]. Table 3 summarizes the comprehensive national actions and the resulted critical impacts on the company's business.

We identify four situation characteristics that mainly affect APC business: curfew and movement restriction, economic sector closure, risk disease transmission, and national call for support. Table 3 shows, among others, that economic sector closure may result in production shut down and the failure to deliver goods and services from local and global vendors. The curfew and movement restrictions might impact an employee's physical presence, documentation delivery, and might cause disrupt operations sequence. The risk of disease transmission might halt production or cause production cuts that affect profits and business continuity. Finally, supporting the national efforts in containing COVID-19 will exert financial pressure on APC. All of which will jeopardies the company business continuity and require RMP to overcome these risks and impacts. Later, in Sects. 9 and 10, these situation characteristics and the impact analysis are discussed in more detail when analyzing challenges (risks) and response analysis.

8   APC Major Challenges

Being one of Jordan's most socially responsible company's, the largest mainstay of the national economy, and extensive labor employer, top management realized the size of APC's national role in maintaining production continuity, serving customers, and ensuring employees' safety day one. Table 4 lists APC's major challenges encountered by different managerial levels and functionalities. Here we map each challenge to the concerned party or department or function to determine related parties, resources, and create communication procedures [2]. This table helps the company's employees find ways to handle different challenges and give a holistic view of the possible recovery plan, RMP, and BC strategies. Each challenge has response activities that should be done by each responsible entity.

9   APC Response Actions to Major Challenges

APC actions in response to COVID-19 incident are directly derived from its core values, and in line with its Risk Appetite Statement that describes the limits and types of risks the company is willing to accept in pursuing its objectives. APC operates with an overall medium risk range, the lowest risk appetite relates to employees' safety, adherence to applicable laws and regulations, and its reputation. Whenever applicable, APC will go beyond the minimum requirements to reduce the levels of any risk that may adversely affect them. APC sets a marginally higher (moderate) risk appetite to its strategic, operational, and financial objectives. It will always attempt to reduce the levels of any risk that may adversely affect them, taking into consideration the cost of the measures taken in relation to their anticipated benefits. The highest risk appetite relates to its innovation, change, and development initiatives, and will encourage its management and employees to provide new ideas and ways of doing business. The response actions to COVID-19 incident are developed both proactively and reactively as challenges arise.

The following is a top-down break down of the response actions to the rising challenge. In what follows, we list APC's response plan at different management levels and across all business functionalities. We list major challenges and their related response actions in Table 5. In the Gap analysis, we conduct a two-step analysis, we assess APC's response adequacy during COVID-19 outbreak, and then, we compare APC actual performance to the desired one.

This process provides a way to identify suboptimal or missing strategies, structures, capabilities, processes, practices, and technologies or skills, to enable top and middle management to set recommendations for a better response if any unfortunate event occurs in the future. Table 5 examines response adequacy during COVID-19.

Table 4 APC major challenges

	Challenges
1. Board of Directors and Executive Management	1.1 Keep operations despite the closure of most economic sectors 1.2 Ensure fulfilling the company CSR and the national role 1.3 Lack of conclusive information about coronavirus in the early days of the outbreak 1.4 Ensure full compliance with governmental directions and requirements 1.5 Guarantee employees and their family's safety and wellbeing
2. Companywide Challenges	1.6 Re-engineer companywide process to match available resources, comply with regulatory requirements, and external environment developments
3. Health and Safety Challenges	1.7 Ensure employees' safety without having any COVID-19 infected cases 1.8 Ensure company readiness to handle and contain any infection spread
4. Operational Challenges	1.9 Lack of workforce enough to operate plants and equipment 1.10 Risk of interrupting off-site activities such as equipment manufacturing and steel fabrication 1.11 Risk of leaving construction sites unsecured 1.12 Risk of delays in project engineering and construction activities 1.13 Risk of extra costs related to contract deferrals 1.14 Risk of losing warranties for delivered equipment 1.15 Risk of delays in inspections, commissioning, and testing due to travel restrictions
5. Financial Challenges	1.16 Pay vendors for supplies of goods and services 1.17 Prepare Letters of Credit (LC) hard copies 1.18 Prepare payroll and deposits
6. Human Resources Challenges	1.19 Ensure manpower availability 1.20 Ensure enough residence and daily needs for employees

(continued)

Table 4 (continued)

	Challenges
7. Logistics Challenges	1.21 Movement restrictions, inbound/outbound goods, and documents
8. Procurement Challenges	1.22 Secure strategic goods/services needed to maintain operations continuity
9. IT Challenges	1.23 Expiration of Microsoft license, and ORACLE ERP support agreements 1.24 Ensure proper remote working and communications without disruptions 1.25 Ensure the effectiveness of IT business continuity and disaster recovery plans 1.26 Provide continuous IT support 1.27 Ensure taking backup tapes and adequate data center conditions
10. Medical Challenges	1.28 Ensure sufficiency of the medical services capabilities and procedures

10 Conclusions

Based on RMP and IIA and the identified gaps in APC response, the following are recommendations and essential lessons learned from the COVID-19 incident to enhance the company response management in such unexpected events. Business continuity is of extreme importance to the organization even though the likelihood of extreme adverse events is low [5].

The actions and procedures taken by APC during the pandemic enrich the company experience in dealing with similar emergencies. The company is developing a procedures manual reflecting the best practices and efficient responses with an accurate mapping to concerned functional departments, ownership of responsibility and the right delegation.

This chapter documents a practical experience that can be used as a guideline for other companies in the playing field. Also, this chapter stresses the importance of planning ahead for uncertainties and the imminent need for incubating risk management thinking in the organization body to ensure business continuity. The following recommendations are from Table 5 (list of gaps bridging needs), RMP, and IIA, and show that APC should:

1. Develop a comprehensive BCP or RMP.
2. Properly set and redistribute workforce numbers to meet business processes and needs.
3. Develop policies, procedures, and infrastructure for a remote working environment. This will require changing APC culture to adapt more to remote work implementation.
4. Develop policies and procedures and infrastructure for a paperless management system.

Table 5 APC response actions to major challenges

Challenge	Key response action	Desired response	Gap (deficiency/weakness/lack of Assets of capabilities)
1. Keep operations despite the closure of most economic sectors stipulated by Law Order no. 1	1.1 Chairman of the Board and CEO conduct an immediate and extensive communication with ministries of labor, Industry, and trade, and other governmental bodies to obtain an exemption to continue operations	Obtain an exemption to continue operations	No Gaps
	1.2 Instruct developing "Emergency Plan and Procedures Manual to control and contain COVID-19	Activation of the BCP and Emergency Plan and Procedures Manual	Absence of a formally developed BCP and Emergency Plan and Procedures Manual
2. Ensure that APC's CSR and the national role is fulfilled	2.1 Support the Kingdom Efforts in its fight against COVID-19 with a total of around 25 million JOD's spent in the first four months of 2020 in areas such as a national emergency fund to counter against COVID-19 and donations to the health sector	Support the Kingdom efforts in its fight against COVID-19	Not Applicable
3 Lack of conclusive information about coronavirus at the early days of the outbreak 4 Ensure full compliance with governmental directions and requirements	4.1 Establish Emergency Committee (EC) as a collaborative cross-functional team composed of representatives from various departments, including Medical Services, Security, Safety, Procurement, and Housing/Utilities Departments	Activation of the BCP and Emergency Plan and Procedures Manual	Absence of a formally developed BCP and Emergency Plan and Procedures Manual

(continued)

Table 5 (continued)

Challenge	Key response action	Desired response	Gap (deficiency/weakness/lack of Assets of capabilities)
5. Ensure employees and their families' safety and wellbeing	5.1 Reduce the number of physically present employees to 450 (22.5% of the total number of employees)	Ensure that the minimum number of employees required to operate the plants are present as per governmental directions	The relatively large number of employees and the high contagiousness rate of COVID-19 make it harder to control and contain
6. Companywide process re-engineers to match available resources, comply with regulatory requirements, external environment developments, and reduce the risk of infection:	6.1 The department identified the employees required to be physically present at the office and those required to work from home	Departments should contact the employees required to be physically present at the office and those required to work from home based on remote working policies and procedures	Lack of remote working policies and procedures
	6.2 Employees are empowered and encouraged to perform the duties of their colleagues who are unable to attend the company premises	Empower and encourage employees to perform the duties of their colleagues who are unable to attend the company premises based on their job rotation history	Lack of job rotation practices
	6.3 Replace paper-based documents with electronic whenever possible	Use a paperless contingency system	Absence of already prepared paperless contingency system
	6.4 Suspend employee's contact and possible infection points like attendance fingerprint devices	Use of face recognition attendance system to replace the fingerprint system	A face recognition attendance system should have been used since day one

(continued)

Table 5 (continued)

Challenge	Key response action	Desired response	Gap (deficiency/weakness/lack of Assets of capabilities)
7. Ensure employees' safety without having any COVID-19 infected cases	7.1 Use personal preventive controls (such as respiratory masks and gloves), install disinfectant dispensers where needed, and keep strategic quantities	Use personal preventive controls (such as respiratory masks and gloves), install disinfectant dispensers where needed, and keep strategic quantities	No Gaps
	7.2 Apply extensive sterilization and sanitization programs for company vehicles, facilities, buildings, and delivered goods	Apply extensive sterilization and sanitization programs for company vehicles, facilities, buildings, and delivered goods	No Gaps
	7.3 Record body temperature for employees and contractors when entering the plants	Record body temperature for employees and contractors when entering the plants	No Gaps
	7.4 Perform a 15-day quarantine for employees who come from infected areas or a family member who traveled recently	Perform a 15-day quarantine for employees who come from infected areas or a family member who traveled recently	Employees come from their widespread residence; this will result in exposing the company to the risk of transmitting the disease
	7.5 Enforce social distancing by stopping physical meetings, replacing it with virtual meetings, and closing socially dense facilities like gyms, garden parks, mosques, and training	Enforce social distancing by stopping physical meetings, replacing it with virtual meetings, and closing socially dense facilities like gyms, garden parks, mosques, and training	No Gaps

(continued)

Table 5 (continued)

Challenge	Key response action	Desired response	Gap (deficiency/weakness/lack of Assets of capabilities)
	7.6 Exclude employees, who may have heart diseases, asthma, or low immunity, from attending work physically	Exclude employees, who may have heart diseases, asthma, or low immunity, from attending work physically	The relatively high employee average age (42.3 years) raises exposure to the risk of infection
	7.7 Deploy field support teams headed by the Quality, Environment, and Safety Directorate (QES) to communicate and ensure compliance with the EC, top management, and governmental instructions	Deploy field support teams headed by the Quality, Environment, and Safety Directorate (QES) to communicate and ensure compliance with the EC, top management, and governmental instructions	No Gaps
8. Ensure Company readiness to handle and contain any infection spread scenarios	8.1 Prepare quarantine rooms in case needed	Prepare quarantine rooms in case needed	No Gaps
9. Lack of workforce sufficient to operate plants and equipment	9.1 Employees are asked to perform additional tasks and activities to ensure production continuity	Employees can perform additional tasks and activities to ensure production continuity	Lack of job rotation practices
	9.2 Physically present, employees are willing to cover more than one shift	Physically present employees cover more than one shift	No Gaps
	9.3 Defer any nonurgent activities and focus on utilizing available resources	Defer any nonurgent activities and focus on utilizing available resources	No Gaps

(continued)

Table 5 (continued)

Challenge	Key response action	Desired response	Gap (deficiency/weakness/lack of Assets of capabilities)
	9.4 Risk of interrupting off-site activities such as equipment manufacturing and steel fabrication	Establish Risk Management Committees to study, issue, and monitor the implementation of Risk Management Plan	No Gaps
	9.5 Risk of leaving construction sites unsecured	Establish Risk Management Committees to study, issue, and monitor the implementation of Risk Management Plan	No Gaps
	9.6 Risk of delays in project engineering and construction activities	Keep projects under steady progress and increase the efficiency of operation for the delayed project	No Gaps
	9.7 Risk of extra costs related to contract deferrals	Renegotiate the contracts and ask contractors to submit a revised schedule	No Gaps
	9.8 Risk of losing warranties for delivered equipment	Reach settlement agreement with equipment suppliers to extend the warranty	No Gaps
	9.9 Risk of delays in inspections, commissioning, and testing due to travel restrictions	Conduct Factory Acceptance Test (FAT) remotely and perform other tests after delivering material to the site	FAT tests at suppliers' factories have not been conducted promptly

(continued)

Table 5 (continued)

Challenge	Key response action	Desired response	Gap (deficiency/weakness/lack of Assets or capabilities)
10. Pay vendors for supplies of goods and services	10.1 Sufficient cash is withdrawn from bank accounts to cover small cash needs	Sufficient cash is withdrawn from bank accounts and added to the contingency cash to cover small cash needs for instant payment	No contingency cash is available to handle contingency payments in crises
	10.2 A full-time employee is assigned to perform urgent transactions and to process wire transfers	Enough full-time employees are assigned to perform urgent transactions and to process wire transfers	No Gaps
	10.3 Authorize payments by the Chief Executive Officers (CFO), and the financial controller remotely	Authorize payments by the Chief Executive Officers (CFO), and the financial controller remotely	No Gaps
11. Prepare Letters of Credit hard copies	11.1 There was no interruption in production and sales during the pandemic. To complete sales transactions, LC's is opened to secure payments; one employee is requested to come to the office to prepare the needed hard copy documents and sign it along with the CFO	There was no interruption in production and sales during the pandemic. To complete sales transactions, LC's is opened to secure payments; one employee is requested to come to the office to prepare the needed hard copy documents and sign it along with the CFO	No Gaps
12. Payroll preparation and depositing	12.1 Prepare salaries and authorizations remotely	Prepare salaries and authorizations remotely	No Gaps
13. Ensure Human Resource availability	13.1 Initiate immediate communications with all departments to determine the required human resources whose physical presence is required and take necessary measures to secure their presence	Actual process mapping defines optimal human resource needs and develop measures to secure their physical presence	This action requires actual process mapping that defines optimal human resource needs and develop measures to secure their physical presence

(continued)

Table 5 (continued)

Challenge	Key response action	Desired response	Gap (deficiency/weakness/lack of Assets of capabilities)
14. Ensure enough residence and daily needs for employees	14.1 Distribute housing among employees while giving priority to operations departments	Distribute housing among employees while giving priority to operations departments	No Gaps
	14.2 Raise food inventory to serve employees' needs for forty days ahead	Raise food inventory to serve employees' needs for forty days ahead	No Gaps
15. Movement restrictions in the face of employees and inbound/outbound goods and documents	15.1 Collaborate with local agents to handle purchased goods transportations, storage, and delivery	Collaborate with local agents to handle purchased goods transportations, storage, and delivery	No Gaps
	15.2 Intensive communications with governmental and third parties to ensure shipping documents are issued	Intensive communications with governmental and third parties to ensure shipping documents are issued	No Gaps
	15.3 Collaborate with all departments to assist in delivering documents among company sites	Collaborate with all departments to assist in delivering documents among company sites	Core operations cover wild areas in Ghore Site, Aqaba Site, which impose some difficulties in managing and monitoring response plans that need to be considered within the BCP
	15.4 Increase company vehicles fleet flexibility in terms of accessible locations and utilization to ensure human resources physical presence	Increase company vehicles fleet flexibility in terms of accessible locations and utilization to ensure human resources physical presence	No Gaps

(continued)

Table 5 (continued)

Challenge	Key response action	Desired response	Gap (deficiency/weakness/lack of Assets of capabilities)
	15.5 Communicate and verify any documents electronically before physical movement	Communicate and verify any documents electronically before physical movement	No Gaps
16. Secure strategic and essential goods/services needed to ensure operations continuity	16.1 Intensive communications with strategic vendors to make sure goods are delivered on time and take any discrepancies into account	Intensive communications with strategic vendors to make sure goods are delivered on time and take any discrepancies into account	No Gaps
	16.2 Cooperate with existing service providers who have movement flexibility to handle local market activities	Cooperate with existing service providers who have movement flexibility to handle local market activities	No Gaps
	16.3 Obtain necessary exceptions to clear goods from ports on time to avoid any delays	Obtain necessary exceptions to clear goods from ports on time to avoid any delays	No Gaps
17. Expiry of Microsoft license and ORACLE ERP support agreements	17.1 Make sure all systems are covered by completing Microsoft license and ORACLE ERP support agreements ahead of time	Make sure all systems are covered by completing Microsoft license and ORACLE ERP support agreements ahead of time	No Gaps
18. Ensure proper remote working and communications without disruptions	18.1 Renew essential software providers licenses and service agreements like Microsoft and Oracle to ensure users access	Renew essential software providers licenses and service agreements like Microsoft and Oracle to ensure users access	No Gaps

(continued)

Table 5 (continued)

Challenge	Key response action	Desired response	Gap (deficiency/weakness/lack of Assets of capabilities)
	18.2 Secure required software and equipment for users whose physical presence is not required	Secure required software and equipment for users whose physical presence is not required	No Gaps
	18.3 Provide online meetings software like Microsoft Teams and Zoom Video Conference	Provide online meetings software like Microsoft Teams and Zoom Video Conference	No Gaps
19. Ensure the effectiveness of IT business continuity and disaster recovery plans	19.1 Test, review, and update plans to ensure all systems are running 24/7	Test, review, and update plans to ensure all systems are running 24/7	No Gaps
20. Provide continuous IT support	20.1 IT engineer is available at the site during the whole period to support users in action	IT engineer is available at the site during the whole period to support users in action	No Gaps
21. Ensure taking backup tapes and adequate data center conditions	21.1 Ensure one person always available with access to Data Center	Ensure one person always available with access to Data Center	No Gaps
22. Ensure proper medical services, capabilities, and procedures	22.1 Ensure that COVID-19 Polymerase Chain Reaction testing (PCR) is available and train the Medical Services Department staff to use them	Ensure that COVID-19 Polymerase Chain Reaction testing (PCR) is available and train the Medical Services Department staff to use them	No Gaps
	22.2 Conduct epidemiological investigations for employees	Conduct epidemiological investigations for employees	No Gaps
	22.3 Prepare separate reception and inspection rooms at the company hospital to deal with COVID-19 suspected cases	Prepare separate reception and inspection rooms at the company hospital to deal with COVID-19 suspected cases	No Gaps

5. Deploy face recognition attendance system.
6. Encourage and promote job rotation practices to equip employees with multiple skills. This will increase flexibility in dealing with contingencies and enhance employee engagement.
7. Secure enough contingency cash amounts on all company premises to avoid cash shortages and to guarantee business continuation.
8. Consider more robust modes of transportation for the product by adopting a more feasible trucking outsourcing model to improve availability and to meet customer's needs.
9. Consider more efficient remote FAT in collaborations with major equipment suppliers for APC.
10. Institutionalize risk management practices at APC to create healthy risk management culture and to include risk management thinking throughout the company structure to enhance risk management maturity level.
11. Document and disseminate the whole risk management actions and RMP in APC during COVID-19.
12. Develop a procedure map with a process mapping.

Acknowledgements We thank all APC's employees for their cooperation and contribution.

References

1. Blyth, M.: Business Continuity Management: Building an Effective Incident Management Plan, pp. 17–23. John Wiley & Sons Inc., New York (2009)
2. Fani, S.V., Subriadi, A.P.: Business continuity plan: examining of multi-usable framework, The fifth information system international conference. Proc. Comput. Sci. **161**, 275–282 (2019)
3. Muflihah, Y., Subriadi, A.P.: A basic element of IT business continuity plan: systematic review. J. Informatika **12**(1), 17–23 (2018)
4. Gallagher, M.: Business Continuity Management: How to Protect Your Company From Danger. Prentice-Hall, pp. 57–128 (2003)
5. Stark, C.: Barriers to Effective Business Continuity Planning. Risk working papers No.2, University of Aberdeen (1999)

The Impact of COVID-19 on Human Resource Practices

Malek B. Elayan

Abstract Recently, it has been noted that COVID-19 prevalence is increasingly affecting organisations operations in varying dimensions. In equal measures, the HR operations have significantly been affected. Through an evaluation of different sources and individual experience, there is a need for the employees and employers to establish relevant strategies in ensuring that they promote continuity in their organisation operations. In the modern business environment, HRM plays a critical function in ensuring that all employees operations and stakeholders are implemented and effective. In this case, they have to play an active role in determining the approach to be followed in the return to work for all employees, flexible working strategies and promoting the health and well-being of their employees. This would assist in imparting the employees with the required skills for harnessing an effective organisation operation. To maximise fully on technology, it would be possible to adopt technology to facilitate effective remote working and online platforms.

Keywords COVID-19 · HRM (human resource management) · HRM practices and strategies

1 Introduction

- The COVID-19 has negatively affected the contemporary world of work quantitatively and qualitatively. This is evidenced by the extent to which organisations globally have invested extensively in people management strategies in the past months. These strategies include remote working, working hours reduction, reduced costs of training and development strategies, and banning on recruitment, among others [1]. Different approaches adopted by the modern organisation in this period are affirmed by [2] study that had focused on Human Resource (HR) function approach on recession impacts and hypothesised that HR plays a critical

M. B. Elayan (✉)
Business Administration Department, Institute of Public Administration, Riyadh, Kingdom of Saudi Arabia
e-mail: Elayanm@ipa.edu.sa

© The Author(s), under exclusive license to Springer Nature Switzerland AG 2021   311
M. T. Alshurideh et al. (eds.), *The Effect of Coronavirus Disease (COVID-19) on Business Intelligence*, Studies in Systems, Decision and Control 334,
https://doi.org/10.1007/978-3-030-67151-8_18

role in implementing different strategies. These are primarily operational in their existence and focus on achieving holistic business performance. Despite various studies having focused on evaluating the role of HR in times of recession and other crisis, [3–5] there exist limited authors who have delved on evaluating the impact of HR in the event of health calamities such as the COVID-19 pandemic. The limited research in this area exists despite the impact of the 101-year old Spanish Influenza that had cost more than 50 million lives is evident. Many organisations globally have failed in integrating strategies that would place them appropriately to tackle the aftermath of the COVID-19. In this chapter, the impact of COVID-19 pandemic on HR practices has been evaluated. To achieve this, a focus on the HR activities at play in the current crisis have been evaluated. Also, the role of top management in supporting the HRM strategies, and proactiveness of HR management in assisting their organisations in absorbing the impacts of the COVID-19 to the HR has been equally evaluated.

2   Research Problem

As noted in [6], as a consequence of the COVID-19 pandemic, more than 72% have been working from home with 40% turnovers being evident. This phenomenon has been common globally with most of the modern organisations being affected by the COVID-19 implications [6]. Based on this view, [7] noted that as a consequence of the prevalence of COVID-19 pandemic, organisations are integrating their mission, values and societal impacts. This is with a focus on their employees, stakeholders, and the community [7]. In specific, modern organisations are supposed to structure their HR practices in a manner that is promoting a climate of leadership and management engagement while providing timely, accurate, assertive, elaborate and consistent communication. This is in the entire organisation levels and as a best practice of the employee's management as they are the most resourceful asset. Taking this problem statement into account, this chapter intends to evaluate the impact of COVID-19 on the HRM practices.

3   Impacts of COVID-19 on HRM Practices

A majority of the issues having a direct impact on the HRM practices as a result of COVID-19 is as a result of people management aspects. In HRM, people management are identified by [8] as being inclusive of strategies set in place by organisation in the management of their workforce and the changing practices. From a general point of view, as a result of COVID-19, the workplaces are now identified as being increasingly agile and responsive owing to the changing phenomenon. This means that organisations must be privy with the need to be compliant with different legal issues and the internal policies and procedures, which could be ignored to offer increased agility and flexibility levels. The importance of this, as noted in [9], is that the agile HRM practices would be ready to proactively respond to the changes in their business environment rather than assuming only an observatory and reactive role. For instance, if organisations continued with their management practices similar to how they were doing them prior to the pandemic, their operations would be more productive and coordinated. In general, the employees now have to learn promptly about supporting an active report working. The phenomenon is also evident in areas of performance, altering the changing benefit scenarios and overall psychological well-being.

4 Current HRM Practices Changes as a Result of COVID-19

The majority of the existing discussions on the impact of COVID-19 on the working practices and HRM are majorly general and primarily focused on defining a new normal. For instance, the discussions argue on the possibility that flexible working or remote working would be a new norm. Despite the validity of these discussions, it is essential to evaluate the impact of the COVID-19 pandemic on the existing working practices, well-being, and HRM in a more specific context. In a study by [10] that had recommended a fuzzy input–output optimisation modeling framework for the organisations operations in the crisis condition, the best practice in operating in such a phenomenon has been provided. The study noted that the HR impact the employee's livelihood and the overall workforce in entirety. The need to systemize the approach of HR on the worker's roles in today's business environment and the future is informed by the likelihood of the emergence of pandemics in the future. This is since all organisations; HR needs to be adequately prepared in mitigating the effects linked to such pandemic's occurrence.

Apart from the people management, business continuity is also critical in their working practices. In particular, the best practice in modern COVID-19 time would involve making sure that organisations are operating through operational technology. Part of this is ensuring that there prevails a high-level communication, being supervised, increased productivity, and performance management. Additionally, the levels of the realignment of the employment gains and redesign of existing policies pertaining to remote working being put into account. [11] report has identified these impacts as including practices to ensure organisations navigate through an active disruption.

4.1 Training and Development Strategies in Remote Working

Globally, the best practice in the period of COVID-19 has been encouraging and ensuring that employees are working from home. This is affirmed by [1] report that had pointed out that during the COVID-19 period, approximately 94% of world employees operate in countries characterised by some degree of workplace closure and measures prioritised. Also, 70% of employees in the modern workplace have ended up working remotely away from their physical workplaces. For organisations to be successful in flexible working, [12] report that had focused on the teleworking and virtual organisations had recommended on the need for ensuring that all employees are provided with skills and knowledge for remote working. These strategies would include the provision of the worker training and HR manager training, mentoring and coaching strategies, and coming up with employee resource groups targeting the remote workers and families. The training would need to scope from establishing a physical office and virtual workspace for time and wellbeing management and guidance to the managers. This is affirmed by the [13] report that had published the best strategy of remote working. Through the use of this strategy, flexible working is put in place that assists an organisation in attracting the top talents, increasing diversity, and improving the overall employee's wellbeing. In the period of COVID-19 and other pandemics, remote working is an effective strategy of supporting the level of business continuity to harness the agile working strategies and other unexpected challenges. The situation is evident from the different organisations are increasingly budgeting resources for re-skilling their workforce. The major challenge for organisations is, however, ensuring that they are designing one-off training programs to their effective one-off training strategies. This challenge cannot be solved by establishing a single program. The best practice would be for the organisations establishing effective ecosystems actively implemented through the adoption of highly powered inventory skills, reskilling strategies, and ensuring that they explore fresh partnerships in active collaboration with both traditional and non-traditional institutions of learning.

4.2 Harnessing Employees Well-being and Future of Workplace

Prior to the emergence of the COVID-19 pandemic, organisations had already focused on enhancing a full disruption in their job's performance, automation strategies, and demographics in their roles. Nevertheless, limited organisations have prioritised on the aspect of well-being as an approach to establishing a future holistic workplace. The importance of the wellbeing of the employees has been made complex by the COVID-19 phenomenon. This is since apart from the mental health issues, the prevalence of COVID-19 has equally contributed to the physical health issues in the modern workforce. The phenomenon is also evident for the employees currently involved in remote learning as they usually are facing mental issues as a consequence of isolation and physical challenges to accommodate their new workspaces. This is since while involved in the work requirements, and such employees could equally be engaged in activities that involve caring for their family, home-schooling roles, and being worried about their immunocompromised or the absent friends and the members of their family. The recommended strategies are affirmed by [14], who had noted that employees engaged in remote working could ensure that they are putting in place distinct strategies for ensuring that all employees are not in any way isolated in their roles and there is a high-level sense of purpose. Additionally, the study had recommended the need for employers to integrate strategies that would ensure that the employees operate through an improved work-life balance [15]. This is by dedicating their time on home working, which is best enhanced by appreciating their financial positioning, setting of varying financial goals, and understanding their trends of using the financial resources.

As a result of the effects of the COVID-19 on the employees' normalcy of doing things, there is a high possibility that their anxiety would be directly affected. As a result, modern organisations must be best positioned in putting in place reliable strategies for mitigating the prevalence of anxiety as a consequence of the coronavirus.

4.3 Re-definition of the Modern Business Practices

The COVID-19 has substantially impacted the living strategies of the populations and doing business. A research conducted by [16]. had noted that these implications would be transitioned beyond the use of Artificial Intelligence and mobile supercomputing. Part of the rethinking process would involve a direct impact on the extent to which the coronavirus presents an opportunity of rethinking the varying assumptions of the products, services, and business model. This is on top of the need to integrate blended training strategies tailored towards arming the employees with relevant skills to assist them in their future practices. As aforementioned, through an increased shift in the workplace strategies, COVID-19 has already substantially altered the extent of modern working in the areas of cooperation between different stakeholders, working remotely, re-skilling strategies, hiring that is informed by existing skills and harnessing the transformation level of sustainable corporate learning. These strategies have been recommended in [17] study that noted them as creativity, which involves establishing and communicating ideas that are sustainable and meaningful. This is since the HRM practices is a critical determinant in the process of controlling and shaping the workplace attitude, behaviours and skills and individuals. Also, HRM determines the most effective policy and practice that can be followed by contemporary organisations in the arrangement and management of work, which is also inclusive of the fundamental structure of the work organisation. It is through such strategies that organisations are best positioned in mitigating the negative implications of the calamities. A similar phenomenon is applicable in the case of COVID-19 where innovation has been prevalent to redefine modern and future workplace practices.

4.4 Reduced Working Hours

As a result of the prevalence of the COVID-19 pandemic, organisations have been prompted to make considerations of requesting their employees to reduce their hours of work to work temporarily. In some countries such as the UK, [18] report points out that the Government has put in place a strategy identified as a partial furlough pay. As a consequence, based on the ordinary employment law, employers through senior management have to agree with the temporary or permanent contractual change. This is a change to a part-time working role, which could be a costly practice if they have to wait until the end of the Furlough scheme. Organisations operating in the COVID-19 period have been required to have precise information pertaining to the rationale of reduction of the working hours and increasing their level of preparedness in responding to all issues of an organisation. The situation would be worse for the employees working for normal hours having to receive less pay and, as such, would receive feelings of demotivation.

5   Role of Top Management in Supporting HRM Practices During and After the Pandemic

It is a core responsibility of leadership in organisations to ensure that their culture is all-inclusive and that employees possess feelings of operating in an environment that support and care for them. As noted in [13], the need for top management coordination is informed by the fact that the HRM practices in the pandemic period tend to affect individuals in different positions along with the hierarchical structure. Since the HRM are the executor of the strategies of transitioning through the current period of change, the senior management would need to put in place the relevant principles and measures for implementing in their workplaces. The core function of the top management is ensuring that they take the primary responsibility of actively navigating their entities within the period of disruption. Also, owing to the increased restrictions put in place by the different governments, the top management roles would ensure that they put in place relevant measures of ensuring that their organisations return to a new norm.

5.1 Planning

It is a responsibility of the top management to support the HR in setting the most appropriate working arrangements during and after the pandemic. This is affirmed by [19] report that had focused on the relevance of the transformational leadership style in enhancing the wellbeing of the employees. The study had recommended that it is the core function of the senior management to work collaboratively with the HRM to ensure they are safeguarding the health and well-being of their employees. In the period of the pandemic, employee's anxiousness is already extensively high with increased expectations of understanding the strategies being put in place by their organisations for retaining their support for the physical and mental health. This is on top of their current views on the flexible and remote working and its effectiveness to the modern employees' operations. This is identified as being a core factor in the decisions and plans being put in place by modern organisations. [18] had recommended that prior rolling out any strategy relating to the mitigation of COVID-19 negative implications, and they must focus on the understanding of they are essential, sufficiently safe, and mutually agreed. It is through this that the HRM can integrate in their practice the aspects of the Government's Job Retention Schemes, health and safety measures in practice, and also flexibility in terms of working times and schedules as a strategy of accommodating distinct issues.

5.2 Legal Considerations

The prevalence of COVID-19 globally has contributed to evidently identification of prevailing gaps in the employment laws that need to be addressed today and in the future. For instance, the increased instances of lay-offs have contributed to the need for understanding if they could be addressed as redundancies. As identified in the law, redundancy is identified to arise in an event the employment contract has already been terminated through an employer strategy as a result of their services becoming superfluous. As evidenced in [20], the legal issues placed to the senior management are similar to the existing legal gap in terms of algorithmic management and gig economy. Apart from the issue with flexible working and return to work after the effects of the pandemic have subsidized, payroll management is also an issue that the senior management must guide the HRM.

Additionally, despite the HRM being required to put in place reliable health, safety, and well-being strategies in their workplace, it would be prerogative of the senior management to decide on the number of employees to return to work after the pandemic. An elaborate identification of the extent of disruption of the contracts of the employees, roles in the workplace, and working strategies with staff culture would have to be considered. As recommended by [16], organisations senior management must similarly partner with the HRM to ensure that they establish strategies for skilling their inventory, reskilling of their employees, and an exploration of private partnerships with the traditional and non-traditional institutions of learning. All these strategies would be intended to establish appropriate recruitment strategies of employees possessing the appropriate skills and capabilities rather than the necessary education.

5.3 Employees Embrace to Change

Inexistence of successful employee buy-in, the plans by the senior management in successfully mitigating the disruption would not be successful. As such, [11] had recommended the need for the top management to ensure that their teams practice high-level empathy. The top management must equally ensure that their employees are demonstrating a detailed appreciation of the crisis implications. This is since there are varying impacts of the COVID-19 to the employees. For instance, some of the employees could have a high-level of serious effects of COVID-19 infections and, as such, being selected to be reluctant to return to their offices. On the other hand, some of the employees could be eager to practice remote work roles but also having caregiving roles that could challenge their abilities to practice their roles.

It is equally vital for the top management to put into the account on the need to streamline the HRM approach to the modalities of work post the pandemic. This is affirmed by [21] report that had focused on COVID-19 impact on modern organisations. In this case, the employees coming back after an extended furlough or a period of remote working could identify the physical layout of their workplace changing with their overall shift being altered. For some employees, a triumphant return to their workplace role could demand a shift in their mindsets, particularly for the employees who work remotely. For the sake of navigating in the distinct changes, the management could liaise with the HRM in ensuring that they appreciate what is required of them and the steps to be adopted in an organisation in mitigating the different implications of their overall well-being.

6 Impact of Technology on HR Practices in COVID-19

A focus on the employee's experience regarding the effects of the COVID-19 pandemic, it is possible to provide suitable information on the effectiveness of stakeholders in improving their roles in flexible employment relationships. To affirm this, [22] had noted that for the different online labour platforms and contracting firms, the experiences of the employees offer sufficient information on addressing the employees demands. It is these insights that provide ample information on aligning their technology infrastructure and other services to achieve high-level sustainability. This is mainly regarding the adoption of modern technology in active support and promotion of work performance and hence an improved workforce quality. Also, through the use of technology, employees increase their awareness of their working conditions in the context of health, work roles, and career development. These aspects lead to an empowered employee in terms of their consciousness on self-care and requesting for the rights of their employees.

7 Conclusion, Implications and Directions for Future Research

In summary, this chapter has established that as the prevalence of COVID-19 continues to affect different countries in distinct ways, HR operations are equally affected in equal measures. The findings in this report have established the need for the employees and employers to put in place relevant approaches for harnessing the continuity of their operations. Apart from the expected return to work by the employees, other strategies include the extent to which the organisations improve flexible working practices and enhanced well-being. The senior management must equally take an active role in ensuring that the HRM is putting the relevant measures of ensuring a successful transition in their workplaces. The demand for a new training protocol has also be established where the employees need to acquire relevant skills for the smooth operations of their organisations. A successful maximising of the use of the technology for facilitating remote working and online services with the health and wellbeing of all employees being prioritised have been identified and recommended. Ensuring the existence of active communication between all stakeholders in an organisation would be useful in enhancing a shift of workforce to the new normal of working from home and ensuring a constant interaction with all stakeholders, which would eventually improve on their experience. These findings hence indicate that despite the COVID-19 pandemic having a significant challenge for the sustainable of modern HRM business environment, there are substantial areas where they will positively impact HRM performance. This includes areas such as reimagining work roles in the contemporary, challenging business environment which could undoubtedly accelerate the future of work. Also, COVID-19 pandemic will positively contribute to improved learning on how to build an improved resilience and efficiencies in organisations operations to assist people to live healthier and assume more sustainable lives.

Despite of the existing studies having evaluated on how COVID-19 has altered the modern HRM environment, there still lacks a higher scope of research in this area. This could be informed by the fact that the pandemic is still a new concept and hence a successful cross-pollination of concepts and knowledge along distinct disciplines is still work on progress. Hence, as a best future practice, scolars need to focus more on gaining relevant knowledge for supporting relevant management practice to promote best HRM practices.

Acknowledgements Thank you for your cooperation and contribution.

References

1. International Labour Organisation. COVID-19 and the World of Work (2020). Online available at https://www.ilo.org/wcmsp5/groups/public/@dgreports/@dcomm/documents/briefingnote/wcms_745963.pdf. Retrieved on 21/6/2020
2. Gunnigle, P., Lavelle, J., Monaghan, S.: Multinational companies and human resource management in Ireland during recession: a retrospective from a highly globalized economy. Thunderbird Int. Bus. Rev. **61**(3), 481–489 (2019)
3. Susaeta, L., Suarez, E., Pin, J.R.: Economic crisis and communication: the role of the HR manager. Bus. Syst. Rev. **2**(2), 278 (2013)
4. Lai, Y., Saridakis, G., Blackburn, R., Johnstone, S.: Are the HR responses of small firms different from large firms in times of recession? J. Bus. Ventur. **31**(1), 113–131 (2016)
5. Lange, T.: Social capital and job satisfaction: The case of Europe in times of economic crisis. Eur. J. Industrial Relations **21**(3), 275–290 (2015)
6. SHRM. SHRM COVID-19 Research Teleworking Realities & Challenges (2020). Online available. https://www.ioe-emp.org/fileadmin/ioe_documents/publications/COVID-19/SHRM_Johnny_C_CV19_IOE_Presentation_for_JT_3.22.pdf. 12/08/2020
7. Walden University. What HR Managers Should Know About COVID-19 (2020). Online available https://www.waldenu.edu/online-masters-programs/ms-in-human-resource-management/resource/what-hr-managers-should-know-about-covid-19. 12/08/2020
8. Brown, D., Hirsh, W., Reilly, P.: Strategic Human Resource Management in Practice: Case Studies and Conclusions–from HRM Strategy to Strategic People Management (2019)
9. Heilmann, P., Forsten-Astikainen, R., Kultalahti, S.: Agile HRM practices of SMEs. J. Small Bus. Manage. 1–16 (2020)
10. Aviso, K.B., Mayol, A.P., Promentilla, M.A.B., Santos, J.R., Tan, R.R., Ubando, A.T., Yu, K.D.S.: Allocating human resources in organizations operating under crisis conditions: a fuzzy input-output optimization modeling framework. Resour. Conserv. Recycl. **128**, 250–258 (2018)
11. PWC. Returning to the Workplace after COVID-19 (2020). Online available at https://www.pwc.com/us/en/services/governance-insights-center/library/covid-19-returning-workplace-boards.html. Retrieved on 21/6/2020
12. Lamond, D., Daniels, K., Standen, P.: Teleworking and virtual organisations: the human impact. The New Workplace, 197 (2003)
13. CIPD. Getting the Most from Remote Working (2020). Online available at https://www.cipd.co.uk/knowledge/fundamentals/relations/flexible-working/remote-working-top-tips. Retrieved on 21/6/2020
14. Bahar, U.: Impact of Emotional Labor on Employee Wellbeing: Mediating Role of Eustress and Moderating Role of Psychological Capital (Doctoral dissertation, Capital University) (2019)
15. AON. COVID-19 Outbreak: Supporting Employee Wellbeing (2020). Online available at https://www.aon.com/getmedia/12ef3be1-0d77-477a-86bd-0ef96372cbd0/Supporting-Employee-Wellbeing-During-The-COVID-19-Outbreak-2020-03-27.aspx. Retrieved on 21/6/2020
16. Forbes. The Impact of the Coronavirus on HR and the New Normal of Work (2020). Online available at https://www.forbes.com/sites/jeannemeister/2020/03/31/the-impact-of-the-coronavirus-on-hr-and-the-new-normal-of-work/#74af39d22b60. Retrieved on 21/6/2020
17. Nagrani, K.: Enterprise Value Creation Through Creative HR Practices. Prof. Komal Nagrani City Premier College, Nagpur Co-Author: Dr. Owais Talib. Enterprise Value Creation Through Creative Management Practices, p. 159 (2019)
18. CIPD. COVID-19: Returning to the Workplace (2019a). Online available at https://www.cipd.co.uk/knowledge/fundamentals/emp-law/employees/workplace-guide-returning-after-coronavirus. Retrieved on 21/6/2020
19. Nielsen, K., Yarker, J., Brenner, S.O., Randall, R., Borg, V.: The importance of transformational leadership style for the well-being of employees working with older people. J. Adv. Nurs. **63**(5), 465–475 (2008)

20. Duggan, J., Sherman, U., Carbery, R., McDonnell, A.: Algorithmic management and app-work in the gig economy: a research agenda for employment relations and HRM. Human Resource Manage. J. **30**(1), 114–132 (2020)
21. Sugandha, S., Rastogi, S.R.: Covid-19 crisis and motivation as a tool to reduce stress in the organization. CLIO Ann. Interdisc. J. History **6**(10), 263–266 (2020)
22. Spurk, D., Straub, C.: Flexible Employment Relationships and Careers in Times of the COVID-19 Pandemic (2020)

Social Media

The Effect of Service Recovery Justices Strategies on Online Customer Engagement Via the Role of "Customer Satisfaction" During the Covid-19 Pandemic: An Empirical Study

R. Al-Dmour, F. AlShaar, H. Al-Dmour, R. Masa'deh⊙, and M. T. Alshurideh⊙

Abstract This research aims at examining how the service recovery justices' strategies influence online customer engagement (OCE) in Jordan's telecom sector via the role of "customer satisfaction" as a mediating variable during the Covid-19 pandemic. Based on the literature review, a conceptual framework was developed to guide this study. A Web survey questionnaire was used to collect the required data from all customers who have been experienced with any failure problem with Telecommunication service providers during the period of the Covid-19 pandemic in Jordan. A convenience sample of 357 respondents was reached. Linear regression and partial correlation analysis were used to validate and test the research hypotheses. The empirical findings of the study confirmed that there is a significant relationship between service recovery justices' strategies and online customer engagement and proved the effective role of customer satisfaction as a mediating variable in this relationship in the telecom service industry: These findings may aid future researchers in their quest to understand the inherent relationships that lie between the constructs' questions and

R. Al-Dmour (✉) · R. Masa'deh
Department of Management Information Systems, School of Business, The University of Jordan, Amman, Jordan
e-mail: rand.aldmour@ju.edu.jo

R. Masa'deh
e-mail: r.masadeh@ju.edu.jo

F. AlShaar · H. Al-Dmour
Department of Marketing, School of Business, The University of Jordan, Amman, Jordan
e-mail: fawzialshaar@yahoo.com

H. Al-Dmour
e-mail: dmourh@ju.edu.jo

M. T. Alshurideh
Department of Marketing, School of Business, The University of Jordan, Amman, Jordan
e-mail: m.alshurideh@ju.edu.jo; malshurideh@sharjah.ac.ae

Department of Management, College of Business Administration, University of Sharjah, Sharjah, United Arab Emirates

© The Author(s), under exclusive license to Springer Nature Switzerland AG 2021
M. T. Alshurideh et al. (eds.), *The Effect of Coronavirus Disease (COVID-19) on Business Intelligence*, Studies in Systems, Decision and Control 334,
https://doi.org/10.1007/978-3-030-67151-8_19

may provide a platform for telecom service providers in their efforts in developing service recovery justices strategies and choosing an appropriate manner to satisfy customers and reinforce online customer engagement.

Keywords Service recovery · Justices strategies · Customer satisfaction · Online customer engagement · Telecom sector · Jordan

1 Introduction

The Covid-19 pandemic has forced online service companies to maintain and build relationships with consumers when their world turns. Businesses are now facing tension between generating sales during a period of severe economic hardship and respect for threats to life and livelihoods that have changed consumer preferences. This tension is very real, especially for telecommunication service companies that provide access to telephone and related communications services. Across the business environment, the telecommunication industry is considered a vital socio-economic engine of contemporary economies [1, 2]. This ought to be justified according to its inputs and contributions to other sectors' evolution; as no one could deny its role in either emergency, disease outbreak, disasters and risks as well as in enabling country-urban migration. Telecommunication service providers over the world play a vital role in ensuring communities remain connected during the coronavirus crisis to support crucial community needs. Support includes measures for those struggling to meet bills, removing data caps on broadband services; and offering new mobile and landline packages—including free calls—to help people stay in touch with each other. Some telecommunication service providers have issued compliance advice with "specific details on automatic compensation and the Broadband Speeds Code of Practice".

The spread of coronavirus is likely to boost digital media consumption across the board as people spend more time at home and communicate in person less. In Jordan, where the effects of the virus are still relatively new, we expect digital media consumption to increase across social media, over-the-top video and online gamin-similar to what we have already seen in other countries such as USA and China. Social networks could be a major beneficiary, as people turn to these platforms to connect with friends and family who may be at a distance or to access news content. Beyond encouraging more virtual conversation between family and friends, the fast-changing nature of the coronavirus could translate to more users tracking real-time news updates via social media. Despite the concerns they may have about tech companies' ability to act as a reliable information source, more than half (65%) of the Jordanian population got news from social media networks. But despite the increased reliability and modern flexibility, Telecommunications networks for physical damage, the associated risk of communication failure remains serious due to the increased dependence on these tools in emergency operations.

During the Covid-19 pandemic, customers experience deep levels of uncertainty, requiring telecoms to be flexible to remain responsive to their needs. Rising demand adds pressure to the network, which affects service levels at call centres. Moreover, increased demand for telecommunication services in a pandemic may cause customers to switch to higher subscription plans or buy additional data packs. Therefore, telecommunications service operators must continuously try to identify customers' most likely need their services and determine how best to support them as well as sort out their problems—for example, by providing flexible payment options or recovery service system. This continuous process requires a well-planned customer service system for recovery that empowers the desired excellence in telecom service operators and retaining their current customers also, acquiring new customers. The practical significance of service recovery is the core tool to deal with cases of dissatisfied customers who avoid complaining to the service providers [2]. However, the minority, which opposes mainly adds up material information to make the journey of improving customer satisfaction much more manageable. Commonly, businesses adopt different recovery strategies in this situation, like apologies, empathic listening, fixing the core problem, remediation and compensation in response to any service breakdown.

Since the effect of service recovery on online customer engagement on telecommunications service sector has not empirically been studied before, this study has come to answer the following question: "Is there any significant relationship between service recovery justices strategies (SRSs) and online customer engagement via customer satisfaction in telecom businesses operating in Jordan during the Covid-19 pandemic?". During this time, people are being forced to change their lifestyles, adjust to working from home and reevaluate their needs. Therefore, companies should understand and keep their customers satisfied and highly engaged. Therefore, the results of this study are expected to be valuable to telecommunications service providers to develop appropriate recovery justices strategies in general and word-of- mouth strategies in particular to enhance and maintain customer satisfaction and loyalty.

2 Jordanian Telecommunication Industry

Jordan's information and communication technology (ICT) sector is a leading sector and sturdy one in the MENA region, supported by about 15 years of industry-friendly policies and a youthful, well-educated and thriving community of digital users. The sector is a significant economic contributor, designating exciting annual growth rates since the year 2000. The ICT sector encounters several obstacles, most prominently substantial tax burden which was increased by the governmental decision to double two essential industry taxes in 2013. Besides, competition between Jordan's biggest three telecom operators has dropped prices, which serves the consumer. However, it reduces the profitability of services. Nevertheless, the sector remains powerful and enduring. It is backed by open-ended government initiatives and a young, tech-savvy population, with the Kingdom's robust network of IT incubators and accelerators required to benefit from new initiatives encouraging start-ups and entrepreneurship. As a result, Jordanians have become among of the most digitally literate people in the region, with internet penetration rising from 50% in 2015 to 76% in the first quarter of 2018 or 6.9 m users. Internet subscription rose by 32.6% in 2015 from 2.46 m users in the first quarter to 2.78 m in the fourth quarter of the same year. Mobile broadband users dominate the subscriber base, accounting for 1.91 m subscribers, or 83% of the total (Jordan ICT and ITC Sector Statistics, 2018).

Three major players dominate the telecom sector in Jordan: Orange Jordan, Zain Jordan and Umniah. Despite a successful liberalization program launched in 1995, Orange Jordan is the only integrated operator in the country, offering mobile, fixed-line, wholesale telecom and internet services. Zain Jordan is the leading mobile provider, with a 40% market share, according to Zain Group's 2014 annual report. Orange Jordan holds a 31% share of the market, while Umniah holds a 29% share. The target population of this research includes the different segments of customers of telecommunication firms who are actively working in the Jordanian market.

3 Literature Review and Theoretical Background

3.1 Service Failure and Repurchase Intention

[3] defined service failure as any real and/or perceived service-related problem that appears during a consumer's experience with the service provider. Failures produce intensive, immediate and emotional reactions which include dissatisfaction, contrary word-of-mouth and complaint behaviours from consumers (either directly or to third parties) [4]. Many studies have reported service failures that have a significant impact on customer repurchase intention [5–7] also supported this view by investigated the effects of a firm's reputation for service quality on customer responses to service failures in airlines and restaurant services in Virginia, USA. The study was based on 322 responses from volunteer customers who frequently used air travel and restaurant services. The outcomes indicated that service failures had a direct negative impact on customer repurchase intentions. This conclusion was also proposed by [8]. However, other studies showed that post-recovery satisfaction has a positive significant impact on customer repurchase intentions [9]. In addition to the effect of service failure on repurchase intention, many service failure studies suggest that the severity of failure and failure criticality lead to negative *WOM communication* by customers [10, 11, 13]. Furthermore, other studies have assured that service failure plays a significant role as a moderator on the relationship between service failure and loyalty [12–14].

Since service failures are incredibly costly for firms [15] and failures are inevitable during service delivery [3], organizations need to respond effectively to failures to maintain their performance and their relationship with customers [16]. To avoid unfavorable outcomes of failures, it is essential to take corrective actions or remedies, known as service recovery. Fitzsimmons [17] has defined service recovery as "a company's resolution of a problem from a dissatisfied customer, converting them into a loyal customer". Also, it was defined as "the action a service provider offers in response to a failure in services" [17]. The concept of service recovery implies the definition of customer satisfaction as well as such a recovery would lead to thought-out and well-planned techniques to turn frustrated customers to satisfied ones with the provided services [18]. The service recovery concept has to be distinguished from complaint management, as the first one mainly stresses on service failure and the action is taken towards it, while the complaint management stresses on customer complaints as an introduction to a service failure. As stated by [19], dissatisfied customers are hesitant to provide complains, where such behaviour would limit the attempts to solve problems and recover services before other customers complain and turn to be dissatisfied. Thus, both of complaint management and service recovery is categorized as customer retention- adapted strategies [3].

Retention strategies, including follow-up and value co-creation among others, positively enhance service recovery effectiveness efforts [20]. Besides, many previous publications that tackled the service recovery issue concluded that a decent recovery positively influences the extent of satisfaction, word of mouth, recommen-dation intention, image, loyalty and trust [20]. Active service recovery could not only eliminate the loss of service failure but also improved the extent of service satisfaction compared with service failures. Arguments about the topic state that decent recovery can rise satisfaction to a more sophisticated level than if nothing had gone wrong in the first place. This is referred to as the service recovery paradox. Many researchers provided evidence on the existence of the service recovery paradox from rational customer expectations through interaction between employees and customers under service failure [21].

Despite its significance, companies pay less attention to managing the complaints of their customers [21]. This was supported by publications, which widely accepted that complaints from dissatisfied customers are not adequately handled. As a result of operating in a service-oriented world economy, companies might be shocked by this apparent disregard of a customer complaint, mainly considering customer dissat-isfaction (i.e., negative word of mouth) worth to be scrutinized at both the short term and the long terms [21, 22]. Businesses are highly projected to incur more costs than to attract new customers. In other words, it has been proposed that dealing with any customer complaint should follow one of the positive approaches to maintain the relationship with current customers who will help generate encouraging communi-cation about the company [19]. Therefore, one of the service recovery benefits is the prevention of customer dissatisfaction, which leads to customer retention. As it is more expensive to gain new customers than to maintain current ones, customers will be more profitable the longer they stay with the company. [4] have suggested a three-dimensional concept of justice with the dimensions of distributive justice, procedural justice and interactional justice. After service recovery is performed, customers evaluate the service recovery strategy through the dimensions of justice [23]. Customers then assess whether they perceive that they have obtained justice out of the service recovery or not, which, in turn, influences the extent of their satisfaction about the provided service(s) [24]. When evaluating the service recovery process, there are three dimensions of justice used: distributive justice, procedural justice and interactional justice. Since the outcome of the perceived justice has an impact one satisfaction, we will use the three dimensions of justice to examine the core of the service recovery process that makes customers dissatisfied. These three dimensions are:

(1) **Distributive Justice**: In a service recovery, the customer evaluates the organization's solution of any projected or actual failure in the provided service(s) through the dimensions of justice [25]. According to [26] distributive justice is the perceived fairness on how resources and rewards are distributed. Hence, a solution of a service failure within service recovery relies mainly on the distributive justice dimension. For distributive justice to be considered high, the compensation should reflect the costs that the customer has incurred due to the service failure [27]. However, there are not only monetary factors that customers weigh in their evaluation of distributive justice. Customers also add values involving non-monetary factors, which include time and effort they spent on complaining, or emotions they experienced during the process, such as anger and embarrassment [28]. Accordingly, distributive justice is evaluated according to the results of service recovery attributes comprehensively, but the one which customers can assess the most is the solution of the service failure. When the focus is on fixing the problem by offering a replacement or compensation, distributive justice is valued as more important to the customer than procedural and interactional justice [26]. Some of past empirical studies stated that perceived distributive justice effects on service recovery evaluation positively [29].

(2) **Procedural Justice**: Within procedural justice, the customer evaluates the procedure and progression of how organizations mainly manage the service recovery process [25]. The customer assesses the organization's policies and whether they are taken into consideration by the company [4]. It is more likely that a process is judged to be fair if the customer is granted a voice and feels that she or he is a part of what will lead to and become the outcome [26]. For this reason, the service recovery attribute of employee empowerment is essential, since employees need to be responsive to the shaped expectations about the customers' service recovery [25]. Furthermore, the employees should have the authority to provide a range of different service recovery solutions so that they can match the service recovery solutions with the customer's expectation [25–27]. Perceived procedural justice is also dependent upon the service recovery speed since the customer makes an assessment based on how quickly the organization provides the service recovery in their perception of procedural justice [27].

(3) **Interactional Justice**: Interactional justice refers to the perceived treatment during encounters from the organization in the service recovery process [24]. Here, the customers evaluate the interactions between them and the organization. Customers make an assessment based on employees who have provided a better service recovery, as well as their behaviour towards them [24, 25]. In the service

recovery context, the organization can give an apology to attempt to enhance the overall situation that the customers have experienced by showing a concern with their needs [27]. Customers expect employees to provide an honest explanation, proper treatment and a genuine effort to solve the issue [30]. The perceived interactional justice is dependent on how customers have been treated during the recovery process concerning factors, such as courtesy, respect, concern and empathy [26]. Thus, how a customer assesses interactional justice depends on the psychological efforts of the main attributes of service recovery [26].

3.2 Online Customer Engagement

The concept of customer/brand relationship has become more complicated with the creation of new platforms such as TripAdvisor, Twitter, Facebook, blogs and video-sharing platforms such as YouTube and Vimeo that facilitate this relationship via multiple devices such as smartphones and desktop computers [31, 32]. The changes in the core of the association between customers and brand can be observed based on how their usage will describe it. 'Participation', 'Interaction', 'Involvement' and 'Engagement' are different concepts that have been used in the past decade to represent the nature of customer behaviour in the online environment. It is crucial to explain the difference between these concepts to understand the reason behind the selection of the extent of customer engagement for this research. The growth of the Internet uses as a communications medium enabled geographically dispersed individuals to gather online. In this regard, [33] introduced many-to-many communication as an important Internet capability. [34] emphasized the Internet's capability to act as a medium that enables customers to access online content to communicate with companies. When discussing the introduction of online communities, the concepts of participation and interaction are the most-used ones. [35] defined the participation as "a virtual community which is a product of the frequency and duration of community visits". This does not differ from other definitions discussed in other articles [36]. Both concepts (participation and interaction) emphasized the behavioural elements of the customers in online communities. [32] stated that 'engagement' includes the extent of emotional and cognitive activity rather than mere behavioural one in brands' community. Thus, [32] defined 'customer brand engagement' as "the level of a customer's motivational, brand-related and context-dependent state of mind characterized by particular concentrations of cognitive, emotional and behavioural activity in brand interactions". After 2010, the concept of engagement has received more attention from researchers and emerged in the marketing literature. The important difference in these studies when compared with the first stream of research is in viewing the concept of engagement as a promising concept, which is expected to provide "enhanced predictive and explanatory power of focal customer behavior outcomes including brand loyalty [32]. The engagement-related terms in this group of studies are in line with the focus of the current research.

An overview of the stream of research using engagement-based concepts that could be seen in the studies after 2010 shows that the engagement-based concepts include customer engagement, which is used in the study by [31], online engagement by [37], customer brand engagement by [32] and community engagement by [38]. As this study follows the theoretical study by [31], the concept of 'customer engagement' is adopted and this term is used in the current study. The following observations from reviewing the conceptualizations of engagement in this group of studies can be made. First, the concept of engagement is represented as a multi-dimensional concept. For example, In [32] study, the concept includes three dimensions: 'cognitive', 'emotional' and 'behavioral'. The concept of engagement exhibits a motivational state that is important for many online offerings. [32] believed that the motivational state occurs by "virtue of an individual's focal interactive experiences with a particular object or agent". Reviewing the conceptualizations of engagement reveals that the individual includes customer and consumer, and the object or agent includes brands and organization.

A newly published research by [39] exclusively and precisely addressed the gap in the need for an empirical study to develop a scale of customer engagement in online brand communities. Although they attempted to bridge the mentioned gap mainly by testing the customer motivations in online brand communities, they also proposed them as different dimensions of the 'customer engagement' construct. However, there is limited literature on customer relationship marketing studies that address customer engagement in an online brand community. The study by [31] that explicitly addressed the concept of customer engagement in an online brand community was adopted to guide the development of the construct scales. It defined customer engagement in the online brand community as follows: "Customer engagement in a virtual brand community involving specific interactive experiences between consumers and the brand, and/or other members of the community. Consumer engagement is a context-dependent, psychological state characterized by fluctuating intensity levels that occur within dynamic, iterative engagement processes" [31].

There are two critical reasons for adopting the selected definition. Firstly, the definition matches the definitions of engagement-related concepts by [32]. The definition goes beyond a mere behavioural aspect; [31] explained that customer engagement includes cognitive and emotional aspects as well. Secondly, this definition shed lights on customer engagement in the context of an online brand community; this directly serves the current research setting. As mentioned, the current study has adapted the "dynamic conceptual model" of the consumer engagement process in an online brand community proposed by the research of [31]. Customer engagement comprises five dimensions: Learning, sharing, co-developing, advocating and socializing.

1. *Learning*: The first dimension is learning defined as "characterizing the vicarious acquisition of cognitive competencies that customers apply to purchase and consumption decision-making" [31]. This dimension is similar to what [31, 39] call 'seeking assistance' as a dimension of engagement in an online brand community. At this level, customers share their questions with others, so that

other customers may help them. [39] defined the dimension of 'seeking assistance' as "the degree to which a community member wants to receive help from fellow community members who share their knowledge, experience, or time with his/her". These two concepts characterize the same dimension of customer engagement in an online brand community in these two studies. The current research is in line with [31] study in adopting the term 'learning' to represent this dimension.

2. *Sharing*: The second dimension of customer engagement is known as sharing. It includes the "sharing of personal relevant information, experience and knowledge through the process of active contributions to the co-creation of knowledge within the online community" [39]. The cognitive and behavioral dimension of consumer engagement is explained in this stage. This dimension is also determined in the study by [39] as 'helping'. This engagement dimension is defined as "the degree to which a community member wants to help fellow community members by sharing knowledge, time or experience" [39]. Both dimensions reflect the same aspect of customer engagement in an online brand community, and the current research uses 'sharing' to represent the second dimension.

3. *Co-developing*: The third dimension of customer engagement in online brand communities is 'co-developing', which was defined as "a process where customers contribute to organizations and/or organizational performance by assisting in the development of a new product, brand, service or brand meaning" [31]. [31] show how customers can contribute to the development of a new product through the engagement process. They have suggested new features for the new product and justified the need for the new product. This dimension is also defined by [39] study, where it is called 'brand influence'. They defined this dimension as "the ardent affection that a community member has for the brand". The current research adopts the term 'co-developing' for this dimension.

4. *Advocating*: When customers are capable to encourage other members to buy a specific brand and recommend a service or product to them, this is known as advocating. [31] suggested that "advocating is an expression of customer engagement". This dimension includes the degree to which a community member wants to express the brand recommendation through engagement in the online brand community.

5. *Socializing*: The last dimension of customer engagement is socializing in an online brand community, which was defined by [31] as "two-way, non-functional interaction through which customers acquire and/or develop attitudes, norm and/or community language". The degree to which a community member is interested in communicating and talking with other community members exists in the socializing dimension.

3.3 Customer Satisfaction

Service recovery presents a critical moment of truth that offers an opportunity for the organization to satisfy and retain its customers [40]. Satisfaction evaluates the consumer's evaluation of a product or service according to the extent to which that product or service has met his/her needs or expectations [41–43]. Failure to meet expectations and needs are assumed to result in frustration with products or services. In the setting of service recovery, [44] proposed that SRS relates to pleasure with objection handling in the customer's evaluation of how the provider reacts failure in providing services. It is a pretty crucial determinant, as it assists in preserving sound connections with clients. Also, it refers to the degree to which customers have positive emotions to a provider's recovery efforts [44]. It leads to better customer's reaction after the consumption of service (i.e., service recovery) [45]. Accordingly, the service provider appraises the level of customer satisfaction and implements that learning to enhance the services, regularly performing a notable competitive advantage over rivals. Accordingly, customer satisfaction assessment is fundamental to enhance customer retentions and encourage positive word of mouth communication.

3.4 Service Recovery Justices Strategies and Online Customer Engagement

Service marketing literature confirms that a firm's profitability is mostly determined by the loyalty of its customers [14]. Yang et al. [46] illustrated that the increase in terms of customer retention is estimated at 5% and this ought to cause a positive growth in profit ranging from 25 to 85%. Consequently, market leaders in at today's business improved their awareness of the role of customer loyalty, as it is considered as the most valuable intangible asset especially when it comes to service firms that desire to stay in a good position in the competitive market. Besides, loyal customers represent a powerful marketing tool and force throughout their word of mouth in case it was positive [46]. It follows that one of the main goals of modern companies is to attract and maintain loyal customers [45].

[47] pointed out that the commitment and quality of services provided by a firm can be tested in response to the case of service failure. Thus, service recovery is deemed a strategic aspect for firms to compete in the marketplace [48]. It has also been recognized as a crucial matter in the past literature of both service quality and marketing eras, especially because service failure is not entirely avoidable due to the intangible nature of services [49]. Additionally, it signifies a tool that would never be eliminated by companies intending to retain satisfied and loyal customers [50]. Moreover, service recovery positively contributes to the positive evaluation of the service provider by its consumers in business to customer (B2C) context, which can develop customer relationships [44] and improve satisfaction, commitment and trust aspects [51]. [52] consider service recovery satisfaction as a primary determinant

of any long-term relationships between customers and their company. Also, it is deemed as one of the foundations of marketing [52] and is an imperative element to the vestige of any business [47]. The more customer satisfaction noticed the higher financial returns and customer retention rate is. However, when customers suffer, the firm may lose the values of all next transactions. It may lose t long-term profits as a result of a negative word-of-mouth by a dissatisfied customer [53]. Accordingly, service businesses have to pay obeisance to service recovery satisfaction because of its role in customer retention. The service recovery unit does not only solve customer concerns but further deals with seeks out and learn from the anticipated service failures [54] even before they were reported. This stresses that recovery implies a test of customer retention and contains conditions where providers do not only have complaint-response package yet also foster a corporate culture where workers are authorised and trained to rectify service failures even before complaints are registered. They refuse the possibility of learning from the lessons and encounters of handling such failures [54] contrary word-of-mouth [40].

4 Conceptual Framework

This research seeks to examine how perceived service recovery justices strategies affect online customer engagement in the Jordanian telecommunication sector and the mediating role of customer satisfaction in this relation. Based on the literature review and related studies such as [31, 43, 55, 56], a conceptual framework was developed to guide this study. It is assumed that the dependent variable" online customer engagement" could be viewed as a function of the independent variables " perceived service recovery justices strategies" through the effect of customer satisfaction as a mediating factor. The suggested hypothetical relationships are diagrammed in Fig. 1.

Therefore, the following hypotheses *are proposed on Jordan's Telecom service sector during Covid-19 pandemic*;

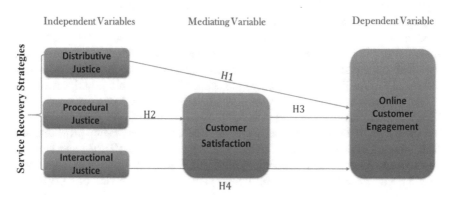

Fig. 1 The research conceptual framework

H1. There is no relationship between the perceived service recovery (SR) justices strategies and online customer engagement.

H2. There is no relationship between perceived service recovery (SR) justices strategies and customer satisfaction.

H3. There is no relationship between customer satisfaction and online customer engagement.

H4. There is no mediation effect of customer satisfaction on the relationship between perceived service recovery (SR) justices strategies and online customer engagement.

5   Research Methodology

To examine the hypotheses of this study, a systematic approach was followed to investigate the effect of perceived service recovery (SR) justice's strategies on online customer engagement via the mediating role of customer satisfaction of Jordanian telecommunication customers. Consequently, a quantitative approach was used to examine the relationships between variables. Moreover, the study relied heavily on collecting primary data from a sample of subjects who were the right group to achieve the study objectives. The target population in this study consists of all Jordanian customers who have been experienced any failure problem with telecommunication services (mobile and Internet) during the pandemic (Covid-19) period in Jordan. The required data were collected from a convenience sample of 357 respondents using a web survey questionnaire. The data was collected during the period 25th March to 10th May 2020.

In this survey, the questionnaire consisted of two sections with the first including the participants' demographic characteristics and the second relating to the variables of the study. The contents of the questionnaire (measures) were mainly selected and adopted from relevant previous studies. With regards to the measures used in this study, the mediating variable "customer satisfaction" was measured using a 5-point scale developed by [57]. Independent variable "service recovery justices' strategies" were measured using a 5-point scale developed by [1, 5, 26, 40]. The dependent variable, "online customer engagement" was measured using a 5-point scale developed by [31]. For validation purposes, the questionnaire contents were modified to the practice of Jordanian culture context based on the results of a pilot study and feedback from five professional academic staff members in this filed.

With regards to respondents' profile characteristics, 77.6% of them were males 80.6% were bachelor degree holders, 67% were from 36 to 45 years old. 84.9% has more than five years' experience with service telecommunication providers and 63.6% were Zain customers, followed by Orange customers (20.7%) and Umniah customers (15.7%). When respondents were asked about the type of problems that they have experienced with the service provider during the period of COVID-19, the majority had technical issues (63.3%), followed by customers who had financial issues (15.1%), service issues "activation/ deactivation of service" (11.2%), customer service issues (8.1%) and finally customers who had other kinds of issues (2.2%).

6 Data Analysis

6.1 Normality and Reliability

Mean, standard deviations, skewness, kurtosis and Cronbach's alpha were used to examine the normality and reliability of the study questionnaires. The descriptive statistics presented in Table 1 indicates positive disposition towards the measured items. The standard deviation (SD) values ranged from 1.040 to 1.139, which indicates a narrow spread around the means. The mean values of all items were greater than the midpoint (3.000) and ranged from 3.01 to 3.670, indicating that the respondents have relatively medium responses to the study's questions. Using skewness and kurtosis, the data was found to be normally distributed. Indeed, skewness and kurtosis were normally distributed, since most of the values were inside the adequate ranges of normality (that is, -1.000 to $+1.000$) for skewness and less than 10 for kurtosis. These outcomes confirm the ability to generalize the study outcomes to customers of the telecommunication sector. Finally, Cronbach's alpha reliability was examined for each factor/construct. Each coefficient greater than 0.60 for adapted and 0.070 for existing scales were considered a reliable indicator of the constructs under study.

6.2 Hypotheses Testing Results

This section discusses the testing of the alternative hypotheses, which are assumed true, but tested for possible acceptance. To answer the questions related to the research problem regarding the influence of service recovery strategies on online customer engagement via the customer satisfaction as a mediating factor, this research has used both simple and multiple linear regression analysis to test the research hypotheses. Before conducting the multiple regression analysis, the severity or degree of multicollinearity was tested by examining using Tolerance and Variance measures. The results of the collinearity statistics of VIF indicate that there is no multicollinearity within the data, which in turn strengthens the model of the research by avoiding the problem of having interchangeable "β" values between independent variables and reducing the bias resulting from type II error. The simple regression analysis technique is employed to test the first three hypotheses. Table 2 summarizes the results of simple regression analysis, with the F-ratio test, for the study hypotheses (H1, H2 and H3). The results indicate that each of these hypotheses is rejected at $\alpha \preceq 0.000$.

Table 1 Descriptive statistics of the scale of measurement

Code	Variables	Mean	SD	Skewness	kurtosis	C A
	Dependent Variable: Online Customer Engagement					0.84
	Co-Developing					
C1	I am motivated to participate in the online community (i.e. social media channels, feedback on the company website) because I can help improve the service provider and its services	3.35	1.046	-0.235-	-0.706-	
C2	I like to know that my comments and suggestions can influence my current telecom firm and its services	3.14	1.075	0.390	-0.589-	
C3	I hope to improve the telecom services through my participation and expression within the online community	3.16	1.109	0.016	-1.026-	
	Sharing (Helping)					
C4	I like participating in the online community because I can use my experience to help others	3.14	1.284	-0.112-	-1.000-	
C5	I like to share my experience and knowledge with others in the online community to help them be more educated about the firm and its services	3.10	1.162	0.779	-0.214-	
C6	I like helping other community members with their questions	3.01	1.305	0.187	-1.090-	
	Leaning (Seeking Assistance)					
C7	I am motivated to participate in the online community because community members can use their knowledge to help me	3.17	1.379	0.103	-1.215-	
C8	I need to be able to use the online community to find answers to my questions about a certain service or the company	3.28	1.275	0.237	-1.086-	
	Socializing					
C9	I use the online community to communicate with people who share the same interests with me	3.06	1.277	-0.081-	-1.098-	
C10	I need to have a conversation with other members online who share the same opinion	3.0.27	1.195	0.417	-0.670-	
	Advocating					
C11	I promote the services through my participation in the online community	3.42	1.186	0.606	-0.503-	
C12	When asked, I recommend the firm or its services to other members of the online community	3.20	1.126	-0.416-	-0.495-	

(continued)

Table 1 (continued)

Code	Variables	Mean	SD	Skewness	kurtosis	C A
	Independent Variables: Recovery Strategies					0.86
	Distributive Justice					
R1	The outcome I received was fair	3.25	1.153	-0.073-	-1.115-	
R2	I got what I deserved	3.28	1.144	-0.123-	-1.068-	
R3	The result of the complaint was right	3.22	1.114	-0.102-	-0.907-	
R4	In resolving the problem, the firm gave me what I needed	3.15	1.125	0.182	-1.088-	
	Procedural Justice					
R5	The firm responded quickly and fairly to my needs	3.31	1.159	-0.219-	-0.914-	
R6	The firm showed adequate flexibility in dealing with my problem	3.24	1.149	-0.074-	-1.114-	
R7	I got a chance to tell them the details of my problem	3.43	1.049	1.072	0.395	
R8	The policies and procedures the firm had in place were adequate for addressing my concerns	3.22	1.142	-0.163-	-1.092-	
	Interactional Justice					
R9	The firm was appropriately concerned about my problem	3.18	1.127	-0.026-	-1.070-	
R.10	The firm put the proper effort into resolving my problem	3.33	1.167	-0.278-	-1.023-	
A.11	The firm's communications with me were appropriate	3.27	1.103	-0.320-	-0.831-	
A.12	I was given a reasonable explanation as to why the original problem occurred	3.16	1.155	0.207	-1.110-	
	Mediating Variable (Customer Satisfaction)					0.89
S1	The content of the service recovery matches exactly with my interests	3.29	1.161	-0.347-	-0.560-	
S2	I am satisfied with my decision to become a member of this firm	3.67	1.088	-0.611-	-0.463-	
S3	The service recovery process provided by the firm meets my needs	3.54	1.125	-0.327-	-0.737-	

Table 2 The results of the first three hypotheses

Hypotheses	R	R Square	Adjusted R Square	F. Test	Sig
H1	0.233	0.054	0.052	20.439	0.000
H2	0.889	0.790	0.790	36.228	0.000
H3	0.211	0.045	0.042	16.536	0.000

Accordingly, it may be concluded that there is a significant relationship between service recovery strategies and online customer engagement, between the service recovery strategies and customer satisfaction and between customer satisfaction and online customer engagement. This result empirically proved that service recovery strategies have a positive direct impact, not only on customer satisfaction but also on online customer engagement.

To test the fourth hypothesis (H4) and to determine whether customer satisfaction acts as a mediator in the relationship between service recovery strategies and online customer engagement, path and regression analysis were conducted. As shown in Table 3, the path of service recovery strategies (SR) with customer satisfaction is significant (*P*-value = 0.000) while the other two regression paths; online customer engagement (OCE) and customer satisfaction (CS); online customer engagement and customer satisfaction are not statistically significant. Accordingly, it can be concluded that there is a full- mediation effect on customer satisfaction in the relationship between service recovery strategies and online customer engagement.

6.3 Discussion

This study aims to examine the effect of service recovery strategies on online customer engagement via customer satisfaction as a mediating variable in the telecommunication sector during the pandemic (Covid-19) period in Jordan. The findings show that service recovery strategies have a positive significant impact on online customer engagement and customer satisfaction, and these results were incompatible with those of prior studies [16, 43, 48]. This conclusion implies that organizations should have a policy of identifying where a severe failure has taken place and to have a clear policy for service recovery. So, when organizations employees are not capable of delivering this essential, acceptable level of service, clients feel not pleased to continue dealing with the business, especially when there are acceptable alternatives. Therefore, the more businesses concentrate on resolving customer questions, answering their queries, responding to their needs quickly, compensating them and following-up service, the better the level of customer satisfaction with the service recovery in these businesses.

The positive significant impact of customer satisfaction on online customer engagement is also found in other researchers' studies [43, 44, 48]. In particular, analysis indicates that service recovery strategies also have an indirect impact on online customer engagement through its impact on customer satisfaction, where customer satisfaction fully mediates the relationship between service recovery strategies and online customer engagement. This result is supported by previous studies, such as [48, 53, 58, 59]. It could be said that such a significant impact of SRS on online customer engagement will decrease when service recovery satisfaction is mediating this relationship. Also, it can be explained in resolving customer problems and offer

Table 3 The results of Hypothesis (H4)

(a) Regression Weights: (Group Number 1-Default Model)

			Estimate	S.E	C.R	P	Label
CS	←	SR	0.958	0.026	36.606	***	
OCE	←	CS	-0.015	0.102	-0.151	0.880	
OCE	←	SR	-0.214	0.110	-1.940	0.052	

(b) Variances: (Group Number 1-Default Model)

	Estimate	S.E	C.R	P	Label
SR	0.868	0.065	13.342	***	
e1	0.212	0.016	13.342	***	
e2	0.788	0.059	13.342	***	

C.R (Critical Ratio); *SE* (standardized estimate)

explanations as required, in addition to compensation and follow-up, will create higher levels of customer satisfaction and, consequently, they will decrease the level of negative customer interaction with company services or with other service users within online communities.

7 Conclusions and Implications

The telecommunication sector in Jordan is besieged with stiff competition and better-informed user publics. Therefore, companies try to create a competitive advantage via discouraging consumers' detrimental actions amidst service failures as well as encouraging customers to grow upward in the loyalty ladder. This research concludes the arguments and test of hypotheses. First, consumers could complain from service failures in the telecom industry while showing readiness to retain their support if their issues are adequately tackled by service providers, thereby obligating them to improve their services. Second, the independent variables "service recovery justices' strategies" were found critically affecting online customer engagement through customer satisfaction as a mediating factor. Therefore, developing appropriate service recovery justices strategies could lead customers to be more satisfied and to display likelihoods to engage in positive online customer engagement that will affect company reputation and performance. Third, the statistical relationship between the independent and dependent variables is significantly mediated by customer satisfaction. Enhanced technical efficiency recovers the means of manipulation of the dimensions of service recovery and eventually provides a significant influence on online customer engagement.

Service recovery covers a comprehensive managerial framework; thus, the current findings expand the current body of knowledge of customer services and, specifically, contribute to the growth of the literature addressing the telecommunication industry. This research contributes to the literature on service recovery by building on and expanding emerging customer assessments of service failures experienced by others. The academic circles are provided with another stream of validated and cross-validated research pieces of evidence as well as an extension of the theory that stimulate further inquiries and perhaps, project learning from mistakes as a sure way of building managerial experience and competitive advantage. The proposed model added the constructs of online customer engagement, which was neglected by previous scholars [43, 53, 55, 57]. Further, the study attempted showing how technical efficiency moderates the relationship between service recovery alternatives justices' strategies and measures of post-complaint behaviour. Some of the study findings, which contradict and/or confirm extant literature, add to the body of the theoretical framework. In cross-validating empirical evidence, customer satisfaction and customer engagement were found to have statistically significant impacts on post-complaint behaviour. To a large extent, these findings enrich the academic strength and stimulate replications to build theory. Practically, the justices strategies of service recovery ought to enhance customer satisfaction and online customer engagement, as the research showed statistically significant relationships between them. Practically, the responsiveness of service quality should be given a high level of attention as customers have many channels, such as (the company's website and different forms of social media) to interact and impact current and prospective customers, which would significantly and positively affect on different dimensions related to customer behaviours and opinions.

Several important managerial implications emerge from this study. First, telecommunication service companies in Jordan should train employees to understand what aspects of perceived justice, the fair distributive treatment, interpersonal communication is important to the customers. If telecommunication service companies can improve these aspects of service recovery, the customers will be satisfied with service recovery that benefits the company in future. Second, to enhance distributive justice, which had a stronger relationship with recovery satisfaction; telecommunication service companies should develop specific monetary compensation guidelines while training employees to quickly and properly react to various service failure situations. Telecommunication service companies should also implement a training program, which clearly illustrates the reactions improving the interactional justice practice through teaching how to properly treat angry and frustrated customers, showing empathy and apology.

Third, the study also showed that customer satisfaction plays a mediating role in the relationship between perceived service recovery justice strategies and online customer engagement in distributive and interactional justice situations. Therefore, telecommunication service companies should make continuous efforts to keep customer's satisfaction and perception in the mind of customers, even in service failure/recovery situations.

8 Limitations and Future Studies

The applications of the study conclusions may be limited by time because of the following points: First, as it focuses on Jordan, to generalize the results, future research is warranted in other countries. Second, it would be interesting to test the study model in other business service contexts. This study is limed to the telecommunication service sector, future studies may focus on other service sectors. Third, also, the researchers have used the cross-sectional surveys, where all data are collected in a particular period, and analysis is restricted to this specified period. Future research can use longitudinal survey where data are gathered over a while.

References

1. Shin, H., Casidy, R., Mattila, A.: Service recovery, justice perception, and forgiveness: the "Other customers" perspectives. Serv. Mark. Quarterly **39**(1), 1–2 (2018)
2. Chang, C., Hung, J.S.: The effects of service recovery and relational selling behavior on trust, satisfaction, and loyalty. Int. J. Bank Mark. **36**(7), 1437–1454 (2018)
3. Ozuem, E.: An exploration of consumers' response. Int. J. Market Res. **59**(3), 33–45 (2016)
4. Alzoubi, H., Alshurideh, M., Kurdi, B., Inairat, M.: Do per-ceived service value, quality, price fairness and service recovery shape customer satisfaction and delight? A practical study in the service tele-communication context. Uncertain Supply Chain Manage. **8**(3), 579–588 (2020)
5. Mostafa, R., Lages, C.R., Sääksjärvi, M.: The CURE scale: a multidimensional measure of a service recovery strategy. J. Serv. Mark. **28**(4), 300–310 (2014)
6. Tarofder, A.K., Nikhashemi, S.R., Azam, S.F., Selvantharan, P., Haque, A.: The mediating influence of service failure explanation on customer repurchase intention through customers satisfaction. Int. J. Qual. Serv. Sci. **8**(4), 516–535 (2016)
7. Hess, E.: Interactional service failures in a pseudo-relationship: the role of organizational attributions, forthcoming. J. Retail. Special Issue Competing Through Serv. **83**(1), 79–95 (2008)
8. Kalamas, E.: Reaching the boiling point: consumers' negative affective reactions to firm-attributed service failures. J. Bus. Res. **61**(8), 813–824 (2008)
9. Vaerenbergh, E.: Customer intentions to invoke service guarantees: do excellence in service recovery, type of guarantee and cultural orientation matter. Managing Serv. Qual. **24**(1), 45–62 (2013)
10. Okazaki, D.: Handbook of Research on International Advertising. Research on International Advertising. Edward Elgar Publisher, Cheltenham (2012)
11. Tyson, E.: Small Business for Dummies, 4th edn. John Wiley and Sons, Hoboken (2012)
12. Chelminski, E.: An examination of consumer advocacy and complaining behaviour in the context of service failure. J. Serv. Mark. **25**(5), 361–370 (2011)
13. Choi, B., Choi, B.J.: The effects of perceived service recovery justice on customer affection, loyalty, and word-of-mouth. Eur. J. Market. **48**(½), 108–131 (2014)
14. Komunda, E.: Remedy or cure for service failure? Effects of service recovery on customer loyalty and satisfaction. Bus. Process Manage. J. **18**(1), 82–103 (2012)
15. Suharto, S., Sulistiyono, S.: The relationship of service quality on consumer satisfaction in the shipyard industry. Modern Appl. Sci. **9**(11), 247 (2015)
16. Wang, G., Oh, S.H., Courtright, A., Colbert. E.: Transformational leadership and performance across criteria and levels: a meta-analytic review of 25 years of research. Group Organization Manage. **36**(2), 223–270 (2011)
17. Fitzsimmons, J.: Service Management: Operations, Strategy and Information Technology, 7th edn. McGraw-Hill College (2011)
18. Grönroos, C.: Marketing services: the case of a missing product. J. Bus. Industrial Market. **13**(4/5), 322–338 (1998)
19. Stauss, B., Seidel, W.: Complaint management. In: Seidel, S.A. (eds.) The Heart of CRM (2005)
20. Gohary, A., Hamzelh, B., Alizadeh, H.: Please explain why it happened! A study of Iranian online shopper. J. Retail. Customer Serv. **31**, 127–142 (2017)
21. Homburg, C., Furst, A.: How organizational complaining. J. Market. **69**, 95–114 (2005)
22. Schminke, M.J., Ambrose, M.L., McMahon, S.R.: Better than ever? Employee reactions to ethical failures in organizations, and the ethical recovery paradox. Organ. Behav. Hum. Decis. Process. **123**(2), 206–219 (2014)
23. Kurdi, B., Alshurideh, M., Alnaser, A.: The impact of employee satisfaction on customer satisfaction: theoretical and empirical underpinning. Manage. Sci. Lett. **10**(15), 3561–3570 (2020)
24. Ellyawati, J., Dharmmesta, B., Purwanto, B.M.S., Herk, H.V.: Perceived justice in service recovery: the study of experimental design on Indonesian customers. Int. J. Bus. Manage. Stud. **2**(2), 511–522 (2013)

25. Bortoli, L.V., Pissutti, C.: Consumer's evaluation about service recovery: the role of social comparison. Ram Revista De De Admindtracao Mackenzie **18**(4), 65–86 (2017)
26. McColl-Kennedy, J.R., Daus, C.S., Sparks, B.A.: The role of gender in reactions to service failure and recovery. J. Serv. Res. **6**(1), 66–82 (2003)
27. Nguyen, D.T., McColl-Kennedy, J.R., Dagger, J.: Strategies of service recovery. J. Market. **53**(2), 31–35 (2012)
28. Tang, X., Chang, E.C., Huang, X., Zhang, M.: Timing and compensation strategies in service recovery. J. Serv. Mark. **32**(6), 755–766 (2018)
29. Miller, J., Christopher, W., Craighead, L., Karwan, K.: Service recovery: a framework and empirical investigation. J. Oper. Manage. **18**(4), 387–400 (2000)
30. McCollough, M.A., Berry, L.L., Yadav, M.S.: An empirical investigation of customer satisfaction after service failure and recovery. J. Serv. Res. **3**(2), 121–137 (2000)
31. Gustaffson, A.: Customer satisfaction with service recovery. J. Bus. Res. **62**, 1220–1222 (2009)
32. Wirtz, J., Mattila, A.S.: Consumer responses to compensation, speed of recovery and apology after a service failure. Int. J. Serv. Industry Manage. **15**(2), 150–166 (2004)
33. Brodie, R.J., Ilic, A., Juric, B., Hollebeek, L.: Consumer engagement in a virtual brand community: an exploratory analysis. J. Bus. Res. **66**(1), 105–114 (2013)
34. Hollebeek, L.: Exploring customer brand engagement. J. Strategic Market. **19**(7) (2014)
35. Hoffman, D.L., Yung, Y.F.: Measuring the customer experience in the online environment. Market. Sci. **19**, 22–44 (2000)
36. ValckK, De., Van Bruggen, G.H., Wierenga, B.: Virtual communities: a marketing perspective. Decis. Support Syst. **47**(3), 185–203 (2009)
37. Dholakiaa, U., Bagozzi, R., Pearo, L.: A social influence model of consumer participation in the network. J. Res. Market. **21**(2004), 241–263 (2004)
38. KuoY, F., Hu, T.L., Yang, S.C.: Effects of inertia and satisfaction in female online shoppers on repeat-purchase intention: the moderating roles of word-of-mouth and alternative attraction. Managing Serv. Qual. Int. J. **23**(3), 168–187 (2013)
39. Calder, E.: An experimental study of the relationship between online engagement and advertising effectiveness. J. Interactive Market. **23**(4), 321–331 (2009)
40. Algesheimer, R., Dholakia, U.M., Herrmann, A.: The social influence of brand community: evidence from European car clubs. J. Market. **69**(3), 19–34 (2005)
41. Baldus, B.J., Voorhees, C., Calantone, R.: Online brand community engagement: scale development and validation. J. Bus. Res. **68**(5), 978–985 (2015)
42. Michel, S., Bowen, D., Johnston, R.: Why service recovery fails: tensions among customer, employee, and process perspectives. J. Serv. Manage. **20**(3), 253–273 (2009)
43. Huang, W.H.: The impact of other-customer failure on service satisfaction. Int. J. Serv. Industry Manage. **19**(4), 521–536 (2008)
44. Ilieska, K.: Customer satisfaction index—as a base for strategic marketing management. TEM J. **2**(4), 327–331 (2013)
45. Al-Dmour, H., Ali, W.K., Al-Dmour, R.H.: The relationship between customer engagement, satisfaction, and loyalty. Int. J. Customer Relationship Market. Manage. (IJCRMM) **10**(2), 35–60 (2019)
46. Cheng, B.L., Gan, C.C., Imrie, B.C., Mansori, S.: Service recovery, customer satisfaction and customer loyalty: evidence from Malaysia's hotel industry. Int. J. Qual. Serv. Sci. **11**(2), 187–203 (2019)
47. Alshurideh, M., Gasaymeh, A., Ahmed, G., Alzoubi, H., Kurd, B.: Loyalty program effectiveness: theoretical reviews and practical proofs. Uncertain Supply Chain Manage. **8**(3), 599–612 (2020)
48. Alshurideh, D.M.: Do electronic loyalty programs still drive customer choice and repeat purchase behaviour? Int. J. Electronic Customer Relationship Manage. **12**(1), 40–57 (2019)
49. Cheung, M., Wai, M.: A customer-dominant logic on service recovery and customer satisfaction. Manag. Decis. **54**(10), 2524–2543 (2016)
50. Yaya, L., Marimon, M., Casadesus, M.: Can ISO 9001 improve service recovery? Industrial Manage. Data Syst. **113**(8), 1206–1221 (2013)

51. Jung, N.Y., Seock, Y.K.: Effect of service recovery on customers' perceived justice, satisfaction, and word-of-mouth intentions on online shopping websites. J. Retail. Consumer Serv. **37**, 23–30 (2017)
52. Kim, G.: The service recovery strategies, customer satisfaction, customer loyalty. Asian J. Qual. **8**(1), 76–86 (2007)
53. Sousa, R., Voss, C.A.: The effects of service failures and recovery on customer loyalty in e-services: an empirical investigation. Int. J. Oper. Prod. Manage. **29**(8), 834–864 (2009)
54. Michel, S., Bowen, D.E., Johnston, R.: Why service recovery fails: tensions among the customer, employee and process perspectives. J. Serv. Manage. **20**(3), 253–273 (2009)
55. Andersson, E., Graselius, H., Nilsson, K.: Service recovery a study of the complaint handling. Eur. J. Mark. **34**(1/2), 156–175 (2005)
56. Nagm, S.M., Dakrory, A.A., Tani, M.I., Khashan, M.A.: The relationship between service recovery and patronage intentions: the mediating role of relationship quality. Int. Bus. Res. **10**(8), 215–231 (2017)
57. Edmondson, B., Johanson, M., Gustafsson, A., Stradvik, T.: The effects of satisfaction and loyalty. Total Qual. Manage. **11**(7), 917–927 (2011)
58. Awan, S., Cheng, Y.: Improve communication quality by understanding customer switching behavior in China's telecom sector. In: Business, pp. 70–83 (2016)
59. Shahriar, S.H., Arafat, S., Khan, M.F., Islam, M.U.: Service recovery strategies versus customers' expectations, race for sustainability: a qualitative study on microbrands. Market. Information Decis. J. **1**(2), 42–53 (2018)

Determinants of E-Word of Mouth on Social Media During COVID-19 Outbreaks: An Empirical Study

M. Al Khasawneh, M. Abuhashesh, A. Ahmad, M. T. Alshurideh⑩, and R. Masa'deh⑩

Abstract In today's digitalization, consumers are highly aware of their surroundings, which make it difficult to convince them. Advertisements are not viewed as truthful; therefore, consumers seek a reliable and a trust worthy source, where other consumers are considered an unfailing source of trust. Therefore, e-word of mouth is one of the most dominant behaviors, in specific on social media platforms due to the increase in technology. E-Word of Mouth on Social Media plays an essential role during COVID-19 Outbreak; its role becomes significant for communication during the quarantine period in epidemic outbreak. Thus, this research studies the key drivers that affect e-word of mouth in social media during COVID 19. The research objective includes the analysis of quantitative data that was collected by distributing an online survey on the key drivers affecting e-word of mouth on social media during COVID 19. With previous literature support, a model was developed that included three variables that aimed to examine the impact of convenience, information quality, and social interaction on e-word of mouth. In this study, it was found that convenience and information quality have a high correlation with e-word of mouth on

M. Al Khasawneh (✉) · M. Abuhashesh · A. Ahmad
E-Marketing and Social Media Department, Princess Sumaya University for Technology (PSUT), Amman, Jordan
e-mail: m.alkhasaawneh@psut.edu.jo

M. Abuhashesh
e-mail: m.abuhashesh@psut.edu.jo

A. Ahmad
e-mail: a.ahmed@psut.edu.jo

M. T. Alshurideh
Department of Marketing, School of Business, The University of Jordan, Amman, Jordan
e-mail: m.alshurideh@ju.edu.jo; malshurideh@sharjah.ac.ae

Department of Management, College of Business Administration, University of Sharjah, Sharjah, United Arab Emirates

R. Masa'deh
Department of Management Information Systems, School of Business, the University of Jordan, Amman, Jordan
e-mail: r.masadeh@ju.edu.jo

M. T. Alshurideh et al. (eds.), *The Effect of Coronavirus Disease (COVID-19) on Business Intelligence*, Studies in Systems, Decision and Control 334, https://doi.org/10.1007/978-3-030-67151-8_20

347

social media during COVID 19. The key findings of this study have both practical and academic implications.

Keywords E-word of mouth · Social media · COVID-19 · Consumer behavior · Convenience · Social interaction · Information quality

1 Introduction

In a world that offers many types of products and information from many sources, consumers nowadays have become ignorant to traditional marketing advertisements. Correspondingly, marketers are realizing the importance and focusing its marketing efforts on e-word of mouth, which is considered a powerful and professional marketing tool with regard to its association with trust. Furthermore, the flow of information is important and can considerably impact the situation during an epidemic disease. The role of e-word of mouth is also essential in any situation related to health protection, especially during COVID 19. Now, with the current COVID-19 outbreak, people are under quarantine, with many countries around the world experiencing epidemic disease outbreaks. During quarantine people have more time to use social media platforms, which could be a source of information and knowledge. E-word of mouth is also a useful source for people in the medical field and businesses for discussions and interactions among each other's. In a simpler way, e-word of mouth is defined as "the intentional influencing of people communications by professional marketing techniques in social media platforms" [1]. Moreover, e-word of mouth can either influence people positively or negatively. E-WOM is one of the most effective type of communication that effects people life and customers buying behaviors [2, 3].

The basic factor that affects E-WOM is the strength of the relationship between those who make a decision and the sources of recommendation, however, knowing who is making a decision on a personal basis is not a major necessity that a source must go through [4]. Moreover, E-WOM is an efficient method for acquiring new customers and sharing information [5], due to the increase of digitalization, e-word of mouth has increased consumers' attainability of unbiased opinions about products and services [6] and made it easier for people to spread their opinions to a large number of audiences [7], therefore, companies have taken that to their advantages.

The availability of online expressions has provided marketers the opportunity to monitor what is being said about their companies [1]. Thus, this gave marketers the benefit to use the power of WOM to create and control E-WOM messages, such as, having the ability to decide whether to keep or delete comments on the company's website. Noting that, it also gives marketers an insight about consumers' attitudes towards the company [8]. In many aspects, interpersonal influence includes a many communication processes; it is electronically contacted, due to higher information availability that can be obtained rather than the traditional usage. Moreover, taking into consideration that E-WOM is not just focused on products, but also can be used for entertaining others, meaning that a company may use it to create an entertaining WOM advertisement, these ads can be amusing and interactive, which could be more or less connected to their brand. In addition, online social presence is important during an epidemic disease, online social presence increased the ability to obtain information about the safety measures and products availability online. This research tends to answer the following questions:

(RQ1): What are the key drivers that affect E-word of mouth on social media during COVID-19 Outbreak?

(RQ2): To what extent does convenience influence E-word of mouth during COVID-19 Outbreak?

(RQ3): To what extent does Information quality influence e-word of mouth during COVID-19 Outbreak?

(RQ4): To what extent does Social interaction influence e-word of mouth during COVID-19 Outbreak?

2   Literature Review

2.1   Social Media During COVID-19 Outbreak

The following research seeks to examine the impact of social media related characteristics on e-word of mouth during COVID-19 outbreak. According to previous studies, there were significant studies that supported word of mouth electronically but few focused on social media platforms. It is noted that the key drivers that affect e-word of mouth were found, but from a narrow aspect. In the twenty first century, social media began boosting with many social networking sites, which highly transformed the interaction of organizations and individuals of common interests [9–11]. This has led to diversified effects whether positive or negative. The main advantages consist of social media being a tool of communication, source of information, sharing of ideas, and a marketing tool [12]. Nowadays, social media and web 2.0 are frequently used. This has developed the consumption of social media in businesses as well. Thus, a change in all types of communications have appeared; business-to-business, business-to-customer, and customer-to-customer. Previous studies have measured various variables on social media such as, interactivity, platform quality and knowledge sharing mechanisms and the outcomes showed that the platform quality and knowledge sharing mechanisms influence online brand communities, customer relationship and integration, however, it varies from one industry to the other. On the other hand, trust, intention to buy, and perceived usefulness were tested using a technology acceptance model and results showed that trust encouraged by social media significantly affect intentions to buy [13]. Moving to another study that has used a theory of planned behavior, psychological ownership theory, safety termination theory and self-concept theory, using the following variables; behavioral attitudes, control variables, self-determination, self-expression, quest for connections, autonomy, and self-concept discrepancy, all these lead to the intention to use blogs. The outcomes were that it highly supports the psychological ownership needs, also, support the privacy concern construct, as well as develop better functions, overall, they study human's behavior in social media in general and research on adoption, privacy, and affective commitment. Noting that multiple researches identified and associated social media and e-word of mouth to the entertainment value that motivates positive influence as well as social media usage [14]. Furthermore, E-WOM provides new directions to people's thoughts and views about any condition and especially during an epidemic outbreak. Currently, the top trend is COVID-19 epidemic disease discussed in news and online social media platforms. In addition, many studies proved that there has been a large increase in the number of people willing to find health and products information on social media [15].

2.2 E-WOM During COVID-19 Outbreak

Electronic word of mouth is a channel of communication where people tend to send both positive and negative messages that are either created by, both product experts and customers that use the service or buys the products [6]. There are different types of e-WOM on social media platforms for example, blog posts, consumer review websites, social media websites where consumers share their experiences and opinions [16, 17].

Considering a research that states that e-WOM can highly affect consumers' purchase intentions and information search. Electronic word of mouth usually takes place on social media platforms or websites where purchase decisions takes place or post- purchase feedback are posted in order to dispel consumer's doubts. Thus, if consumers take the e-WOM messages into consideration, it can directly drive them to a purchase action [18].

Moreover, the diffusion of information occurs in a short time by e-WOM on social media websites. It includes a diversity of ways to exchange knowledge and information asynchronously [19]. Also, an increase of social profiles providing health information was also observed in the recent years, especially during COVID19 epidemic disease. People using social media felt great confidence in sharing their views and data collection during COVID-19 epidemic outbreak quarantine. Health promotions broadcast by the media are used to promote awareness [15], adding to the availability of the e-WOM communications that are accessible at any time when archived. Measurability is another variable of e-WOM, the quantity of e-WOM communications are presented in volume in addition to its persistence [17, 20].

The Possibility of anonymity is a distinct feature of the communication via internet. Unknown people or people with little relationship with one another can communicate by the E-WOM. However, positive information is mainly given than negative when communication occurs between strangers however; the interaction is in power and encourages discussions and sharing of feedbacks [21, 22].

2.3 Convenience

Convenience was associated with 'Portability' which is defined as the ability to use a product in any given location [23], furthermore, [24] defined it as, highly distributed products that require a short period of time, mental and physical effort. Moreover, [25] interpreted convenience in e-commerce as the scope of which customers perceive a website as simple, user-friendly and sensual. For the current research, Convenience is defined as the capability of using a product at any given time or place. Convenience as a general view is a necessary factor to consumer behavior [26] which became an important benefit for consumers that are time-constrained, thus, it is increasingly becoming one of the marketer's interest [27]. "Convenience" designated consumers' effort and time used when purchasing a product, rather than a product's attribute or characteristic. Concentrating on the energy, time, and opportunity that consumers waste when buying goods and services, researchers started to perceive convenience as a feature that minimize a product's nonmonetary price [28, 29]. A previous study confirmed the effectiveness of service convenience on consumer behavior influencing e-word of mouth further. Consumer satisfaction is considered an important element that influences consumer behavioral intentions. Therefore, the convenience of services should be taken into concern ensuring consumer satisfaction which generates positive e-word of mouth. In details, the perception of service convenience causes consumers to communicate positive e-word of mouth and recommendations when meeting their expectations [30]. Andaleeb and Basu [31] stated that there is a significant relationship between convenience and behavioral intentions, customer satisfaction, and consumer switching behavior, as well as, customer retention and perception. Furthermore, consumer's behavioral intentions are influenced by their perceptions concerning service convenience, which means, communicating positive E-WOM. Service convenience (i.e. effort and time) has to be understood within the frame of consumers' activities that they undergo when performing the purchasing process and using a service [27]. Even though this engagement of service convenience is based on the consumption process received matters regarding linear cumulation of understanding of convenience across stages, it is the only up to date empirical model that is considered. Therefore, an increase in different views of convenience has a positive effect on outcome [30]. Based on this, studies approve that convenience is an important service attribute that influences the evaluation of consumers' while encountering a service, and that behavioral intentions results in consumer satisfaction [32] which leads to positive e-word of mouth [33]. Thus, the following hypothesis is raised:

H1: Convenience has a significant positive effect on E- word of mouth during COVID-19 Outbreak.

2.4 Information Quality

Information quality has been defined as a group of characteristics, such as accuracy, completeness, inquiry, adequacy and the right time [34]. Furthermore, it is also defined as the degree in which information provides value for its users and to the organization as a whole [35]. In advance, [36] viewed information as the data that is subject to processing for the purpose of giving it value and making it beneficial for the organization and for the users, information as data and facts that we get through observation, experiment and education, as they are also data and facts superior to ideas and opinions. For the current research, we used [34] definition, which is Information quality is a group of characteristics, such as accuracy, completeness, inquiry, adequacy, and the right time. Conversely, the most important aspect of health promotions is increasing people's patience and keeping them calm during the quarantine period to avoid anxiety. The Chinese government improved public endurance by their emotional awareness through message sharing, which encouraged people to share their views. It is a common for people to first think that social media information is only rumors during an outbreak; as such, people were at risk of being ignorant of health information shared by the government or by individuals. The quality of information shared among people and government official play an important role for people to trust messages posted and shared on social media through E-WOM.

When acquiring information, people face motives that lead them to talk more: (1) what making risky decisions, complex, necessary, ridden by uncertainty or (2) alternative sources of information are not trust worthy or unavailable [37]. Information is an essential part in social media for gathering information quality which affects shopping transactions [38]. It has also become important for the relationship between suppliers and consumers, due to the increase of trust and loyalty from the presence of efficient information, which therefore, leads to a better understanding and consumer fulfillment. Social media platforms can captivate, and retain consumers by giving information of value [39]. Furthermore, it provides access to all types of consumers to attain information, as well as experiences, [40] which create an importance for marketers to use it [41]. Moreover, [42] stated that if information quality has a positive effect on the satisfaction of consumers, then, consumers will be more satisfied with accurate and competent information. Based on this, it can be concluded that information quality leads to consumer satisfaction and a pleasant consumer satisfaction generates positive e-word of mouth [33]. Therefore, the following hypothesis is raised:

H2: Informational quality has a significant positive effect on E-word of mouth during COVID-19 Outbreak.

2.5 Social Interaction During COVID-19 Outbreak

It is well known that social interaction refers to the talk that occurs between individuals when they meet, it occurs when two or more personalities, groups, or social systems that mutually influence one another [43]. Since it is very broad, social interaction as "A situation where the behaviors of one actor are consciously reorganized by, and influence the behavior of, another actor, and Vis versa" [44]. It is also the way a reference group's actions affect a person's behavior [45], as well as being the basic element in the way of how societies work [46]. For the current research, we defined Social interaction as a situation where two individuals interact and influence one another. Furthermore, Online Social Presence in Epidemic Outbreak According to social presence theory, social presence is about intimacy, feeling of closeness, familiarity, immediacy, and urgency to exchange information and motives in society. Online social presence is important during epidemic outbreak quarantine periods. It not only plays an important role in the coordination of society but also in the creation of motives. Online social presence is predicted by online streaming, mediating communal television pleasure. Among the magnitudes of social presence, online social presence indirectly plays a role in the mediating direction. Findings encouraged innovative marketing policy through which participation can be optimistic by refining presence fundamentals [15].

There are two main types concerning social interactions, one is focused, which is the essence of interaction that occurs in a group of people who has similar goals, they might be familiar with each other's, either in the past, or starting to know one another for the first time, during the interaction. The other type is unfocused, under the content that people are mainly unaware of an interaction, and they tend to be unfamiliar of one another, with different goals [43].

Online communities provide opportunities for nurturing relationships and friendships, resulting from shared values, beliefs, and interests [47]. Participation and membership in a group virtually can become a main concern in a social life on an individual. Virtual community members tend to engage in exchanging e-WOM [48] which justifies the importance of e-WOM from the perspective of marketing. E-WOM appears in groups that are characterized by strong connections and relationships, such as close friends and family; while e-WOM out of group rises within people of weaker connections, such as groups in social networks that aim for reaching the public. E-WOM is a social wonder that appear in group settings [49]. The more interaction between consumers in a group occur, the higher their usage of e-WOM is in order to reflect their own knowledge and build an expert's reputation about products [50, 51].

From a marketing perspective, E-WOM derives social interaction between consumers, where it can significantly influence consumer's decisions and behaviors affecting customer retention [52], as well as social learning through observations where it can be a part of social interaction. Moving to impact of social interaction on word-of-mouth, generally word of mouth requires the presence of sustained and specific social relationships between consumers. The need of word of mouth within consumers has increased dramatically with the changing environments and radical communication technologies that exist nowadays [53, 54]. Based on this, the following hypothesis is proposed:

H3: Social interaction has a significant positive effect on E-word of mouth during COVID-19 Outbreak.

3   Methodology

3.1   Survey Design

A sample survey was used to collect data from the study population and in specific, a non-probability sample. The objective of the survey was to collect data and gather them to transform them into suitable and valid information. A random sample of individuals was collected and tested to obtain information that influence consumer behavior and preference. The survey's questions were built on the base of previous items of scales that were chosen due to the highest correlation to examine tested variables.

3.2   Survey Method

A survey was collected from 190 consumers using Survey Monkey, an online survey tool. The purpose of using an online survey rather than using a traditional survey due to higher accuracy and higher response rates by reaching target audiences fast, getting real time results for easier and more fast conduction of market research at a fraction of the usual cost, known to be much easier, and is more readily for analyzation at any time possible [55]. For the purpose of this research, the sampling method used is the non-probability sampling which is known as the selection of a small fraction from the overall population.

4   Data Analysis

4.1   Sample Profile

We used three demographic factors (Gender, Age, and Education) which resulted in having a higher female response rate (62.6%) almost twice higher than male response rate (37.4%). As for the age, the highest response rate was from the age group {18–24} which made up (46.3%) responses. Furthermore, the education factor resulted in having the Bachelor Degree as the highest response rate (66.8%).

Table 1 Exploratory factorial analysis

Items	Factors			
	1	2	3	4
SI1		0.640		
SI2		0.728		
SI3		0.633		
SI5		0.698		
IQ2	0.642			
IQ3	0.638			
IQ4	0.767			
IQ5	0.791			
IQ6	0.665			
WM1				0.636
WM2				0.774
WM3				0.688
CON1			0.773	
CON2			0.726	
CON3			0.782	

Extraction Method: Principal Component Analysis
Rotation Method: Varimax with Kaiser Normalization
SI = Social Interaction; IQ = Information Quality; WM = Word of mouth; CON = Convenience

4.2 Exploratory Factorial Analysis

Exploratory Factorial Analysis (EFA): is defined as the process of examining to which point k observed variables measure m hypothetical variables in order to measure indirectly the un-observable by taking measures of the observed variables.

Using the following criteria (i.e. item loadings > 0.50; cross loadings < 0.30); this led to delete the items below (Table 1).

4.3 Reliability Analysis

Cronbach Alpha: It's a measurement that accommodates of the internal consistency of a scale, and is signified as a number between 0 and 1. Moreover, internal consistency is characterized as the range to which all the items in a scale measure the same concept. Table 2 shows that the Cronbach alpha of each variable is above 0.70.

Table 2 Reliability analysis-Cronbach alpha

Social Interaction	(α) Item Level	(α) Variable Level
SI1	0.595	0. 757
SI2	0.0498	
SI3	0.567	
SI5	0.558	
Information Quality		
IQ2	0.694	0.885
IQ3	0.611	
IQ4	0.791	
IQ5	0.784	
IQ6	0.733	
Word of mouth		
WM1	0.623	0.817
WM2	0.703	
WM3	0.686	
Convenience		
CON1	0.724	0.869
CON2	0.743	
CON3	0.787	

4.4 Convergent Validity

In order to study the relationship between a measure and another measure it's important to take into account two main validities; which are convergent as shown in the table below (Table 3) and discriminant (in Table 4). When explaining convergent validity, it is to examine the extent to which the scale score is in correlation with scales of related constructs, and in order to examine the convergent we have used the Average Variance Extracted (AVE) method. Average Variance Extracted (AVE) provides an indication of the average percentage of the variation that is analyzed by the measuring items for latent construct (should be above 0.50), (Refer to Eq. 1).

Table 3 Convergent Validity

Variables	AVE
Convenience	0.794
Information Quality	0.686
Social Interaction	0.580
Word of mouth	0.732

Table 4 Discriminant Validity

	Convenience	Information Quality	Social Interaction	Word of mouth
CON1	**0.8762**	0.5211	0.4231	0.4605
CON2	**0.8884**	0.6121	0.5579	0.4716
CON3	**0.9087**	0.5721	0.4579	0.4716
IQ2	0.5062	**0.8051**	0.5474	0.4931
IQ3	0.3766	**0.7266**	0.4787	0.4216
IQ4	0.5726	**0.8749**	0.4683	0.5061
IQ5	0.5363	**0.88**	0.5031	0.5685
IQ6	0.6262	**0.8451**	0.517	0.5497
SI1	0.4671	0.5146	**0.7819**	0.3705
SI2	0.3024	0.3508	**0.7122**	0.36
SI3	0.4638	0.5126	**0.7868**	0.4133
SI5	0.3992	0.4592	**0.7619**	0.3745
W1	0.4581	0.5587	0.3996	**0.8371**
W2	0.4488	0.4858	0.4205	**0.863**
W3	0.4405	0.5349	0.4615	**0.8667**

$$AVE = \sum K^2 / n \qquad (1)$$

* K = factor loading of every item and n = number of items in a model.
Table 3 shows that the AVE of each variable is above 0.50.

4.5 Discriminate Validity

On the other hand, discriminant validity is the unit of measurement to examine the scale scores that are uncorrelated with scales o unrelated constructs. Cross Loadings represent the main candidates for removal from subsequent analysis with the main purpose of improving the fit model. The items of each variable loaded significantly higher on its own construct compared to other constructs.

4.6 Hypothesis Testing

SmartPLS was used to verify the hypotheses of the current study (See Fig. 1). SmartPLS advantages: modeling the dependent and independent variables.

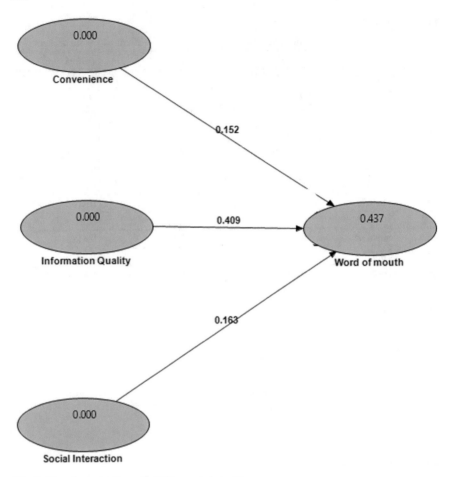

Fig. 1 Hypotheses testing with PLS reported results

H1: convenience has a significant and positive effect on COVID-19 e-word of mouth. The findings provide support for this hypothesis ($\beta = 0.152$; Sig $= p < 0.05$), thus the hypothesis is accepted.

H2: Informational quality has a significant and positive effect on COVID-19 e-word of mouth. The findings provide support for this hypothesis ($\beta = 0.409$; Sig $= p < 0.05$), thus the hypothesis is accepted.

H3: Social interaction has a significant and positive effect on Covid 19e-word of mouth. The findings provide support for this hypothesis ($\beta = 0.163$; Sig $= p < 0.05$), thus the hypothesis is accepted.

All the results are reported in Table 5.

Table 5 The Results of the Hypotheses

Hypotheses	Path coefficient	t-value	Sig. Level	R^2	Decision
ConvenienceWord of mouth	0.152	2.027	$p < 0.05$	0.437	Accept
Information quality Word of mouth	0.409	5.557	$p < 0.05$		Accept
Social interaction Word of mouth	0.163	2.259	$p < 0.05$		Accept
Samples = 190					

5 Discussion and Conclusion

The objective of the current study is to examine the key drivers of E-word of mouth on social media during COVID-19 outbreak. To achieve this objective, a research model was developed based on a comprehensive revision of related literature. The current proposed research model included three independent variables related to social media characteristics and one dependent variable represented by E-word of mouth on social media during COVID-19 Outbreak. In particular, the empirical results of the current research strongly support the newly proposed model in predicting consumers' E-word of mouth on social media influencers during Covid 19. Taking this point further, the overall explanatory power of the current research model had an R-square of 43.7% for E-word of mouth on social media during COVID-19 indicating that the tested model and the three independent variables had the capability of explaining a relatively moderate percentage of variation of word of mouth on social media during COVID-19 outbreak. In addition, the current study found that consumer E-word of mouth on social media during COVID-19 was significantly impacted by information quality, social interaction, and convenience in their order of influencing strength.

Results of Hypothesis One (Convenience has a significant positive effect on e-word of mouth during COVID-19 outbreak). Convenience was hypothesized to having a positive effect on e-word of mouth on social media during Covid 19. This positive effect was assumed to have an increased behavior in spreading positive e-word of mouth [30]. This relationship was supported suggesting that convenience exerted a positive effect on e-word of mouth, due to the level of significance shown in Table 5. This justifies that consumers who are involved with convenience, whether it was in the time they donated, or the effort they put [63] has led to consumer's satisfaction [32] which furthermore drives behavior of spreading e-word of mouth regarding COVID-19 [33]. Furthermore, this result is consistent with the latest assertion's presented by [15] who stated that people under quarantine had more free time to use social media as the most convenient source of message and news sharing of topics related to health issues such as COVID-19 related information. Moreover, social media platforms makes it convenient for consumers to attain information, as well as experiences [40] which creates an understanding why it is important for marketers to use it [41]. An increase in different views of convenience has a positive effect and outcome which will then have the influence on e-word of mouth [30]. As it was demonstrated previously where the relationship was supported by the results of the hypothesis, therefore, convenience does have a significant and positive effect on e-word of mouth during COVID-19 outbreak.

Results of Hypothesis two (Information quality has a significant and positive effect on e-word of mouth on social media during COVID-19 outbreak). This hypothesis was developed to test the effect of social media information quality on e-word of mouth during COVID-19 outbreak. The findings provided support for this hypothesis and showed consistency in which there is a relationship between information quality and e-word of mouth during Covid 19. This finding aligns with [38] study as it was stated that gathering information quality on social media, results in an influence in transactions. Furthermore, [42, 64] found that information quality has a positive impact on consumer's satisfaction, which leads to the generation of positive e-word of mouth [33]. It is noted that the role of social media has become increasingly more essential to collect up to date and high quality of COVID-19 related information [15]. Consistent with this view, it was found that peoples' tendencies to share and spread online information related to health issues are higher on social media compared to offline discussions.

Results of Hypothesis three (Social interaction has a significant and positive effect on e-word of mouth). The relevancy between social interaction and word of mouth during COVID-19 was hypothesized to having a positive relationship. The results of the data analysis showed that the hypothesis was accepted as the level of significance was ($p < 0.05$), which showed consistency with Operman's (2000) study who stated that social interaction enabled by social media platforms lead consumers towards the dissemination of E-word of mouth. Online communities provide the development of relationships and friendships, resulting from shared values, interests, and beliefs [47, 53]. Participation and membership in virtual groups may become a main part of an individuals' social life as it is known that members tend to engage in word of mouth [48] which justifies e-WOM's importance during COVID-19 outbreak. E-WOM appears in groups that are represented by strong connections and strong relationships, such as close friends and family; while e-WOM out of group occurs within people with weaker ties, such as in social networking groups that aim for reaching the mass public. Since e-WOM is a social experience that exist in group settings [49], higher interactions within consumers in a group, leads to a higher use of e-WOM to reflect their abilities and provides opinions about products as experts [50]. Social interaction is important during epidemic outbreak quarantine periods as it has a significant role in the coordination of online communities as well as the creation of motives such as escape and socializations motives [56]. Additionally, peoples' motives to share e-word of mouth in quarantine during epidemics such as COVID-19 Virus outbreak would be highly impacted by social interaction [15].

5.1 Implications

The current research highlighted and added valuable information concerning the key drivers related to social media characteristics that affect e-word of mouth on social media during COVID-19 Virus outbreak. Issues raised by the findings show that need to examine key drivers that affect word of mouth. As a result, several aspects have implications for the theory and practice. The findings show the need for marketers, health authorities to recognize the importance of e-word of mouth, from a theoretical perspective, these findings add to the understanding of consumer behavior (e-word of mouth) during epidemic periods.

5.2 Theoretical Implications

The current research contributes to theory within the area of consumer behavior. From a theoretical point of view, the research provides advanced information on the key drivers that affect e-word of mouth during COVID-19 outbreak. Thus, the study adds to widen our knowledge of the key drivers affecting e-word of mouth on social media. Therefore, in doing so, the research used theories in the context of consumer behavior, which extended to the use of an already existing theory, such as, The Theory of Planned Behavior (TPB) Model. For several years, previous researchers used the theory of planned behavior (TPB) for the sake of suggesting that human behavior is highly affected by the preferences and attitudes of people, their norms, as well as, their control of behavior and intentions to act [57]. With time, this theory extended, and had variables added to it, such as; social conflict [58] social group, norms, and identity [59]. These variables showed a constant effect on different cultures [60]. Thus, it is noted that consumers tend to be affected by their social surroundings, that affect their intention to act and behave, and of such behaviors, is e-word of mouth.

It is also evident that the model has been applied in different contexts, such as the choice of transportation mode [61], choice of holiday, the choice of travelers' on destinations, the e-word of mouth influence on the choice of restaurants as well as, the attitudes of tourists towards travelling [62]. However, limited studies have applied the TPB model in the context of E-word of mouth on social media during epidemic outbreak periods.

5.3 Practical Implications

The values this research addsin the marketing area is the ability for it to be practiced in reality. In this sense, the value given by this research, in terms of the Theory of Planned Behavior (TPB) is that it widens our understanding of the key determinants related to social media characteristics on E-WOM during COVID-19 Virus Outbreak.

Social media can have an enormously positive influence on health-related knowledge and behaviour and that social media users are able to access and analytically evaluate significant health-related content shared by other users. During Epidemic outbreaks and quarantine periods, as people have more free time to socialize and engage in social media related activities, the convenience of obtaining information related to Covid 19, COVID-19 information quality presented on social media and the ability to participate and interact on social media are all have significant positive impact on E-word of mouth on social media during COVID-19 Virus outbreak periods. Therefore, all these factors should be taken into consideration by policy makers, health authorities and practitioners by paying more attention to managing peoples' participation and interaction on social media during epidemic outbreaks periods. Taking this point further, social media can be a competent source of information and an effective media for staying up-to-date with the huge volume of medical knowledge. Therefore, it is recommended that social media users should be given sufficient knowledge and reliable epidemic related information and educate them on how to critically evaluate health information in the case of an epidemic or pandemic on social media before disseminate it to their family and friends.

5.4 Limitations

Throughout this study, several limitations have been faced; first, the distributed survey was 252 survey samples in total, but only 190 were completed. Not only that, but also the percentage of females who have completed the survey outweighed the percentage of males by 62%. Moreover, the age segment that scored the highest percentage was between 18 and 24 years old with an estimate amount of 95 individuals within the scope of the survey. Secondly, limited researches on electronic word of mouth during epidemic outbreaks have been found on the basis of this topic, as most of them were used traditionally and were outdated, which made it harder to reach and analyze most recent publications on social media.

5.5 Future Recommendations

The limitations noted above, provide a basis for future research, as well as the possibility of studying and measuring this topic according to other variables, such as, engagement, perceived value. Since this research was conducted on the basis of quantitative approach, it is recommended to utilize qualitative as well in order to support the study further. Adding to that, analyzing a higher number of responses, for the sake of gaining more reliable information is needed.

6 Conclusion

Only few studies concerning the key determinants of social media related characteristics on E-word of mouth during COVID-19 Virus outbreaks were conducted previously. The results of the study verified that convenience, information quality, and social interaction on social media have a positive impact on E-WOM during COVID-19 outbreaks. E-Word of mouth is growing rapidly in popularity, not just in word but also in concept. More marketers should consider taking time to learn consumer behaviors such as their E-WOM activity. This research highlighted the main key drivers related to social media that affect e-word of mouth during epidemic periods in which can add a glimpse of what derives people to use electronic word of mouth and how they act in crisis situations.

Acknowledgements I would like to express my deep and sincere gratitude to Princess Sumaya University for Technology (PSUT) and University of Jordan for their support and for providing the research team for all the necessary facilities.

References

1. Kozinets, R., Valck, K., Wojnicki, A., Wilner, S.: Networked narratives: understanding word-of-mouth marketing in online communities. J. Mark. **74**, 71–89 (2010)
2. Yang, S., Hu, M., Winer, R., Assael, H., Chen, X.: An empirical study of word-of-mouth generation and consumption. Mark. Sci. **31**(6), 952–963 (2012)
3. Boguszewicz-Kreft, M., Kuczamer-Kłopotowska, S., Kozłowski, A., Ayci, A., Abuhashesh, M.: The theory of planned behaviour in medical tourism: international comparison in the young consumer segment. Int. J. Environ. Res. Public Health **17**(5), 1626 (2020)
4. Argan, M.: Word-of-Mouth (WOM) as a tool of health communication: a case study of Turkey. Health MED **6**(1), 216–221 (2012)
5. Garnefeld, I., Helm, S., Eggert, A.: Walk your talk: an experimental investigation of the relationship between word of mouth and communicators' loyalty. J. Serv. Res. **14**(93), 93–107 (2010)
6. Hennig-Thurau, T., Walsh, G.: Electronic word of mouth: motives for and consequences of reading customer articulations on the Internet. Int. J. Electronic Commerce **8**(2), 51–74 (2004)
7. Dellarocas, C.: The digitization of word of mouth: promise and challenges of online feedback mechanisms. Manage. Sci. **49**(10), 1407–1424 (2003)
8. Goyette, I., Ricard, L., Bergeron, J., Marticotte, F.: E-WoM scale: word-of-mouth measurement scale for e-services context. Canadian J. Admin. Sci. **27**(1), 5–23 (2010)
9. Edosomwan, S., Prakasan, S.K., Kouame, D., Watson, J., Seymour, T.: The history of social media and its impact on business. J. Appl. Manage. Entrepreneurship **16**(3), 79–91 (2011)
10. Abuhashesh, M.Y.: Integration of social media in businesses. Int. J. Business Soc. Sci. **5**(8) (2014)
11. Alsamydai, M.J., Al Khasawneh, M.H.: Antecedents and consequences of E-Jordanian consumer behavior regarding Facebook advertising. Int. J. Bus. Manage. Res. **3**(4), 41–59 (2013)
12. Siddiqui, S., Singh, T.: Social media its impact with positive and negative aspects. Int. J. Comput. Appl. Technol. Res. **5**(2), 71–75 (2016)
13. Hajli, M.: A research framework for social commerce adoption. Information Manage. Comput. Security **21**(3), 144–154 (2013)

14. Abuhashesh, M., Al-Khasawneh, M., Al-Dmour, R., Masa'deh, R.: The impact of Facebook on Jordanian consumers' decision process in the hotel selection. IBIMA Bus. Rev. 1–16 (2019a)
15. Yasir, A., Hu, X., Ahmad, M., Rauf, A., Shi, J., Nasir, S.: Modeling impact of word of mouth and e-government on online social presence during COVID-19 outbreak: a multi-mediation approach. Int. J. Environ. Res. Public Health **17**(8), 2954 (2020)
16. Bickart, B., Schindler, R.M.: Internet forums as influential sources of consumer information. J. Interactive Mark. **15**(3), 31–40 (2001)
17. Cheung, C.M., Thadani, D.R.: The impact of electronic word-of-mouth communication: a literature analysis and integrative model. Decis. Support Syst. **54**(1), 461–470 (2012)
18. Erkan, I., Evans, C.: The influence of eWOM in social media on consumers' purchase intentions: an extended approach to information adoption. Comput. Hum. Behav. **61**, 47–55 (2016)
19. Hung, K., Li, S.: The influence of eWOM on virtual consumer communities: social capital, consumer learning, and behavioral outcomes. J. Advertising Res. **47**(4) (2007)
20. Ahmad, A., Abuhashesh, M., Obeidat, Z., AlKhatib, M.: E-WOM and airline e-ticket purchasing intention: mediating effect of online passenger trust. Manage. Sci. Lett. **10**(12), 2729–2740 (2020)
21. Kaijasilta, N.: The conceptualization of electronic Word-of-Mouth (EWOM) and company practices to monitor, encourage, and commit to EWOM-a service industry perspective. Master's thesis, Aalto University (2013)
22. Abuhashesh, M., Mohammad, S.J., Khasawneh, M.A.: The attitude of Jordanian customers towards virtual stores. Int. J. Islamic Mark. Branding **4**(1), 59–75 (2019)
23. Gehrt, K.C., Yale, L.J.: The dimensionality of the convenience phenomenon: a qualitative reexamination. J. Bus. Psychol. **8**(2), 163–180 (1993)
24. Copeland, M.T.: Relation of consumers' buying habits to marketing methods. Harvard Bus. Rev. **1**(2), 282–289 (1923)
25. Srinivasan, S.S., Anderson, R., Ponnavolu, K.: Customer loyalty in e-commerce: an exploration of its antecedents and consequences. J. Retail. **78**(1), 41–50 (2002)
26. Chen, M.C., Chang, K.C., Hsu, C.L., Yang, I.C.: Understanding the relationship between service convenience and customer satisfaction in home delivery by Kano model. Asian Pacific J. Mark. Logistics **23**(3), 386–410 (2011)
27. Farquhar, J.D., Rowley, J.: Convenience: a service perspective. Mark. Theor. **9**(4), 425–438 (2009)
28. Kelley, E.J.: The importance of convenience in consumer purchasing. J. Mark. **23**, 32–38 (1958)
29. Etgar, M.: The household as a production unit. In: Sheth, J.N. (ed.) Research in Marketing, vol. 1, pp. 79–98. JAI Press, Greenwich, CT (1978)
30. Khazaei, A., Manjiri, H., Samiey, E., Najafi, H.: The effect of service convenience on customer satisfaction and behavioral responses in bank industry. Int. J. Basic Sci. Appl. Res. **3**(1), 16–23 (2014)
31. Andaleeb, S.S., Basu, A.K.: Technical complexity and consumer knowledge as moderators of service quality evaluation in the automobile service industry. J. Retail. **70**(4), 367–381 (1994)
32. Chang, Y.W., Polonsky, M.J.: The influence of multiple types of service convenience on behavioral intentions: the mediating role of consumer satisfaction in Taiwanese leisure setting. Int. J. Hospitality Manage. **31**, 107–118 (2012)
33. Ennew, C.T., Banerjee, A.K., Li, D.: Managing word of mouth communication: empirical evidence from India. Int. J. Bank Mark. **18**(2), 75–83 (2000)
34. Long, L.: Management Information System. Prentice-Hall International Inc., USA (2011)
35. Chaffey, D., Wood, S.: Business Information Management: Improving Performance Using Information Systems. Pearson Educational Limited, Haelow, England (2005)
36. O'Brien, J., Marakas, G.: Management information systems, McGraw –Hill compamies, USA. In: Pollard, C., Sipior, J., Leidner, D., Lai, L., Cheung, C., Cristobal, D. (2010) Information Technology for Management. John Wiley & Sons, Asia (2011)
37. Berger, J.: Arousal increases social transmission of information. Psychol. Sci. **22**(7), 891–893 (2011)

38. Xu, H., Koronios, A.: Understanding information quality in e-business. J. Comput. Information Syst. **45**(2), 73–82 (2005)
39. Kim, H., Niehm, L.S.: The impact of website quality on information quality, value, and loyalty intentions in apparel retailing. J. Interactive Mark. **23**, 221–233 (2009)
40. Senecal, S., Nantel, J.: The influence of online product recommendations on consumers' online choices. J. Retail. **80**(2), 159–169 (2004)
41. Do-Hyung, E., Jumin, L., Ingoo, H.: The effect of on-line consumer reviews on consumer purchasing intention: the moderating role of involvement. Int. J. Electronic Commerce **11**(4), 125–148 (2007)
42. Kim, S., Stoel, L.: Dimensional hierarchy of retail website quality. Information Manage. **41**(5), 619–633 (2004)
43. Bardis, P.: Social science. Int. Honor Soc. Soc. Sci. **54**(3), 147–167 (1979)
44. Memon, S.A., Hadikusumo, B.H., Sunindijo, R.Y.: Using social interaction theory to promote successful relational contracting between clients and contractors in construction. J. Manage. Eng. **31**(6) (2014)
45. Scheinkman, J.: The new palgrave dictionary of economics, 2nd edn. In: Durlauf, S., Blume, L. (eds.), Palgrave Macmillan (2008)
46. Diekmann, A.: Social Interaction and Society: Perspectives of Modern Sociological Science (2016). https://ethz.ch/content/dam/ethz/special-interest/gess/chair-of-sociology-dam/documents/sis2016/Booklet%20SIS2016.pdf
47. McKenna, K., Green, A., Gleason, M.: Relationship formation on internet: what's the big attraction? J. Soc. Issues **58**(1), 9–31 (2002)
48. Alon, A., Brunel, F.F., Siegal, W.S.: Word-of-mouth and community development stages: towards an understanding of the characteristics and dynamics of interpersonal influences in internet communities. Adv. Consum. Res. **29**(1), 429–430 (2002)
49. Abrantes, J., Seabra, C., Lages, C., Jayawardhena, C.: Drivers of in-group and out-of-group electronic word-of-mouth (eWOM). Eur. J. Mark. **47**(7), 1067–1088 (2013)
50. Wojnicki, A.: Word-of-mouth and word-of-web: talking about products, talking about me. Adv. Consum. Res. **33**(1), 573–575 (2006)
51. Algharabat, R.S., Rana, N.P., Alalwan, A.A., Baabdullah, A.M.: Investigating the impact of social media commerce constructs on social trust and customer value co-creation: a theoretical analysis. In: Digital and Social Media Marketing, pp. 41–55. Springer, Cham (2020)
52. Opermann, M.: Tourism destination loyalty. J. Travel Res. **39**(1) (2000)
53. Prahalad, C.K., Venkatram, R.: Co-opting customer competence. Harvard Bus. Rev. **78**(1), 79 (2000)
54. Hendrayati, H., Pamungkas, P.: Viral marketing and e-word of mouth communication in social media marketing. In: 3rd Global Conference On Business, Management, and Entrepreneurship (GCBME 2018), pp. 41–48. Atlantis Press (2020)
55. Fricker, R.D., Schonlau, M.: Advantages and disadvantages of internet research surveys: evidence from the literature. Field Methods **14**(4), 347–367 (2002)
56. Guan, W.J., Ni, Z.Y., Hu, Y., Liang, W.H., Ou, C.Q., He, J.X., Du, B., et al.: Clinical characteristics of coronavirus disease 2019 in China. N. Engl. J. Med. **382**(18), 1708–1720 (2020)
57. Armitage, C.J., Conner, M.: Efficacy of the theory of planned behaviour: a meta-analytic review. Br. J. Soc. Psychol. **40**(4), 471–499 (2001)
58. Olsen, S.O.: Antecedents of seafood consumption behaviour: an overview. J. Aquat. Food Prod. Technol. **13**(3), 79–91 (2004)
59. Terry, D.J., Hogg, M.A., White, K.M.: The theory of planned behaviour: selfidentity, social identity, & group norms. Br. J. Soc. Psychol. **38**, 225–244 (1999)
60. Tuu, H.H., Olsen, S.O., Thao, D.T., Anh, N.T.: The role of norms in explaining attitudes, intention and consumption of a common food (fish) in Vietnam. Appetite **51**, 546–551 (2008)
61. Banberg, S., Ajzen, I., Schmidt, P.: Choice of travel mode in the theory of planned behaviour: the roles of past behaviour, habit, and reasoned action. Basic Appl. Sci. Psychol. **25**, 175–188 (2003)

62. Sparks, E., Pan, G.W.: Chinese outbound tourists: understanding their attitudes, constraints and use of information sources. Tourism Manage. **30**, 483–494 (2009)
63. Brown, D.J.: Decentralization and School-Based Management. Psychology Press (1990)
64. Cry, I.: CAFTA Blocked for Ignoring Indigenous Rights (2008)

The Impact of Social Media Marketing Communications on Consumer Response During the COVID-19: Does the Brand Equity of a University Matter?

Ahmad Aljumah⊙, Mohammed T. Nuseir⊙, and Muhammad Turki Alshurideh⊙

Abstract The main objective of the current study is to investigate the impact of social media marketing communication on the consumer response to University in UAE during COVID-19. In addition to that, the study has also examined the mediating role of brand equity of university in the relationship between social media marketing communication and consumer response during the COVID-19. The schedule for the academic session of an academic year, their starting and ending dates, schedule for vacations all vary from country to country; as a result, the conditions were not similar and varied due to the same reason. However, few countries suspended their classic room lectures from March or April 2020 up to further notification. We employed a combination of inferential and descriptive analyses to carry out the data analysis. For this purpose, the PLS-SEM approach was integrated, which is a second-generation technique for structural equation modeling. PLS-SEM is a relatively new approach and provides reliable results when coupled with SEM models. The response rate is above 50%. Customer-Based Brand Equity illustrated the significant indirect impacts of these significant indirect effects on the association among Customer-Based Brand Equity (CBBE) and social media marketing communications. As a result, this research further added the perception of Customer-Based Brand Equity (CBBE) in the perception of UNIVERSITY perception through endorsing

A. Aljumah
Management Department, Emirates College of Technology, P.O. Box 41009, Abu Dhabi, United Arab Emirates
e-mail: aljumah37@gmail.com

M. T. Nuseir (✉)
Department of Business Administration, College of Business, Al Ain University, Abu Dhabi Campus, P.O. Box 112612, Abu Dhabi, United Arab Emirates
e-mail: mohammed.nuseir@aau.ac.ae

M. T. Alshurideh
Department of Management, College of Business Administration, University of Sharjah, Sharjah, United Arab Emirates
e-mail: malshurideh@sharjah.ac.ae; m.alshurideh@ju.edu.jo

Department of Marketing, School of Business, The University of Jordan, Amman, Jordan

© The Author(s), under exclusive license to Springer Nature Switzerland AG 2021
M. T. Alshurideh et al. (eds.), *The Effect of Coronavirus Disease (COVID-19) on Business Intelligence*, Studies in Systems, Decision and Control 334,
https://doi.org/10.1007/978-3-030-67151-8_21

the Customer-Based Brand Equity (CBBE), which estimates the brand sustainability, explicit consumers associations, functional brand image, perception, hedonic brand image and experiences with UNIVERSITY brands by brand awareness. This research work also positioned the two levels of social media communication, such as UGC and FCC, into marketing communication through endorsing the degrees of social media advertising, social media promotions, and social media interactive marketing. Hence, the research was capable of discussing the various roles of these social media marketing communications on the consumer response and the growth of Customer-Based Brand Equity (CBBE). Moreover, Customer-Based Brand Equity was illustrated to have significant influences on consumer responses.

Keywords Social media marketing · Media promotions · COVID-19 · Consumer response · Brand equity · Communications

1 Background

In most companies developing and managing brands, equity is of high preference, and therefore due to this reason, brands are the significant resources for any company [1]. According to analysis, there is a significant role of brand equity; academic institutions employ their maximum efforts to determine different variables associated with brand equity that would impact and support in the growth of brand equity and understanding the benefits associated and gained with growth in brand equity, most notably how it benefits them in terms of the viewpoint of their consumer[2]. [3] recommended that there are three different types of constructs for the research study of brand equity. This research work can be investigated by these constructs that are the organizational perspective, from the customer perspective, and the financial perspective [4].

One of the derivatives for research and estimate for determining the brand equity is Consumer-Based Brand Equity (CBBE) as it reflects the consumer's mindset and consumer's perspectives. Moreover, there are two significant features for the research study of Consumer-Based Brand Equity (CBBE) [5]. Firstly to acknowledge the variables that are associated with it and participate towards the growth of Consumer-Based Brand Equity (CBBE), such as marketing communications, hence the estimation regarding marketing practices and marketing communication, for instance, interactive marketing, advertising, sales promotion, word of mouth (WOM) and 20 further support to increase the growth of Consumer-Based Brand Equity (CBBE) [6].

The second feature is to examine and investigate the impact of Consumer-Based Brand Equity (CBBE) on consumer behavior and its reactions in the form of brand preference and purchase intention [7]. Thus, the significance and impact of Consumer-Based Brand Equity (CBBE) on customer behavior and attitudes. [8] recommended that the brand understandings, which is shown by the brand image and brand awareness that is the depiction of Customer-Based Brand Equity (CBBE).

Hence, it illustrated how brand relations and brand knowledge impact the behaviors of the consumer. Hence, Consumer-Based Brand Equity (CBBE) explained how consumer behavior and consumer perception of a brand are affected [9].

The findings of earlier discussions and research have suggested encouraging behavior and reactions from the consumer's side, which has been a significant issue for the brand managers and the scholars because the managers apply aggressive estimation towards the growth of their brand resources, for instance, their brand equity [10]. The development of efficient brand equity plays a significant role. It is essential in terms of strategy for sustaining the satisfaction level among consumers, most importantly, the enhancement of favorable consumer reactions in a highly competitive market and acceptance [11].

Branding provides benefits to the companies, which the main reason behind it, and that is why it is among the list of consumer's preferences. It offers the organization differentiating competencies and competitive benefits [12]. According to the [13], the definition of brand equity is creating differential features of a brand; brand equity points out the variance of brands among the similar category of products and services for their consumers. Thus, the features of brand equity of a brand invite consumers towards the brand through the development of the brand image, which ultimately impacts the consumer reactions. Consumer responses are the influence of consumers' overall estimation about a brand Consumer-Based Brand Equity (CBBE) and the data derived by the appeals of marketing communications [13].

The research work related to Consumer-Based Brand Equity (CBBE) started developing since the initial years of the 2000s and even regularly further afterward as the marketing communications environment, which changed continuously [14]. These researches consistently illustrated that one of the points which showed that there had been some rapid changes and developments in the situation of marketing communications that included developments and emergence of social media, for instance, Twitter, Facebook, Micro-blogs and YouTube that have now become most significant forums for the development and sustainability of brand equity and their success [15].

By the brand profiles and fan pages that are developed on social media, brand managers are provided by the infinite offers for posting and sharing their information in the form of messages, photos, videos, and comments regarding their organization and brands [16]. However, few empirical research through their results, and findings proved that social media communications are the outstanding traditional media communications among the emerging Consumer-Based Brand Equity (CBBE) [17].

The highly infecting types of disease as declared based on its nature. According to the report, the disease is contagious and highly infected, and the disease can be transferred between individuals through the respiratory droplets and various other types of contact types, for instance, mouth, hand, and nose [18]. In the same way, in various fields, this pandemic influenced the educational process. Hence, the governments made certain decisions which focused the integrated targets of delaying the transmission of COVID-19 through avoiding the face to face interaction as in teaching and through maintaining the social distance. Consequently, the virtual classrooms have been introduced instead of the classical classroom by the implementation and utilization of complementary technologies to facilitate the settings. Few countries announced an extraordinary halt towards the proceeding of the academic session in order to gain more time for further preparation and develop a better form through the environment of distance learning.

In UAE, the educational institutions and universities were instructed by their Ministry of Education (MOE) to stop in-class teaching. They decreased their preference from going to campus by substituting it with the digitalized systems of distance learning. Initially, the Ministry of Education (MOE) announced the spring holidays for two weeks, which started from March 29th and was afterward further extended. The students were then instructed to pursue their academic activities and lectures through distant learning from home for the next 14 days. Afterward, the Ministry of Education (MOE) announced a further extension of the online learning until the session end for the entire academic year 2019 till the end of 2020.

On the other hand, for the process of distance learning, there are various systems and settings available for planning and conducting online-learning. However, the most important part is an examination or assessment that would be an important and challenging task to be assumed in distance learning [19]. However, various academic institutions and universities were anxious regarding how to examine and evaluate students based on their academic performance and their assessment with justice and integrity. The results of universities in lockdown situation possibly further prolonged till the end of this academic year however the highlighted issues in their mind was regarding assessment and concerning the grading that was one of the vital policy task and challenge. Ministry of Education (MOE) announced the regulations and rules that governed all the policies, which gave them a standardized strategy and outlined guidance for overall university campuses of UAE in which they discussed all concerns and questions which were developed in mind by the administrations of these campuses of universities.

As per the above-discussed research, the acknowledgment regarding consumers on consumer-bases brand equity in the perception of university brands, mainly as customers of university products, was derived through a problematic situation of decision making during buying of any product [7]. However, university consumers are outstandingly proactive and strongly taking an interest in decision making about their favorite brands. These customers would depend upon the features and various aspects related to the resources of university brands to make a more straightforward process regarding their decision making [7]. Contrary to the earlier research work, this research presented the certified model of consumer-based brand equity through the perception of university-industry [20]. This research work also investigated various influences of social media marketing communications on the consumer-bases brand equity of consumer responses for university brands.

2 Hypothesis Development

Generally, social media has been recognized as the most influential media with infinite offerings regarding brand communication concerning emerging and Consumer-Based Brand Equity (CBBE) enhancement [21]. Additionally, the capability to distribute on-time information and impact on the consumer viewpoint are the significant reason for which social media is attaining further consideration from brand managers and marketers [22].

Although the utilization of social media mainly for reasons of marketing communications is still growing, there has been an excessive result regarding the influence of social media on brand equity. It has also been determined through various research works that have assumed various constructs in attaining the association among brand equity and social media communications [23]. [24] investigated the influence of social media and traditional media communication on Consumer-Based Brand Equity (CBBE) in three industries belonging to different product service segments such as pharmaceuticals, telecommunications, and tourism.

Their research work engaged UGC and FCC both to estimate the influence on the Consumer-Based Brand Equity (CBBE) by social media communications. Advertising was utilized for the illustration of traditional media marketing communications [25]. The scholars investigated the influence of a functional brand image, hedonic brand image, and brand awareness on the brand purchase intention and the brand attitude as measurements for brand equity [26]. The findings from the survey, which was conducted online with the sample size of 393 consumers, discovered that user-generated, advertising, and 68 firm-created are the most important influence on consumer-based brand equity (CBBE). Their results showed that two types of brand images that are hedonic and functional have a positive impact on social media communications.

With the agreement of research findings of researchers[27, 28] also gave his research finding that backed the earlier research that assumed that both types of social media communications such as UGC and FCC have a positive impact on the brand equity (for instance hedonic brand image, functional brand image, and brand awareness). As a result, their brand equity impacts brand purchase intention. [28] reported that a standardized survey was conducted online through Facebook for data gathering. For determining the model fit, the structural equation modeling in AMOS 21.0 was used.

The research acknowledged that user-generated and firm created both are the kinds of social media communications that have a positive impact on brand equity. Brand Equity was also recognized to have a significant impact on brand purchase intention. Generally, the study suggested that the company must target through social media as it is an effective mechanism for the development of brand equity and effecting purchase intension.

In various research work, the scholars [29] added the hypothesized influence of social media communication on brand equity by using an approach which is called mixed method. The research acknowledged that the communications which are associated with the brand on social media could be categorized into three categories, such as consumer-brand communication, consumer-to-consumer communication, and brand-consumer communication. The structure generated in this research demonstrated that social media communications have a positive influence on the degrees of brand awareness, cheerful messaging about brand and brand loyalty, brand equity, and positive associations.

[30] recommended through a case study based on a survey through he gave a comparative and systemic analysis of earlier research work to generate a conceptual model that how influence on brand equity development is carried through social media engagements of consumers. The results and approaches illustrated in their research work, which offered perception into how organizations can misuse the fan page of a brand to grow viral brand awareness, the development of the strong brand image, impacts of consumer's decision making, and relationships as well. On the other hand, the 70 empirically the research clarifies the influence of FCC as a kind of social media communications on the development of substantial brand equity between online customers.

However, the results illustrate that the research survey conducted between Twitter and Facebook users by the scholars [30] who reported a significant influence on brand equity by UGC. The research also engaged the SEM to generate a structural model that illustrated the empirical research associations among consumer brand perception and UGC as one of the essential social media communications elsewhere. It was mentioned as a Consumer-Based Brand Equity (CBBE).

The results by scholars [31] reported that customer's associations by comments on social media pages of brands, feedbacks, and comments are the illustrations of UGC. These types of associations were claimed to have a positive influence on brand equity. Similarly, the researchers [31, 32] theorized that there is a positive impact on brand equity by the UGC. Their research work recognized that the hypothesis was acknowledged from the structural model features, which were described by the analysis of SEM. In the same way, the scholars [33] assumed the constructs of brand equity, such as brand image, brand awareness, brand loyalty, perceived quality, and brand association. Their research also claimed the indirect influence of brand equity of constructs on the influence of another type of UGC, which is called electronic word-of-mouth on the decision of consumers.

H1: Social Media Advertising has a significant impact on brand equity.

H2: Social Media Sales Promotion has a significant impact on brand equity.

H3: Social Media Interactive Marketing has a significant impact on brand equity.

The findings of the research work pointed out the efficient brand communications which are deliberately distributed to develop an encouraging concept of brand equity in the consumer's point of view in mind. However, the potential that brands are not incorporated in the mindsets of consumers is enhanced with the marketing communications [34]. Thus, impacting the method of consumer responses and brand decision making in the forms of brand preferences and purchase intention [35].

As per the activities regarding social media marketing and communication efforts due to the utilization of opportunities of social media for broadcasting the perception regarding the brand association, which is offered to both consumers and brand managers to interact with brand information, comment, post, interact and share [36]. This opening supported to enhance the consumer's commitments and associations with online brands and responses, and afterward, the impact on consumer behaviors [37].

The research developed an influence on consumer responses by social media marketing communications. Although, various research works have investigated the influence of consumer response on a fragmentary basis by social media marketing communications, which afterward tend towards the instable results [38]. In short, there has been 76 short number of researches that are associating the influence of social media marketing communications towards consumer response further than their purchase intention.

Although the researchers [39] claimed that the activities regarding marketing communication, for instance, the brand advertisements suggested favorable and positive consumer response in the form of brand preference and purchase intention. The scholars [39] recommended that the influence on brand equity by the UGC and FCC are both further prolonged in terms of consumer purchase intention [40]. Thus, by the communication appeals of FCC and UGC both on consumers' purchase intention, a positive impact of brand perception and brand equity is developed. According to this research, the increasing fame of social media communications for contacting the customers, the researchers [41] claimed that there is a significant impact on brand equity and purchase intention by UGC.

The association of this research gives insight into the fact what consumers have in mind about the brand and how social media would influence the consumer's intentions towards the purchase of the brand. Their results are the same as the findings claimed by the [42], which showed that social media marketing efforts could generate an encouraging response from the luxury brand customers; these responses consist of brand preference. [39] claimed a positive influence of a functional brand image, hedonic brand image, and brand awareness by the brand attitude on the brand purchase intention.

In the same way, the research work conducted by the researchers [39, 42] who reported that the impact on brand equity by social media communications generates a positive influence on brand purchase intention [31]. The engagement of these results is that brand purchase intentions show the behavioral performance of consumer's perception developed by the FCC and UGC77 both. In order with the theoretical perception of double attitude, which extended that more experience towards the online advertisement can be implicit and explicit either, the researcher [43] reported about the experimental research work, which was conducted to engage the impact on purchase intention and consumer attitude by the online advertising. One of the other experimental research work reported by the researchers [38] who recognized that consumer perception about the brand image is associated positively with the brand purchase intention. In contrast, consumer perception of brand image is not linked positively with the brand attitude.

H4: Social media advertising has a significant impact on consumer responses.

H5: Social media sales promotion has a significant impact on consumer Responses.

H6: Social media interactive marketing has a significant impact on consumer responses.

The majority of earlier research work has shown that it is essential to target the management and development of the brand [43]. Along with various other significant purposes and intentions, the impact of consumer responses significantly and positively is one of the components and the critical apprehension for the successful growing brand equity. The reason behind it that the strong, positive, and successful features of a brand impact the consumer's preferences and consumer's purchase intentions. On the other hand, the worth of a brand can be judged through the behavior of consumers in the form of consumer's loyalty towards the purchase of the brand and the consumer purchase preferences [44]. The structure of this research work is complete with the research that illustrated the positive association among various factors of brand equity on various degrees of consumer responses.

Mostly the earlier research was targeting the significant factors of Consumer-Bases Brand Equity concerning consumer responses, which targeted the accomplishment of purchase intention. [45] recommended that brand preference and purchase intention are the significant variables for estimating the worth and success of brand equity on consumer responses.

The various research work has empirical results that suggested the significance of brand presence and purchase intention as a degree of consumer responses towards the growth of brand equity for their success. The results of these researches are associated with the opinions suggested by the researcher [41], who claimed that customer response towards the brand values success through buying the brands and creating liking among them through a positive perception in their minds. In the same way, [31] recommended that beyond this, there are two different kinds of product involvements such as low 79 involvement (such as household cleaners) and the high involvement (such as hotels)- the successful brand equity tend towards the purchase intention and brand preference.

[44] revealed the association among overall brand equity, brand equity dimensions, and the consumer's responses from two various countries from Europe, such as Spain and the United Kingdom. [31] estimated the consumer responses with purchase intention that is linked with the premium price, brand preference, and brand extension. The results of these scholars illustrated a significant impact on consumer responses by brand equity, particularly the brand preferences and purchase intention. In the same way, researchers [8] estimated the impacts on consumer response dimensions by brand equity. It was recommended that the high level of value regarding brand equity tends towards a positive response from the customer sides, such as brand preference and purchase intention. [27] reported that brand preference and purchase intention are the two significant consumer responses, most importantly, when employment with high involvement brands, for instance, the university brands.

H7: Consumer-Based Brand Equity has a significant impact on consumer responses.

H8: Consumer-Based Brand Equity mediates between Social media advertising and consumer responses.

H9: Social media sales promotion mediates between Social media advertising and consumer responses.

H10: Social media interactive marketing mediates between Social media advertising and consumer responses.

3 Methodology

In the present research, a survey was conducted by distributing 186 questionnaires to the respondents. Oversampling was done to avoid sampling error and non-response bias [46]. For this purpose, we administered 186 questionnaires to the targeted respondents. Moreover, it has been argued by [47] that if the sample size is small, then there is a higher tendency of error, while there is higher susceptibility to obtain accurate results with large sample size. In this regard, oversampling will likely facilitate in avoiding any potential loss arising from damages and non-cooperative subjects [48]. In particular, oversampling is usually done by the researchers because it ensures that the results will remain unaffected from any non-response rate and non-response bias. The response rate obtained in this research is consistent with [49] argument, according to which it is acceptable to obtain a 50% response rate in social surveys.

Once the data collection process is completed, we employed a combination of inferential and descriptive analyses to carry out the data analysis. For this purpose, the PLS-SEM approach was integrated, which is a second-generation technique for structural equation modeling. PLS-SEM is a relatively new approach and provides reliable results when coupled with SEM models. The SEM models involve a set of latent variables that may have a cause and effect relationships among them. According to [50], the PLS-SEM approach serves as a flexible and powerful tool to establish statistical models and predict the nature of relationships between the variables. Thus, for the measurement and structural model development, we adopted the Smart PLS path modeling in this study. The measurement or outer model is estimated by determining and explaining the validity and reliability of the model constructs. On the other hand, the structural model is generally used for establishing the correlations and the effects of constructs on other model constructs' by performing a bivariate correlation analysis and regression analyses.

4 Results

In this study, a two-stage method was employed to assess the PLS-SEM path model and report its results [51]. The two steps are as follows: (1) assessing the outer or the measurement model, and (2) assessing the inner or the structural model (Fig. 1).

Each construct's outer loadings are observed to ascertain the individual item reliability [52]. According to [50] suggested rule of thumb, all those items that fall within 0.70 to 0.99 must be retained in the study (Table 1).

By assessing the measurement model, we are meant to observe a set of criteria, such as the internal consistency reliability, individual item reliability, convergent validity, content validity, and discriminant validity [51]. In order to determine discriminant validity, the indicator loadings and the cross-loadings are compared. Therefore, when all the indicator loadings turn out as greater in comparison to the cross-loadings, sufficient discriminant validity is achieved (Table 2).

The average variance extracted (AVE) calculations and the values of the coefficients are presented in Table 3, which are ranging from 0.50 to 0.86. Since all the AVEs of latent variables are above 0.50, therefore convergent validity is said to be achieved. Obtaining satisfactory outcomes for the convergent validity, composite reliability, item loadings, and AVE coefficients signify that items are representing their unique latent constructs, which resulted in achieving adequate convergent validity.

In contrast, by the term discriminant validity, we mean that those measures which are assumed as unrelated are not related. As recommended by [52], the discriminant validity can be ascertained by calculating the AVE's square root values for each construct. These square root values are placed diagonally in the correlation matrix. [50] suggest that the square root AVEs must be greater in value than the squared correlation estimates, as it would lead to the achievement of desirable discriminant validity. In addition, the diagonal elements must be relatively greater in value as compared to the off-diagonal elements or coefficients.

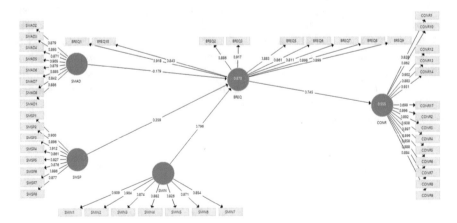

Fig. 1 Measurement model

Table 1 Outer loadings

	BREQ	CONR	SMAD	SMIN	SMSP
BREQ1	0.918				
BREQ10	0.843				
BREQ2	0.886				
BREQ3	0.917				
BREQ5	0.883				
BREQ6	0.861				
BREQ7	0.811				
BREQ8	0.898				
BREQ9	0.899				
CONR1		0.828			
CONR10		0.862			
CONR12		0.902			
CONR13		0.883			
CONR14		0.931			
CONR17		0.700			
CONR2		0.896			
CONR3		0.892			
CONR4		0.938			
CONR5		0.897			
CONR6		0.896			
CONR7		0.858			
CONR8		0.888			
CONR9		0.884			
SMAD2			0.879		
SMAD3			0.893		
SMAD4			0.871		
SMAD5			0.905		
SMAD6			0.879		
SMAD7			0.885		
SMAD8			0.842		
SMIN1				0.909	
SMIN2				0.904	
SMIN3				0.874	
SMIN4				0.882	
SMIN5				0.826	
SMIN6				0.871	

(continued)

Table 1 (continued)

	BREQ	CONR	SMAD	SMIN	SMSP
SMIN7				0.854	
SMSP1					0.900
SMSP2					0.896
SMSP3					0.912
SMSP4					0.861
SMSP5					0.827
SMSP6					0.876
SMSP7					0.886
SMSP8					0.877
SMAD1			0.886		

Table 2 Reliability

	Cronbach's Alpha	rho_A	Composite reliability	Average Variance Extracted (AVE)
BREQ	0.963	0.965	0.969	0.775
CONR	0.977	0.978	0.979	0.769
SMAD	0.958	0.960	0.965	0.775
SMIN	0.949	0.949	0.958	0.765
SMSP	0.958	0.959	0.965	0.774

Table 3 Validity

	BREQ	CONR	SMAD	SMIN	SMSP
BREQ	0.880				
CONR	0.745	0.877			
SMAD	0.703	0.719	0.880		
SMIN	0.724	0.710	0.686	0.875	
SMSP	0.747	0.735	0.738	0.698	0.880

After assessing the outer model, the structural model is then determined. By employing 5000 benchmarks resamples, a bootstrapping procedure is applied to measure the path coefficients' significance [53]. The structural model estimates, including the moderating variable, are shown in Fig. 2 and Tables 4 and 5.

The results in Table 4 indicate that all the direct paths are significant at a p-value of less than 0.05.

The results in Table 5 indicate that all the mediating paths are significant at a p-value of less than 0.05.

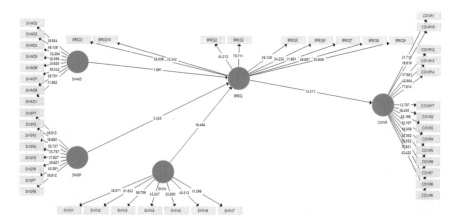

Fig. 2 Structural model

Table 4 Direct relationships

	(O)	(M)	(STDEV)	(|O/STDEV|)	P Values
BREQ → CONR	0.745	0.745	0.056	13.311	0.000
SMAD → BREQ	0.179	0.184	0.096	1.861	0.032
SMAD → CONR	0.133	0.137	0.073	1.831	0.034
SMIN → BREQ	0.796	0.793	0.048	16.488	0.000
SMIN → CONR	0.593	0.589	0.047	12.593	0.000
SMSP → BREQ	0.359	0.368	0.102	3.520	0.000
SMSP → CONR	0.267	0.275	0.082	3.255	0.001

Table 5 Mediating relationships

	(O)	(M)	(STDEV)	(|O/STDEV|)	P Values
SMAD → BREQ → CONR	0.133	0.137	0.073	1.831	0.034
SMIN → BREQ → CONR	0.593	0.589	0.047	12.593	0.000
SMSP → BREQ → CONR	0.267	0.275	0.082	3.255	0.001

In PLS-SEM structural model estimation, R-square is an important criterion and is usually termed as the coefficient of determination (R^2) [50]. It shows how much variation in the model's endogenous construct is describable by the model's exogenous constructs [54]. However, the acceptable range for the value of R^2 is determined by the research context [51]. Thus, the minimum acceptable range of R^2 suggested by Falk and Miller (1992) is 0.10. In addition, other R^2 values, such as 0.67, are proposed as substantial, 0.33 as moderate, and 0.19 as weak (Table 6).

Following the [55] suggestion, this study performed a predictive relevance test through a blindfolding approach. Although, in PLS-SEM analysis, this method serves

Table 6 R-Square

	R Square
BREQ	0.878
CONR	0.555

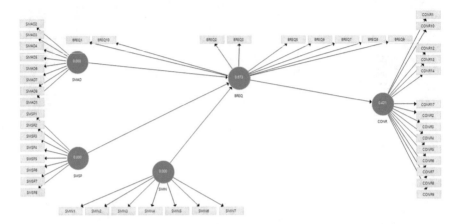

Fig. 3 Blindfolding

as an additional tool for checking the goodness-of-fit. Blindfolding procedure is not only used for testing the research model's predictive relevance; rather, [56–62] argued that an important condition for applying blindfolding procedure is that the endogenous variables must have an operational reflective measurement model (Fig. 3).

[50] also states that in the reflective measurement model, an unobservable concept or a latent variable may bring about change in the observable indicators. Therefore, we intentionally applied a blindfolding method to this research because all endogenous constructs were reflective in nature.

	SSO	SSE	Q^2 (=1 − SSE/SSO)
BREQ	1953.000	639.532	0.673
CONR	3038.000	1759.799	0.421

5 Conclusion

The research work reported semi-structured discussions to confirm the point of view regarding Customer-Based Brand Equity (CBBE) in the perception of university brands. This research work also investigated the performance of social media marketing communications in the emerging Customer-Based Brand Equity (CBBE) and the associations on consumer response by Customer-Based Brand Equity (CBBE). The current research work has also investigated the indirect impact on the association between consumer response and social media communications by the Customer-Based Brand Equity (CBBE).

The conclusion of this research shows that Customer-Based Brand Equity (CBBE) is a valid perception in the framework of university brands. The significant variables or degrees that were depicted by the Customer-Based Brand Equity (CBBE) in the framework of university brands are brand sustainability, brand awareness, hedonic brand image, and functional brand image. Moreover, this research work illustrated that brands associated with the framework that is arranged by the brands or company owners and the consumers both on social media are reduced through various marketing communications, which consist of sales, word-of-mouth, advertising, interactive marketing, and promotions.

The research work has been based on the earlier research of marketing communications similar to other studies such as [62–67], the research work of brand equity, and the theoretical framework of CSRM to theorize the direct association among consumer response, CBBE, and social media marketing communications. These marketing communications were reported to have a significant association among the consumer response and the Customer-Based Brand Equity (CBBE) of university brands. According to this study, the results of this research work maintained the overall hypotheses presented in this research work. Although, the association among social media advertising, social media advertising, consumer response, CBBE, social media interactive marketing was reported as negative.

References

1. Banerjee, P., Saini, G.K., Kalyanaram, G.: The role of brands in recruitment: mediating role of employer brand equity. Asia Pacific J. Human Resources **58**(2), 173–196 (2020)
2. Iglesias, O., Markovic, S., Singh, J.J., Sierra, V.: Do customer perceptions of corporate services brand ethicality improve brand equity? Considering the roles of brand heritage, brand image, and recognition benefits. J. Bus. Ethics **154**(2), 441–459 (2019)
3. Hepola, J., Karjaluoto, H., Hintikka, A.: The effect of sensory brand experience and involvement on brand equity directly and indirectly through consumer brand engagement. J. Product Brand Manage. **26**(3), 282–293 (2017)
4. Lakatos, V., Erdey, L., Szűcs, N., Rózsa, A., Nagy, A.: Establishment of the customer perspective of the balanced scorecard system at sports enterprises through the example of a Hungarian football club. Appl. Stud. Agribusiness Commerce **13**(1–2), 5–16 (2019)
5. Tasci, A.D.: Testing the cross-brand and cross-market validity of a consumer-based brand equity (CBBE) model for destination brands. Tourism Manage. **65**, 143–159 (2018)
6. Sijoria, C., Mukherjee, S., Datta, B.: Impact of the antecedents of electronic word of mouth on consumer-based brand equity: a study on the hotel industry. J. Hospitality Mark. Manage. **28**(1), 1–27 (2019)
7. Raji, R.A., Mohd Rashid, S., Mohd Ishak, S.: Consumer-based brand equity (CBBE) and the role of social media communications: qualitative findings from the Malaysian automotive industry. J. Mark. Commun. **25**(5), 511–534 (2019)

8. Çifci, S., Ekinci, Y., Whyatt, G., Japutra, A., Molinillo, S., Siala, H.: A cross-validation of consumer-based brand equity models: driving customer equity in retail brands. J. Bus. Res. **69**(9), 3740–3747 (2016)

9. Algharabat, R., Rana, N.P., Alalwan, A.A., Baabdullah, A., Gupta, A.: Investigating the antecedents of customer brand engagement and consumer-based brand equity in social media. J. Retailing Consumer Serv. **53** (2020)

10. Rana, J., Paul, J.: Consumer behavior and purchase intention for organic food: a review and research agenda. J. Retailing Consumer Serv. **38**, 157–165 (2017)

11. Gong, X., Cheung, C.M., Zhang, K.Z., Chen, C., Lee, M.K.: Cross-side network effects, brand equity, and consumer loyalty: evidence from mobile payment market. Int. J. Electronic Commerce **24**(3), 279–304 (2020)

12. Gebauer, H., Saul, C.J., Haldimann, M., Gustafsson, A.: Organizational capabilities for pay-per-use services in product-oriented companies. Int. J. Prod. Econ. **192**, 157–168 (2017)

13. Datta, H., Ailawadi, K.L., Van Heerde, H.J.: How well does consumer-based brand equity align with sales-based brand equity and marketing-mix response? J. Mark. **81**(3), 1–20 (2017)

14. Alvarado-Karste, D., Guzmán, F.: The effect of brand identity-cognitive style fit and social influence on consumer-based brand equity. J. Product Brand Manage. (2020). Vol. ahead-of-print No. ahead-of-print

15. Zarantonello, L., Grappi, S., Formisano, M., Brakus, J.: How consumer-based brand equity relates to market share of global and local brands in developed and emerging countries. Int. Mark. Rev. **37**(2), 345–375 (2020)

16. Ogle, D.L., Tenkasi, R.R.V., Brock, W.B.B.: The social media presence of organization development: a social network analysis using big data. In: Research in Organizational Change and Development. Emerald Publishing Limited, Vol. 28, pp. 1–41 (2020)

17. Hansen, A., Machin, D.: Media and Communication Research Methods. Macmillan International Higher Education (2018)

18. Liu, L.T., Li, J., Peng, G.Y., Li, J., Wu, T., Yu, Z.Y., Zhao, C.Y.: Analysis of similarities and differences between coronavirus disease 2019 and severe acute respiratory syndrome. World J. Traditional Chin. Med. **6**(2), 145–151 (2020)

19. Rodrigues, H., Almeida, F., Figueiredo, V., Lopes, S.L.: Tracking e-learning through published papers: a systematic review. Comput. Educ. **136**, 87–98 (2019)

20. Porto, R.B.: Consumer-based brand equity: benchmarking the perceived performance of brands. Revista Brasileira De Marketing **18**(4), 51–74 (2020)

21. Adetunji, R.R., Rashid, S.M., Ishak, M.S.: Social media marketing communication and consumer-based brand equity: an account of automotive brands in Malaysia. Jurnal Komunikasi: Malaysian J. Commun. **34**(1), 1–19 (2018)

22. Godey, B., Manthiou, A., Pederzoli, D., Rokka, J., Aiello, G., Donvito, R., Singh, R.: Social media marketing efforts of luxury brands: influence on brand equity and consumer behavior. J. Bus. Res. **69**(12), 5833–5841 (2016)

23. Lamberton, C., Stephen, A.T.: A thematic exploration of digital, social media, and mobile marketing: research evolution from 2000 to 2015 and an agenda for future inquiry. J. Mark. **80**(6), 146–172 (2016)

24. Pelamo, L.: Brand Equity Building Through Social Media Marketing: Case: EasySoda Finland Oy (2018)

25. Andrews, J.C., Shimp, T.A.: Advertising, Promotion, and Other Aspects of Integrated Marketing Communications. Nelson Education (2017)

26. Schivinski, B., Dabrowski, D.: The effect of social media communication on consumer perceptions of brands. J. Mark. Commun. **22**(2), 189–214 (2016)

27. Bruhn, M.: Relationship Marketing. Verlag Franz Vahlen (2012)

28. Schivinski, B.: Effects of social media communication on brand equity and brand purchase intention. Ph.D. Interdisc. J. **2**, 157–162 (2011)

29. Zailskaite-Jakste, L., Kuvykaitė, R.: Communication in social media for brand equity building. Econ. Manage. **18**(1), 142–153 (2013)

30. Zailskaite-Jakste, L., Kuvykaite, R.: Consumer engagement in social media by building the brand. In: Proceedings in EIIC-1st Electronic International Interdisciplinary Conference (No. 1) (2012)

31. Schivinski, B., Dabrowski, D.: The impact of brand communication on brand equity through Facebook. J. Res. Interactive Mark. **9**(1), 31–53 (2015)

32. Abu Zayyad, H.M., Obeidat, Z.M., Alshurideh, M.T., Abuhashesh, M., Maqableh, M., Masa'deh, R.E.: Corporate social responsibility and patronage intentions: the mediating effect of brand credibility. J. Mark. Commun., 1–24 (2020)

33. Severi, E., Ling, K.C., Nasermoadeli, A.: The impacts of electronic word of mouth on brand equity in the context of social media. Int. J. Bus. Manage. **9**(8), 84–96 (2014)

34. Duffett, R.G.: Influence of social media marketing communications on young consumers' attitudes. Young Consumers **18**(1), 19–39 (2017)

35. Deng, X., Xu, Y.: Consumers' responses to corporate social responsibility initiatives: the mediating role of consumer–company identification. J. Bus. Ethics **142**(3), 515–526 (2017)

36. Cheung, M.L., Pires, G.D., Rosenberger, P.J., III.: Developing a conceptual model for examining social media marketing effects on brand awareness and brand image. Int. J. Econ. Bus. Res. **17**(3), 243–261 (2019)

37. Iglesias, O., Markovic, S., Rialp, J.: How does sensory brand experience influence brand equity? Considering the roles of customer satisfaction, customer affective commitment, and employee empathy. J. Bus. Res. **96**, 343–354 (2019)

38. Lindgren, S.: Digital Media and Society. Sage (2017)

39. Mirabi, V., Akbariyeh, H., Tahmasebifard, H.: A study of factors affecting on customers purchase intention. J. Multidisciplinary Eng. Sci. Technol. (JMEST) **2**(1), 267–273 (2015)

40. Poturak, M., Softic, S.: Influence of social media content on consumer purchase intention: mediation effect of brand equity. Eurasian J. Bus. Econ. **12**(23), 17–43 (2019)

41. Raji Ridwan, A., Mohd Rashid, S., Ishak, M.S.: User-generated contents in Facebook, functional and hedonic brand image and purchase intention. SHS Web Conf. **33**, 1–6 (2017)

42. Pentina, I., Guilloux, V., Micu, A.C.: Exploring social media engagement behaviors in the context of luxury brands. J. Advertising **47**(1), 55–69 (2018)

43. Huang, Z.J., Cai, L.A.: Modeling consumer-based brand equity for multinational hotel brands—when hosts become guests. Tourism Manage. **46**, 431–443 (2015)

44. Chiyesu, E.K., Lusaka, Z., Chibomba, M.K.: To Understand Consumer's Preference and Consumer's Buying Behavior of Soft Drinks in Kitwe

45. Satvati, R.S., Rabie, M., Rasoli, K.: Studying the relationship between brand equity and consumer behavior. Int. Rev. **1–2**, 153–163 (2016)

46. Basheer, M., Siam, M., Awn, A., Hassan, S.: Exploring the role of TQM and supply chain practices for firm supply performance in the presence of information technology capabilities and supply chain technology adoption: a case of textile firms in Pakistan. Uncertain Supply Chain Manage. **7**(2), 275–288 (2019)

47. Hafeez, M.H., Basheer, M.F., Rafique, M., Siddiqui, S.H.: Exploring the links between TQM practices, business innovativeness, and firm performance: an emerging market perspective. Pakistan J. Soc. Sci. (PJSS) **38**(2), 485–500 (2018)

48. Hameed, W.U., Nawaz, M., Basheer, M.F., Waseem, M.: The effect of AmanahIkhtiar Malaysia (AIM) on microenterprise success in Sabah State Malaysia. Dialogue (1819–6462) **14**(2), 223–238 (2019)

49. Muneer, S., Basheer, M.F., Shabbir, R., Zeb, A.: Does information technology expedite the internal audit system? Determinants of internal audit effectiveness: evidence from Pakistani banking industry. Dialogue (1819–6462) **14**(2), 144–157 (2019)

50. Hair, J.F., Ringle, C.M., Sarstedt, M.: Partial least squares: the better approach to structural equation modeling? Long Range Plan. **45**(5–6), 312–319 (2012)

51. Henseler, J., Ringle, C.M., Sinkovics, R.R.: The use of partial least squares path modeling in international marketing. In: New Challenges to International Marketing. Emerald Group Publishing Limited (2009)

52. Fornell, C., Larcker, D.F.: Structural Equation Models with Unobservable Variables and Measurement Error: Algebra and Statistics (1981)
53. Junoha, M.Z.B.H.J.M., Bin Hidthiir, M.H., Basheer, M.F.: Entrepreneurial financial practices in Pakistan: the role of access to finance and financial literacy. Int. J. Innov. Creativity Change 7(9), 210–231
54. Basheer, M.F., Hafeez, M.H., Hassan, S.G., Haroon, U.: Exploring the role of TQM and supply chain practices for firm supply performance in the presence of organizational learning capabilities: a case of textile firms in Pakistan. Paradigms 12(2), 172–178 (2018)
55. Geisser, S.: A predictive approach to the random effect model. Biometrika 61(1), 101–107 (1974)
56. Sattler, H., Völckner, F., Riediger, C., Ringle, C.M.: The impact of brand extension success drivers on brand extension price premiums. Int. J. Res. Mark. 27(4), 319–328 (2010)
57. Al Kurdi, B., Alshurideh, M., Salloum, S.A.: Investigating a theoretical framework for e-learning technology acceptance. Int. J. Electrical Comput. Eng. (IJECE) 10(6), 6484–6496 (2020)
58. Al Kurdi, B., Alshurideh, M., Salloum, S., Obeidat, Z., Al-dweeri, R.: An Empirical Investigation into Examination of Factors Influencing University Students' Behavior towards Elearning Acceptance Using SEM Approach 14(2), 19–41 (2020)
59. Kurdi, B., Alshurideh, M.: Employee retention and organizational performance: Evidence from banking industry. Manage. Sci. Lett. 10(16), 3981–3990 (2020)
60. Kurdi, B., Alshurideh, M., Alnaser, A.: The impact of employee satisfaction on customer satisfaction: theoretical and empirical underpinning. Manage. Sci. Lett. 10(15), 3561–3570 (2020)
61. Alshurideh, M., Gasaymeh, A., Ahmed, G., Alzoubi, H., Kurd, B.: Loyalty program effectiveness: theoretical reviews and practical proofs. Uncertain Supply Chain Manage. 8(3), 599–612 (2020)
62. Alzoubi, H., Alshurideh, M., Kurdi, B., Inairat, M.: Do perceived service value, quality, price fairness and service recovery shape customer satisfaction and delight? A practical study in the service telecommunication context. Uncertain Supply Chain Manage. 8(3), 579–588 (2020)
63. Ammari, G., Alkurdi, B., Alshurideh, A., Alrowwad, A.: Investigating the impact of communication satisfaction on organizational commitment: a practical approach to increase employees' loyalty. Int. J. Mark. Stud. 9(2), 113–133 (2017)
64. Alshurideh, M.T., Shaltoni, A.M.: Marketing communications role in shaping consumer awareness of cause-related marketing campaigns. Int. J. Mark. Stud. 6(2), 163–168 (2014)
65. Alshurideh, M., Al Kurdi, B., Abumari, A., Salloum, S.: Pharmaceutical promotion tools effect on physician's adoption of medicine prescribing: evidence from Jordan. Modern Appl. Sci. 12(11), 210–222 (2018)
66. Al-Dmour, H., Al-Shraideh, M.T.: The influence of the promotional mix elements on Jordanian consumer's decisions in cell phone service usage: An analytical study. Jordan J. Bus. Admin. 4(4), 375–392 (2008)
67. Al-Dmour, H., Alshuraideh, M., Salehih, S.: A study of Jordanians' television viewers habits. Life Sci. J. 11(6), 161–171 (2014)

Customers Online Engagement with Social Media Influencers' Content Related to COVID 19

M. Al Khasawneh, M. Abuhashesh, A. Ahmad, R. Masa'deh◉, and M. T. Alshurideh◉

Abstract The purpose of the current research is to investigate the impact of social media influencers on customer engagement in Jordan with content related to COVID19. Built on a literature review a conceptual framework has been developed in order to guide the research. A quantitative approach was used to collect data through Google Forms. The research questionnaire completed by a convenience sampling technique. A structured online questionnaire was distributed to a sample of 278 respondents. Smart PLS 2.0 was used to test the study's hypotheses. The main findings revealed that the reason for a positive relationship between customers trust in social media influencers and customer participation is when customers find an influencer to be trustworthy they tend to participate more. Practically, the findings of this research revealed that the impact of social media influencers on customers' participation with content related to COVID 19 on social media will give governments and companies a chance to better understand the importance of social media influencers in contributing to the success of social media-based marketing

M. Al Khasawneh (✉) · M. Abuhashesh · A. Ahmad
E-Marketing and Social Media Department, Princess Sumaya University for Technology (PSUT), Amman, Jordan
e-mail: m.alkhasaawneh@psut.edu.jo

M. Abuhashesh
e-mail: m.abuhashesh@psut.edu.jo

A. Ahmad
e-mail: a.ahmed@psut.edu.jo

R. Masa'deh
Department of Management Information Systems, School of Business, The University of Jordan, Amman, Jordan
e-mail: r.masadeh@ju.edu.jo

M. T. Alshurideh
Department of Management, College of Business Administration, University of Sharjah, Sharjah, United Arab Emirates
e-mail: malshurideh@sharjah.ac.ae; m.alshurideh@ju.edu.jo

Department of Marketing, School of Business, The University of Jordan, Amman, Jordan

© The Author(s), under exclusive license to Springer Nature Switzerland AG 2021
M. T. Alshurideh et al. (eds.), *The Effect of Coronavirus Disease (COVID-19) on Business Intelligence*, Studies in Systems, Decision and Control 334,
https://doi.org/10.1007/978-3-030-67151-8_22

385

campaigns through encouraging customers' engagement. The current study provided a significant theoretical contribution to our knowledge due to it is originality.

Keywords Social media · Social media influencers · COVID 19 · Brands · Customers' engagement · Trust

1 Introduction

Social media covers all of the interaction between individuals and the way customers engage with each other's by sharing information, knowledge, and opinions through using web-based applications [1]. [2] added that social media has become a huge part of our everyday life's activities. In the marketing side, social media gives brands the chance to set the tone, tell a story, and build impressions. Therefore, social media has gained a popularity among customers. In order for a company to succeed in social media, it has to listen to the customers and engage with them and be a part of the conversation, the most important function of social media for brands is the fact that it brings the brand closer to the customer. Also, social media are a potentially useful tool for the effective communication of emerging infectious diseases outbreak updates and essential medical information to the public. Thus, Corona virus had impacted people life and ways of living. Infectious with Coronavirus (COVID-19) has become public health sever issue over the world causing severe acute respiratory syndrome, a novel coronavirus has recently emerged from China who made the assessment that COVID-19 can be characterized as a pandemic, it is therefore of utmost importance to prevent any further spread in the public and healthcare settings.

Social media platforms are being studied by health researchers and marketers in addition to mobilized for a variety of purposes. The general strategies and guidelines are social distancing, staying home, treating the patients and contact tracing [3, 4]. Other countries are, however, taking stricter measures to contain the outbreak, such as lockdown and mass testing. Therefore, people are forced to stay at home to stop the spread of corona virus disease. For that reason, people are spending more time on social media platforms to be informed and updated about the current pandemic corona virus outbreak. Accordingly, marketer took advantages of the new life style people living during quarantine in order to use it as a new marketing strategy to reach more target customers while staying at home. [5] argued that social media is a hybrid component of the promotion mix where it combines characteristics of traditional IMC tools while focusing on e-word-of-mouth. It is also a hybrid in that it is a mix between mixed technology and media origins that enable immediate, real-time communication, and uses multi-media formats through using numerous platforms such as Facebook, YouTube, blogs, Instagram, and etc. [6, 7]. The emergence of a really educated and an increasingly suspicious and demanding customer population facilitates the acceptance of social media in the marketplace.

One of the most popular social media platforms is Instagram which is a rising photo and video-sharing platform owned by Facebook was launched in 2010 [8]. [9] proved that the best social media platform for engagement is Instagram, as it was shown the engagement on Instagram is 58 times better than Facebook and 120 times better than Twitter. While [10] mentioned that celebrities are recently seen endorsing products on Instagram. Taking this point further, social media gave customers a chance to get closer to the celebrities and influencers who made customers trust their opinions. As in the marketing world there is always a new trend; therefore, social media had always been a trend for marketers to use in their marketing campaigns. Hence, the current trend and challenge now is the use of social media influencer to market a brand. Thus, companies promoted their brands through social media by employing online personalities known as social media influencers who endorse companies' brands and influence followers' perceptions [2, 11]. The current study tends to answer the following question:

What is the impact of social media influencers on customers' engagement and participation with Content related to Covid 19?

2 Literature Review

A number of relevant studies within the social media literature agreed that social media sites are considered an important medium for engaging with customers [12–14]. They also confirmed that firm-generated brand stories on social media focus on creating and building up customers' relationships with brands by giving a theme for engagement between customers and firms. Moreover, [15] suggested that social media sites provide an opportunity for brands to engage and communicate with existing customers and create a more intimate relationship with them in addition to building a relationship with potential customers. Thus, the emergence of social media, creating an emotional connection with customers has become easier where social media enables customers not to only engage with brands but also create their own content [16, 17]. Similarly, [12] found that the most commonly social media sites are used for communication, maintaining social ties and formulating new connections between brands and customers. Also, the main purpose of brands is trying to maintain a strong relationship with their valued customers. This study emphasized the importance of social media sites as a promising means of publicity, which every brand must hold, as it is seen that social media is a medium to a healthy and direct relation between brands and their customers in an online environment, this offers customers the ability to communicate, influence and retain a stronger position towards brands [18, 19].

Whereas, [20, 21] proposed that social media has also influenced customer from information acquisition to post-purchase information passing such as product reviews which may affect customers feelings towards a product or a brand in terms of trust. On the other hand, [22] proposed that even a small amount of negative reviews from posts by a few individuals may have a significant impact on customer attitudes and might give them the impression that these brands aren't trustworthy. Social media has also strengthened the social ties between people, which in return have a significant influence on customer's trust in a brand or a product. Taking this point further, a study conducted by [23] who found that people are willing to buy a product or a service if their friends or family have bought it or recommended it.

While conducting further research, it was found that the influencer who represents the brand is the reason customers trust a particular brand [19]. Customers nowadays avoid messages through the traditional media (TV, print, and radio) and consider them low on trust. On the other hand, Peer-created content or user-generated content is considered to be more trustworthy, since social media is accessible by everybody it gives the chance to opinion leaders or so called "social media influencers" to stand out and influence their followers [24]. This has been previously confirmed by [8] who proposed that following an influencer on social media sites has a positive result on trust. According to the results of this study, Instagram is the superior social media platform for developing brand awareness. On average, 60% of all the respondents claim that they have been aware of a brand only because of an influencer's endorsement on social media. Only 50% of Twitter users, 10% less than the average, claim to have been aware of a brand only based on a celebrity endorsement. 78% of Instagram users, 18% higher than the average claim to have been aware of a brand only based on influencer's endorsement on social media [8, 25]. In contrast, another study conducted by [26], 27 found a positive relationship between the endorsement campaigns success and the attractiveness, credibility and trustworthiness of the influencer. The same positive relationship was also concluded between the impact and effect of the influencer's campaign and the match between a product and the influencer endorsement [28]. These results have been confirmed by [29], 30 by concluding that both source characteristics (i.e. trust and credibility) are necessary to ensure the effectiveness of influencer's endorsements on social media sites.

2.1 Social Value During COVID 19

[31] defined social value as changing the social system by creating customer equity. The effect of social value could be very visible on a person's engagement on social media, as the social value perceived can encourage a person to engage and contribute more with brands and other people on social media, this is derived from their perceived value and how much people get affected by them. Social value measures the non-monetary effect of organizations and programs; this includes the prosperity of individuals and societies, social capital and surrounding ambience [32, 33].

Furthermore, Infectious with Coronavirus (COVID-19) has become public health sever issue over the world causing severe acute respiratory syndrome coronavirus, thus, to enhance public awareness, social media platform and other internet sources can provide a rapid method of surveillance that predicts the real time burden of disease and hence can guide preventive and curative strategies. Nowadays social media is an important platform to propagate information regarding the risk of infection and its control mechanisms. In addition, social media can be a valuable tool in providing values in distinct climatic and socio-economic context [34].

Social media communication is an increasingly utilized outlet for people to freely create and post information that is disseminated and consumed worldwide through the Internet [35]. Social media can create a sense of anonymity, allowing for unadulterated personal expression when compared to traditional face-to-face meetings, especially among young people and about intimate matters. In this respect, social media provide an additional informal source of data. Despite of the growing body of literature examining social media in health contexts; including issues of public health communications, promotion and surveillance [35], they provide limited insight into how the utility of social media might vary depending on the particular public health objectives governing an intervention—for example, to which extent social media platforms would contribute to enhance the public health awareness and prevention during epidemics disease transmissions. Thus, it is hypothesized that:

H1: *Social Value has a significant positive impact on customer engagement with content related to Covid 19.*

2.2 Opinion Passing

[36] described opening passing as having customers who will initiate and pass along positive and helpful information regarding your product or service. According to [37]38, one of the most distinctive features of internet-based is e-WOM that can be generated through online social networks implying the act of sharing opinions on a product, service or brands or social topics such as COVID 19 across multiple social media platforms. [39] individuals with a high level of opinion passing are known as opinion leaders and they may have huge influence on attitudes and behaviors of others. [40] agreed that the opinion-passing concept describes a way of acquiring new customers through encouraging engagement and opinion passing among customers. [41] stated that opinions passed by customers about their experience which they have observed through consuming a product or a service can help others in their decision-making process. [42] proposed that social media increases sharing opinions on social topics and has a direct effect on customer's engagement. Thus, it is hypothesized that:

H2: *Opinion passing has a significant positive impact on customer engagement with content related to Covid 19.*

2.3 Credibility

Source credibility is the extent to which the recipient of a message sees the communicator as having the relevant expertise and is trustworthy. Source credibility has been considered as a key factor in triggering and further enhancing customers' engagement and communication effectiveness. [43] also confirmed that perceived social media message's credibility was among the first theories that was empirically inspected and was found to have substantial influence on customers' attitudes towards the social media messages. Thus, credibility factors should be incorporated into technology adoption studies. In addition, influencer's credibility plays a significant role in determining the extent to which the individual believes in the honesty and truthfulness of the disseminated information through social media platforms [41–44]. Thus, it is suggested that:

H3: *Credibility of social media influencers has a significant positive impact on customer engagement with content related to Covid 19.*

2.4 Engagement

According to [45]46, engagement consists of the formation of a continuous bond between a brand and its customers based on a continual effort from the brand to encourage the interest of its customers through interaction, shared values, experiential contents and rewards. [47] suggested that the level of customer participation with brands and the influencers on social media is affected by engagement. [48] added that once the customer has full trust for the social media influencer this will lead to a positive impact on the engagement with them. Whereas, [49] proposed that customer's willingness to engage with brands is affected by the level of trust they hold towards the influencers representing the brand.

The observed linkage among participation and commonality also highlights the importance of implementing a social media strategy that offers numerous opportunities for two-way communications between people of the same interests to enhance active engagement among customers on social media [50, 51]. The current study will provide the connection and the direct impact engagement has on the trust and the participation of the customer. Thus, it is hypothesized that:

H4: *Engagement with social media influencers has a significant positive impact on customer trust with content related to Covid 19.*

H5: *Engagement with social media influencers has a significant positive impact on customer participation with content related to Covid 19.*

2.5 Trust

In a recent study, [52] defined trust as the extent to which customers are confident of the communicator's intent to transmit unbiased objective and valid information. It also reflects the confidence level of target audience in the objectiveness and truthfulness of the information provided by the source. As discussed in another recent study by [53]. Trust is an essential part of the relational approach and it is the key to the expansion of the brand-user relationship customers are trustworthy of the information posted by the brand and information posted by other customers online, this encourages participation between the customers and the brand, and between customers themselves where [54] stated in (p. 309) that "without trust in brands customers cannot enter in the loyalty set". [55] believed that customer's loyalty is derived from the customer's decision to consume a service or product over and over in the future because of their trust and belief that it is superior to its competitors. While, [8] findings concluded that credibility and trustworthiness are vital characteristics for celebrity/ influencer to be effective on social media. The most essential factors that determine how reliable a source is our trust and expertise [7, 26, 56, 57]. For example, [26] stated that trust involves customers trusting the source as individuals (this kind of trust is important in personal selling and therefore is not fully applicable to celebrity endorsements) and finally, he mentioned process-based trust as the trust that is built over long period of time.

[58] defined it as the process by which people are enabled to become actively and genuinely engaged instead of being passive observers. In a more recent study, [50] described participation as one of the most important characteristics of social media because it allows users to be more engaged in the conversation enabling them to share their experiences and knowledge and aiding in their decision-making process. On the other hand, [59] described customer participation as the input or resource integrated by customers it also includes psychological, physical and even emotional offerings during the process of producing or delivering services. [48] suggested that influential content, such as emotional content has a positive impact on participation. Studies like [60–64] showed that social media may be used to generate a growth in the customer's participation (comments and feedbacks to the brand). Meanwhile, [65] concluded that the engagement of the customers has a positive impact on the customer participation (comments and feedbacks) on social media. Thus, it is hypothesized that:

H6: *Trust in social media influencers has a significant positive impact on customer participation with content related to Covid 19.*

3 Methodology

3.1 Data Collection and Sample

The current research uses a quantitative approach to test the study's model. Among the different quantitative data collection methods, an online survey instrument was utilized to assess the key factors influencing customer participation in an online context. This design is in line with prior research in this area [66–68]. Similar to previous research, non-probability convenience sampling was employed to collect the required data for the study.

For the purpose of testing the hypotheses created for this research, data was collected online through Google Forms. The survey was distributed on Facebook which generated 222 responses. The sampling method used in this study was convenience sampling. This method was chosen based on the limited time-frame. The number of respondents to the survey was 278 respondents; however, only 80% of the respondent filled out the survey completely. Thus, the respondents' number was reduced to 222.

4 Data Analysis

Two statistical software were used to validate and test the proposed model, namely: SPSS 21 and SmartPLS. The purpose of using SPSS was to conduct an exploratory factorial analysis; while that of Smart PLS was to obtain Cronbach alpha of the scales, establish convergent and discriminant validity, and test the study's hypotheses [69].

4.1 Exploratory Factorial Analysis (EFA)

An EFA was first performed to ensure the researchers of the current study used the principal component analysis technique [70] and applied a Varimax rotation to initially-extracted factors. The recommendations of [71] were also followed in order to evaluate the factorial solutions obtained from SPSS. 20 Items that either had communality (<0.6), loading (<0.50) and cross loadings (>0.3) were removed. This process resulted in reducing the items from 35 to 32. The three items that were deleted were those that measured the construct of 'Engagement'. These items were: "I really like to talk about these brands with others"; "I am always interested in learning more about these brands"; and "I would be interested in products with this brand's name on it". The remaining 39 items of the seven factors loaded significantly on their respective factors ($0.551–0.841$), and items communalities were ≥ 0.60 (see Table 1) indicating satisfactory factorability for all the items. Additionally, the p-values for the Bartlett's test for Sphericity was significant (0.000), and the Kaiser-Meyer-Okline (KMO) measure of sampling adequacy was (0.914), which is well above the cut-off-point of 0.60, suggesting satisfactory factorability for all the items.

4.2 Reliability Analysis

SmartPLS (2.0) was used to assess Cronbach alpha (α) and composite reliability (CR). Cronbach alpha is an index that is concerned with the homogeneity of items within a scale, whereas, CR focuses on the notion that items have different loadings, and therefore, it does not underestimate the internal consistency reliability of latent

Table 1 Items communalities

Items communalities	Extraction
Social value1	0.636
Social value2	0.660
Social value3	0.738
Social value4	0.727
Opinion passing1	0.800
Opinion passing2	0.847
Opinion passing3	0.742
Engagement1	0.633
Engagement2	0.666
Engagement3	0.728
Engagement4	0.768
Engagement5	0.772
Engagement6	0.731
Engagement7	0.823
Engagement8	0.673
Engagement9	0.645
Trust1	0.766
Trust2	0.820
Trust3	0.794
Participation1	0.725
Participation2	0.717
Participation3	0.735
Participation4	0.751
Participation5	0.637
Credibility1	0.627
Credibility2	0.620
Credibility3	0.656
Credibility4	0.671
Credibility5	0.758
Extraction Method: Principal Component Analysis	

variables in PLS path models as (α) does. The results of (α) ranged from 0.70 to 0.942, and all scales exceeded the threshold value of 0.70 [72]. On the other hand, CR ranged from 0.817 to 0.951, and all scales were above the cut-off-point of 0.80 [73]

4.3 Convergent Validity

Convergent validity (CV) refers to the extent to which two or more measures of the same concept covary highly. CV was assessed via average variance extracted (AVE), which measures the level of variance captured by a construct versus the level due to measurement error. Table 2 shows that the AVE of the seven scales exceeded the cut-off-point of 0.50, ranging from 0.662 to 0.806.

5 Model Testing

It is worth noting that PLS is considered more rigorous when assessing a model compared with regression analyses [74]. It also allows for testing the structural relationships between latent constructs in one model simultaneously. When testing a model using Smart PLS, four issues need to be reported: path coefficients (β); R^2 values (0.67, 0.33 and 0.19 for endogenous constructs, which are regarded as legend, moderate and weak, respectively), effect size f^2 (scores of 0.02, 0.15 and 0.35 indicate small, moderate and large effects, respectively) (78 Cohen, 1988), and t-values (0.5 > 1.96 ≤ 2.58 suggest a significance level of 0.05, > 2.58 ≤ 3.26 suggest a significance level of 0.01, and > 3.26 suggest a significance level of 0.001) [75]. However, it should be noted that the decision of whether to accept/reject hypotheses depends on the significance level (t-values) of the path coefficients.

Using the above procedures, the six hypotheses of the current study were tested as follows. With regard to H1, which indicates that social value has a significant positive impact on customer engagement, was accepted. The path coefficient from social value to customer engagement was significant with moderate effect size ($\beta = 0.384; f^2 = 0.019; t \geq 0.001$). With respect to H2, which speculates that opinion passing has a significant positive effect on customer engagement in social media, was accepted. The path coefficient from opinion passing to customer engagement was significant and with small effect size ($\beta = 0.195; f^2 = 0.05; t \geq 0.01$). Concerning H3, which indicates that credibility has a significant positive effect on customer engagement in social media, it was accepted ($\beta = 0.221; f^2 = 0.12; t \geq 0.001$). The

Table 2 AVE values

Variables	Cronbach alpha	Composite reliability	AVE
Social values	0.838	0.892	0.673
Opinion passing	0.867	0.919	0.791
Engagement	0.942	0.951	0.683
Trust	0.880	0.926	0.806
Participation	0.886	0.916	0.687
Credibility	0.872	0.907	0.662

Table 3 Analysis of the Hypotheses

Hypotheses	Path coefficients	R^2	t-values	Sig. Level	f^2	Accepting/Rejecting Hypotheses
Social values → Engagement	0.384	0.515	5.212	>0.001	Moderate 0.19	Accepted
Opinion passing → Engagement	0.195	0.515	2.911	>0.01	Small 0.05	Accepted
Credibility → Engagement	0.221	0.515	3.768	0.001	Medium 0.12	Accepted
Engagement → Trust	0.326	0.106	4.901	>0.001	–	Accepted
Engagement → Participation	0.420	0.260	7.366	>0.001	Large 0.57	Accepted
Trust → Participation	0.183	0.260	3.168	>0.01	Small 0.03	Accepted

analysis further revealed that variance in customer engagement that was explained by the three independent variables was relatively high ($R^2 = 0.515$).

Moreover, as far as H4 is concerned, the hypothesis was accepted. The path coefficient from customer engagement to trust was significant ($\beta = 0.326; t \geq 0.001$). The variance in trust that was explained by customer engagement was very weak ($R^2 = 0.106$). With regard to H5, the path coefficient from customer engagement to participation was significant and with a strong effect size ($\beta = 0.420; f^2 = 0.57; t \geq 0.001$). The variance explained in participation by customer engagement was relatively weak ($R^2 = 0.260$). The last hypothesis was accepted (H6). This is because the path coefficient between trust and participation was significant and with small effect size ($\beta = 0.183; f^2 = 0.03; t \geq 0.01$). The variance explained in participation by trust was relatively weak ($R^2 = 0.260$) as shown in Table 3 (Fig. 1).

5.1 Discussion

The current study developed a model of social media influencers related factors influencing consumers' engagement, trust and participation with content related to Covid 19, and the empirical results strongly support the model in predicting consumers' engagement, trust and participation with social media influencers on their Covid 10 related content. In particular, the overall explanatory power of the current research model had an R-square of 51.1% for consumer engagement and 10.7% for consumer trust towards social media influencers Covid 19 related content, and 26% for consumer participation with social media influencers on their Covid

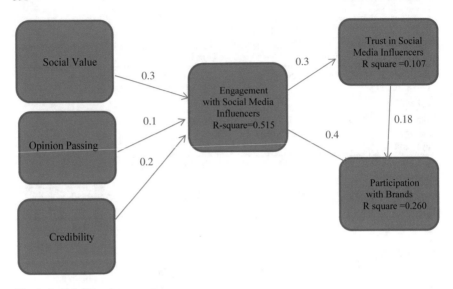

Fig. 1 Study's Hypotheses testing

19 related content, indicating that the tested model had the capability of explaining a relatively moderate percentage of variation of consumer engagement, trust and participation with social media influencers content on Covid 19. In addition, the current study found that consumer engagement with social media influencers was significantly impacted by social value, credibility and opinion passing in their order of influencing strength. The findings show that consumer's trust in social media influencers can be explained by consumer engagement. In addition, participation in Covid 19 related topic presented by social media influencers can be explained by consumer trust and engagement with social media influencers.

Customers' social value was hypothesized to have a significant positive impact on customers' engagement with social media influencers, and findings of the current study support for this hypothesis. It is assumed that the reason for a positive relationship between social value and customer engagement with social media influencers is due to the ability of social value to encourage customers with special norms to engage with social media influencers who may have the same norms as them, in addition to sharing the same beliefs between social media influencers and customers which will result in a better engagement with the brand through the social media influencer of this brand. This study has shown consistency with previous literature, by proving that social value has a significant positive impact on customers' engagement with social media influencers [31]. To the extent of our knowledge no studies were found to confirm the positive relationship between the social value and engagement with social media influencers.

Customers' opinion passing was hypothesized to have a significant positive impact on customers' engagement with social media influencers, and the findings provided support for this hypothesis. It is assumed that the reason for a positive relationship between opinion passing and customer engagement with social media influencers due to the capability of the customers to pass opinion and word of mouth between each other online which will result in engagement between the customers and social media influencers. This study has shown consistency with previous literature, by proving that opinion passing has a significant positive impact on customers' engagement with social media influencers [38–76]. In addition to, [42] who proposed that social media increases sharing opinions on products and has a direct effect on engagement between customers and brands as a single person can take multiple roles of opinion transmitters or leaders.

A significant positive relationship between influencers' credibility and customers' engagement with brands on social media was hypothesized in the study, and the findings provided support for this hypothesis. It was proved that the reason for a positive relationship between influencers' credibility and customer engagement with social media influencers is the ability of social media influencers to affect the customers' attitudes. Similarly, a study conducted by [77] confirmed that influencers who were thought to be credible impact customers' engagement with social media influencers which in return will impact their participation with this brand, in addition to these credible influencers ability to shape customers' perceptions of the brand. This study has shown consistency with previous literature, by proving that credibility has a significant positive impact on customers' engagement with social media influencers [43–77], unlike [78] who did not confirm such a direct relationship between social media influencer's credibility and engagement.

Customers' engagement with social media influencers was hypothesized to have a significant positive impact on customers trust in social media influencers, and findings of this study provided support for this hypothesis. It is assumed that the reason behind the positive relationship between customers' engagement and their trust in social media influencers is the engagement that shows a high level of transparency which in return will positively affect trust. This study has shown consistency with previous literature, by proving that customers' engagement with social media influencers has a significant positive impact on customers trust in social media influencers. [45], 79 confirmed that engagement is viewed as an important concept to provide power for customers that will result in increasing participation of customers with brands. Also, [8] findings concluded that trustworthiness is a vital characteristic for celebrity/influencer to be effective on social media where trustworthy celebrities drive people to participate more with the brand.

Customers' engagement with social media influencers was hypothesized to have a significant positive impact on customers' participation with brands, and the findings provided support for this hypothesis. It is assumed that the reason for a positive relationship between customers' engagement and customer participation with brands because people are motivated to participate more with the brand when they see how the brand engages with customers. This study has shown consistency with previous literature, by proving that customers' engagement with social media influencers has a significant positive impact on customer's participation with brands. For example, [45] suggested that brands could build up a strong relationship with their customers and motivate them to take part in the conversation by actively engaging with them. Also,

[59] described customer participation as the input or resource integrated by customers it also includes psychological, physical and even emotional offerings during the process of producing or delivering services and it is encouraged when customers witness how the brand engages with customers who do participate.

Customers' trust in social media influencers was hypothesized to have a significant positive impact on customers' participation with brands, and findings provided support for this hypothesis. It is assumed that the reason for a positive relationship between customers trust in social media influencers and customer participation with brands is because when customers find an influencer to be trust worthy they tend to participate more. Previous literature has shown consistency with the findings of this study, by proving that customers trust in social media influencers has a significant positive impact on customers' participation with brands [58]. Whereas, [52] demonstrated that the confidence level of target audience in the objectiveness and truthfulness of the information provided by the source which in return can positively impact customers' participation.

5.2 Theoretical Implications

The current study agreed with literature in regards to customers' participation with Covid 19 related contents through social media influencers. From a theoretical perspective, the study presented new information highlighting a relationship with minimal coverage in previous literature. The model of the impact of social media influencers on customers' engagement, trust and participation with Covid 19 related content presented by influencers will provide researchers with the information needed to examine the effectiveness of social media influencers on customers' participation with social topics, particularly on social media. Thus, the model's importance has been improved by determining the factors that influence customers to participate.

The current study focused on variables such as social value and opinion passing that pervious literature did not focus on. These variables were considered to be important for the current study's outcomes that measured the customers' engagement with social media influencers which will therefore affect their participation with community related content that influencers endorse.

5.3 Practical Implications

The importance of this study lies in its ability to be applied in practice. For that, the findings of this study the impact of social media influencers on customers' participation with content related to Covid 19 on social media will give businesses a chance to better understand the importance of social media influencers in contributing to the success of social media-based marketing campaigns through encouraging customer participation. The findings will aid practitioners in understanding how (social value, opinion passing, and credibility) will affect engagement and trust, which will in return have an effect on participation. An important aspect of featuring social media influencers in the campaign is the credibility of that influencer, which is a crucial characteristic for any social media influencer and was found to have a significant positive impact on customer participation.

Also, there are some aspects related to the customers such as how willing they are to be a part of the conversation because in some cultures people are reluctant to participate especially when it's an open discussion held on social media. Social value was found to have a significant positive impact on customer engagement and ultimately participation, practitioners can benefit from that through choosing an influencer who shares the same norms and values with the target audience.

Opinion passing was also found to have a significant positive influence on engagement, trust and participation because some people like to give their opinion and peer-opinion is usually found to be trustworthy by customers, what policy makers can do to benefit from this is providing a podium for customers to speak their mind and take their feedback in mind and develop the product or service accordingly and this will make the customers feel like their opinion matters and it drives them to keep on passing their opinion.

Another vital aspect is the impact of engagement on trust and ultimately participation, when a company or a brand engages with customers and responds to their comments it positively affects participation because people tend to participate more when they feel that somebody is listening and engaging with them and brands should always keep that in mind when implementing a social media campaign strategy.

Trust was also found to have a significant positive influence on participation and this is very logical because customers tends to participate more openly when they trust the influencer delivering the social media message. The main thing that can be done is to make sure that the influencer that is being featured in the campaign is someone who is trusted among the target audience and that can be anticipated when looking at people's comments on social media messages provided by these influencers. Our recommendation to benefit from having a social media influencer discuss social issues is to first study the target audience and their preferences well before implementing anything. Using effective strategies to involve social media influencers in topics related to health issues could, therefore, be a promising strategy in enhancing individual's engagement and participations in such vital topics such as Covid 19 pandemic.

5.4 Limitations

The current study faced some limitations due to some factors. First, there was only a limited amount of time to conduct this study. The fact that this study was a cross-sectional study was also a huge limitation, as the sample of people who answered the survey at a specific point in time; this implies that their answers might have been affected by a transient state of mind or feeling that could have affected their answers in a certain way. It is assumed that if people were asked the same questions at a different point in time, their answers might have been different and therefore this study's results could be highly affected by this limitation.

5.5 *Future Studies*

There are a number of recommendations that can be suggested to any future studies. The first of which is to attempt to add the attractiveness and likeability variables to the model presented in this study. Doing so will allow future studies to improve on the research results presented here. Another suggestion would be to apply the model in a different country because this study was only applied on Jordan. Additional recommendation is to apply this study on a larger sample size, this will improve the validity of the research and the results presented.

5.6 *Conclusion*

Social media covers all of the interaction between individuals online the way they engage, share information, knowledge and opinions through using web-based applications. The study provided important information on how specific factors will affect customers' engagement with social media influencers which will in return affect their trust in social media influencers and participation with brands, moreover, the findings of this study provides a different theoretical and practical implications.

The research model developed in this study provided a significant theoretical contribution to our knowledge due to it is originality. The model tested contradicted and emphasized relationships between variables. This study enriched the consensus by providing further understanding of factors influencing the impact of social media influencers on customer participation with content related to Covid 19. Engagement and trust were proven in the current research to have significant positive impact on customers' participation with Covid 19 related content. Lastly, the current study contributes to companies by providing a new understanding of what drives individuals to participate with influencers on social media. In addition, the study emphasizes a result of increased trust and engagement, which is increased customer participation.

References

1. Safko, L., Brake, D.: The Social Media Bible: Tactics, Tools and Strategies for Business Success. John Wiley & Sons, Hoboken, NY (2009)
2. Abuhashesh, M., Al-Khasawneh, M., Al-Dmour, R., Masa'deh, R.: The impact of Facebook on Jordanian consumers' decision process in the hotel selection. IBIMA Bus. Rev. (2019), Article

ID 928418, https://doi.org/10.5171/2019.928418

3. Kampf, G.: Antiseptic Stewardship: Biocide Resistance and Clinical Implications. Springer International Publishing, Cham (2018)

4. Chan, J.F.W., Yuan, S., Kok, K.H., To, K.K.W., Chu, H., Yang, J., Tsoi, H.W., et al.: A familial cluster of pneumonia associated with the 2019 novel coronavirus indicating person-to-person transmission: a study of a family cluster. The Lancet **395**(10223), 514–523 (2020)

5. Glynn, M., Faulds, D.: Social media: the new hybrid element of the promotion mix. Bus. Horiz. **52**, 357–365 (2009)

6. Arora, A., Bansal, S., Kandpal, C., Aswani, R., Dwivedi, Y.: Measuring social media influencer index-insights from Facebook, Twitter and Instagram. J. Retail. Consum. Serv. **49**, 86–101 (2019)

7. Ahmad, A., Abuhashesh, M., Obeidat, Z., AlKhatib, M.: E-WOM and airline e-ticket purchasing intention: mediating effect of online passenger trust. Manage. Sci. Lett. **10**(12), 2729–2740 (2020)

8. Jatto, O.: Consumer Attitude Towards Celebrity Endorsements on Social Media (2014). https://esource.dbs.ie/bitstream/hadle/10788/2192/mba_jatto_o_2014.pdf?sequen

9. Elliot, N.: Instagram is the King of Social Engagement (2014). https://go.forrester.com/blogs/14-04-29instagram_is_the_king_of_social_engagement/

10. Alford, R.: Brands Need to be More Original on Instagram (2013). https://www.campaignlive.co.uk/article/brands-need-original-instagram/1182680

11. Schomer, A.: Influencer Marketing: State of the Social Media Influencer Market in 2020 (2019). https://www.businessinsider.com/influencer-marketing-report?r=US&IR=T, Accessed 29th Feb 2020.

12. McKenna, K.Y., Green, A.S., Gleason, M.E.: Relationship formation on the internet: what's the big attraction? J. Soc. Issues **58**(1), 9–31 (2002)

13. Gensler, S., Völckner, F., Liu-Thompkins, Y., Wiertz, C.: Managing brands in the social media environment. J. Interact. Market. **27**(4), 242–256 (2013)

14. Lee, J.A., Eastin, M.S.: I like what she's endorsing: The impact of female social media influencers' perceived sincerity, consumer envy, and product type. J. Interact. Advert. **20**(1), 76–91 (2020)

15. Mangold, G., Faulds, D.: Social media: the new hybrid element of the promotion mix. Bus. Horiz. **52**, 357–365 (2009)

16. Rust, R.T., Zeithaml, V.A., Lemon, K.N.: Driving Customer Equity. The Free Press, New York (2000)

17. Britt, R.K., Hayes, J.L., Britt, B.C., Park, H.: Too big to sell? a computational analysis of network and content characteristics among mega and micro beauty and fashion social media influencers. J. Interact. Advert. (accepted), 1–25 (2020)

18. Abuhashesh, M., Mohammad, S.J., Khasawneh, M.A.: The attitude of Jordanian customers towards virtual stores. Int. J. Islam Market. Brand **4**(1), 59–75 (2019)

19. Singh, J., Crisafulli, B., Quamina, L.T., Xue, M.T.: To trust or not to trust: the impact of social media influencers on the reputation of corporate brands in crisis. J. Bus. Res. (2020). https://doi.org/10.1016/j.jbusres.2020.03.039

20. Hanna, R., Rohm, A., Crittenden, V.: We're all connected: the power of the social media ecosystem. Bus. Horiz. **54**, 265–273 (2011)

21. Jiménez-Castillo, D., Sánchez-Fernández, R.: The role of digital influencers in brand recommendation: examining their impact on engagement, expected value and purchase intention. Int. J. Inf. Manage. **49**, 366–376 (2019)

22. Schlosser, A.: Posting versus lurking: communicating in a multiple audience context. J. Consum. Res. **32**(2), 260–265 (2005)

23. Goolsbee, A., Zittrain, J.: Evaluating the costs and benefits of taxing internet commerce, forthcoming. Nat. Tax J. (1999)

24. George, S.L.: Using Social Media Influencers to Promote Qatar as a Tourist Destination (2020). https://hdl.handle.net/10576/12643

25. Hwang, K., Zhang, Q.: Influence of para-social relationship between digital celebrities and their followers on followers' purchase and electronic word-of-mouth intentions, and persuasion knowledge. Comput. Hum. Behav. **87**, 155–173 (2018)
26. Egan, J.: Marketing Communications. Thomson Learning, London, UK (2007)
27. Abu Zayyad, Z., Obeidat, Z., Alshurideh, M., Abuhashesh, M., Maqableh, M., Masa'deh, R.: Corporate social responsibility and patronage intentions: the mediating effect of brand credibility. J. Market. Commun. (accepted) (2020). https://doi.org/10.1080/13527266.2020.1728565
28. Yuan, S., Lou, C.: How social media influencers foster relationships with followers: the roles of source credibility and fairness in parasocial relationship and product interest. J. Interact. Advert. (accepted), 1–42 (2020)
29. Cunningham, N., Bright, L.: The tweet is in your court: measuring attitude towards athlete endorsements in social media. Int. J. Integr. Market. Commun. **4**(2), 73–87 (2012)
30. Subhadip, R., Pansari, A.: Owner or endorser?: investigating the effectiveness of celebrity owners of sports teams as endorsers. Int. J. Sports Market. Sponsorship **15**(2), 89–106 (2014)
31. Libai, B., Muller, E., Peres, R.: Sources of Social Value in Word-of-Mouth Programs (2010). https://www.msi.org/reports/sources-of-social-value-in-word-of-mouth-programs/
32. Wood, C., Leighton, D.: Measuring Social Media Value: The Gap Between Policy and Practice (2010). https://www.demos.co.uk/files/Measuring_social_value_-_web.pdf
33. Luong, A.: All that glitters is gold: the regulation of hidden advertisements and undisclosed sponsorships in the world of beauty social media influencers. William Mary Bus. Law Rev. **11**(2), 565 (2020)
34. Gluskin, R.T., Johansson, M.A., Santillana, M., Brownstein, J.S.: Evaluation of Internet-based dengue query data: Google dengue trends. PLoS Neglected Tropical Diseases **8**(2) (2014)
35. Laranjo, L., Arguel, A., Neves, A.L., Gallagher, A.M., Kaplan, R., Mortimer, N., Lau, A.Y., et al.: The influence of social networking sites on health behavior change: a systematic review and meta-analysis. J. Am. Med. Inf. Assoc. **22**(1), 243–256 (2015)
36. Hennig-Thurau, T., Gwinner, K.P., Walsh, G., Gremler, D.D.: Electronic word-of-mouth via consumer-opinion platforms: what motivates consumers to articulate themselves on the internet? J. Interact. Market. **18**(1), 38–52 (2004)
37. Dellarocas, C.: The digitization of word of mouth: promise and challenges of online feedback mechanisms. Manage. Sci. **49**(10), 1407–1424 (2003)
38. Norman, A.T., Russell, C.A.: The pass-along effect: investigating word of-mouth effects on online survey procedures. J. Comput.-Mediated Commun. **11**(4) (2006)
39. Feick, L.F., Price, L.L.: The market maven: a diffuser of marketplace information. J. Mark. **51**(1), 83–97 (1987)
40. Shirky, C.: Linux for the end user-phase 1. Linux J. (74es), 8 (2000)
41. Hughes, C., Swaminathan, V., Brooks, G.: Driving brand engagement through online social influencers: an empirical investigation of sponsored blogging campaigns. J. Mark. **83**(5), 78–96 (2019)
42. Shu-Chuan, C., Kim, Y.: Determinants of consumer engagement in electronic word-of-mouth (eWOM) in social networking sites. Int. J. Advert. Rev. Market. Commun. **30**(1), 47–75 (2011)
43. Al Samydai, M.J., Al Khasawneh, M.H.: Antecedents and consequences of e-Jordanian consumer behavior regarding Facebook advertising. Int. J. Bus. Manage. Res. **3**(4), 41–59 (2013)
44. Owusu, R.A., Mutshinda, C.M., Antai, I., Dadzie, K.Q., Winston, E.M.: Which UGC features drive web purchase intent? A spike-and-slab Bayesian Variable Selection Approach. Internet Res. (2016)
45. Gambetti, R., Graffigna, G., Biraghi, S.: The grounded theory approach to consumer–brand engagement. Int. J. Market. Res. **54**(5), 659–687 (2012)
46. Lou, C., Yuan, S.: Influencer marketing: how message value and credibility affect consumer trust of branded content on social media. J. Interact. Advert. **19**(1), 58–73 (2019)
47. Koh, J., Kim, D.: Knowledge sharing in virtual communities: an e-business perspective. Expert Syst. Appl. **26**(2), 155–166 (2004)

48. Mosavi, A., Ghaedi, M.: A survey on the relationship between trust, customer loyalty, commitment and repurchase intention. Afr. J. Bus. Manage. **6**(36) (2012)
49. Moorman, C., Deshpande, R., Zaltman, G.: Factors affecting trust in market research relationships. J. Market. **57**, 81–101 (1993)
50. Chan-Olmsted, S., Cho, M., Lee, S.: User perceptions of social media: a comparative study of perceived characteristics and user profiles by social media. Online J. Commun. Media Technol. **3**(4), 149 (2013)
51. Silva, M.J., Farias, S.A., Grigg, M.K., Barbosa, M.D.: Online engagement and the role of digital influencers in product endorsement on Instagram. J. Relat. Market. **19**(2), 133–163 (2020)
52. Smith, C.: Here's Why Instagram Demographics are so Attractive to Brands (2014). https://www.businessinsider.com/instagram-demographics-2013-12
53. Rizan, M., Warokka, A., Listyawati, D.: Relationship marketing and customer loyalty: do customer satisfaction and customer trust really serve as intervening variables? J. Market. Res. Case Stud. Article ID 724178 (2014)
54. Ahmed, Z., Rizwan, M., Ahmad, M., Haq, M.: Effect of brand trust and customer satisfaction on brand loyalty in Bahawalpur. J. Sociological Res. **5**(1) (2014)
55. Jones, T., Sasser, W., Earl, W., Jr.: Why satisfied customers defect. Harvard Bus. Rev. **73**(6), 88–99 (1995)
56. Ohanian, R.: Construction and validation of a scale to measure celebrity endorsers' perceived expertise, trustworthiness, and attractiveness. J. Advert. **19**(3), 39–52 (1990)
57. Choi, S.M., Rifon, N.J.: Who is the celebrity in advertising?. Understanding dimensions of celebrity images. J. Popular Culture **40**(2), 304–325 (2007)
58. Chen, J., Xu, H., Whinston, A.B.: Moderated online communities and quality of user-generated content. J. Manage. Inf. Syst. **28**(2), 237–268 (2011)
59. Rodi, A.R., Kleine, S.S.: Customer Participation in Services Production and Delivery in Swart: Handbook of Serv Market and Manage. Sage Publications, California (2000)
60. Goh, K.Y., Heng, C.S., Lin, Z.: Social media brand community and consumer behavior: quantifying the relative impact of user-and marketer-generated content. Inf. Syst. Res. **24**(1), 88–107 (2013)
61. Kumar, V., Bhaskara, V., Mirchandani, R., Shah, M.: Creating a measurable social media marketing strategy: increasing the value and ROI of intangibles and tangibles for hokey pokey. Market. Sci. **32**(2), 194–212 (2013)
62. Miller, A.R., Tucker, C.: Active social media management: the case of health care. Inf. Syst. Res. **24**(1), 52–70 (2013)
63. Rishika, R., Kumar, A., Janakiraman, R., Bezawada, R.: The effect of customers' social media participation on customer visit frequency and profitability: an empirical investigation. Inf. Syst. Res. **24**(1), 108–127 (2013)
64. Li, X., Wu, L.: Measuring effects of observational learning and social-network word-of-mouth (WOM) on the sales of daily-deal vouchers. SSRN Working Paper (2014)
65. Lee, D., Hosanagar, K., Nair, H.: The Effect of Social Media Marketing on Consumer Engagement (2009). https://www.semanticscholar.org/paper/The-Effect-of-Social-Media-Marketing-Content-on-Lee-Hosanagar/d6e1dc1be10ef988ffe018c3f323499541b8506a
66. Shih, H.P., Jin, B.H.: Driving goal-directed and experiential online shopping. J. Organ. Comput. Electron Comm. **21**(2), 136–157 (2011)
67. Rose, S., Clark, M., Samouel, P., Hair, N.: Online customer experience in e-retailing: an empirical model of antecedents and outcomes. J. Retail. **88**, 308–322 (2012)
68. Shobeiri, S., Mazaheri, E., Laroche, M.: Shopping online for goods vs. services: where do experiential features help more? Int. J. Consum. Stud. **39**, 172–179 (2015)
69. Hair, J.F., Ringle, C.M., Sarstedt, M.: Pls: indeed a silver bullet. J. Market. Theor. Pract. **19**(2), 139–151 (2011)
70. Rietveld, T., Van Hout, R.: Statistical Techniques for the Study of Language and Language Behaviour. Mouton de Gruyter, Berlin-New York (1993)
71. Costello, A., Osborne, J.: Best practices in exploratory factor analysis: four recommendations for getting the most from your analysis. Pract. Assess Res. Eval. **10**(7), 1–9 (2005)

72. Nunnally, J.C.: Psychometric Theory. McGraw-Hill Book Company (1978)
73. Crossley, J., Davies, H., Humphris, B., Jolly, G.: Generalisability: a key to unlock professional assessment. Med. Educ. **36**, 972–978 (2002)
74. Mintu-Wimsatt, A., Graham, J.: Testing a negotiation model on Canadian anglophone and Mexican exporters. J. Acad. Market. Sci. **32**(3), 345–356 (2004)
75. Chin, W.: The partial least squares approach to structural equation modelling. In: Marcoulides, G. (Ed.), Modern Methods for Business Research. Lawrence Erlbaum Associates, Mahwah, NJ, pp. 295–358 (1998)
76. Zhu, L., Yin, G., He, W.: Is this opinion leaders review useful?: peripheral cues for online review helpfulness. J. Electron Commerce Res. **15**(4), 267–280 (2014)
77. Dwivedi, A., Johnson, L., McDonald, R.: Celebrity endorsement, self-brand connection and consumer-based brand equity. J. Prod. Brand Manage. **24**(5), 449–461 (2015)
78. Spry, A., Pappu, R., Cornwell, B.T.: Celebrity endorsements, brand credibility and brand equity. Eur. J. Market. **45**(6), 882–909 (2011)
79. Avnet, T., Higgins, E.: Locomotion, assessment, and regulatory fit: value transfer from 'How' to 'What.' J. Exp. Soc. Psychol. **39**, 525–530 (2003)

Printed in the United States
by Baker & Taylor Publisher Services